Springer Series on
SIGNALS AND COMMUNICATION TECHNOLOGY

Signals and Communication Technology

Wireless Network Security
Y. Xiao, S. Shen, and D. Du (Eds.)
ISBN 0-387-28040-5

Wireless Ad Hoc and Sensor Networks
A Cross-Layer Design Perspective
R. Jurdak
ISBN 0-387-39022-7

Cryptographic Algorithms on Reconfigurable Hardware
F. Rodrgues-Henriquez, N.A. Saaqib, A. Diaz Perez, and C.K. Koc
ISBN 0-387-33956-6

Multimedia Database Retrieval
A Human-Centered Approach
P. Muneesawang and L. Guan
ISBN 0-387-25267-X

Broadband Fixed Wireless Access
A System Perspective
M. Engels and F. Petre
ISBN 0-387-33956-6

Distributed Cooperative Laboratories
Networking, Instrumentation, and Measurements
F. Davoli, S. Palazzo and S. Zappatore (Eds.)
ISBN 0-387-29811-8

The Variational Bayes Method in Signal Processing
V. Šmídl and A. Quinn
ISBN 3-540-28819-8

Topics in Acoustic Echo and Noise Control
Selected Methods for the Cancellation of Acoustical Echoes, the Reduction of Background Noise, and Speech Processing
E. Hänsler and G. Schmidt (Eds.)
ISBN 3-540-33212-x

EM Modeling of Antennas and RF Components for Wireless Communication Systems
F. Gustrau, D. Manteuffel
ISBN 3-540-28614-4

Interactive Video
Methods and Applications
R. I Hammoud (Ed.)
ISBN 3-540-33214-6

Continuous Time Signals
Y. Shmaliy
ISBN 1-4020-4817-3

Voice and Speech Quality Perception
Assessment and Evaluation
U. Jekosch
ISBN 3-540-24095-0

Advanced ManMachine Interaction
Fundamentals and Implementation
K.-F. Kraiss
ISBN 3-540-30618-8

Orthogonal Frequency Division Multiplexing for Wireless Communications
Y. (Geoffrey) Li and G.L. Stüber (Eds.)
ISBN 0-387-29095-8

Circuits and Systems
Based on Delta Modulation
Linear, Nonlinear and Mixed Mode Processing
D.G. Zrilic ISBN 3-540-23751-8

Functional Structures in Networks
AMLn—A Language for Model Driven Development of Telecom Systems
T. Muth ISBN 3-540-22545-5

RadioWave Propagation for Telecommunication Applications
H. Sizun ISBN 3-540-40758-8

Electronic Noise and Interfering Signals
Principles and Applications
G. Vasilescu ISBN 3-540-40741-3

DVB
The Family of International Standards for Digital Video Broadcasting, 2nd ed.
U. Reimers ISBN 3-540-43545-X

Digital Interactive TV and Metadata
Future Broadcast Multimedia
A. Lugmayr, S. Niiranen, and S. Kalli
ISBN 3-387-20843-7

Adaptive Antenna Arrays
Trends and Applications
S. Chandran (Ed.) ISBN 3-540-20199-8

Digital Signal Processing with Field Programmable Gate Arrays
U. Meyer-Baese ISBN 3-540-21119-5

Neuro-Fuzzy and Fuzzy Neural Applications in Telecommunications
P. Stavroulakis (Ed.) ISBN 3-540-40759-6

continued after index

Wireless Network Security

YANG XIAO, XUEMIN SHEN,

and DING-ZHU DU

Springer

Editors:
Yang Xiao
Department of Computer Science
University of Alabama
101 Houser Hall
Tuscaloosa, AL 35487

Xuemin (Sherman) Shen
Department of Electrical & Computer Engineering
University of Waterloo
Waterloo, Ontario, Canada N2L 3G1

Ding-Zhu Du
Department of Computer Science & Engineering
University of Texas at Dallas
Richardson, TX 75093

Wireless Network Security

Library of Congress Control Number: 2006922217

ISBN-10 0-387-28040-5 e-ISBN-10 0-387-33112-3
ISBN-13 978-0-387-28040-0 e-ISBN-13 978-0-387-33112-6

Printed on acid-free paper.

9 8 7 6 5 4 3 2 1

springer.com

CONTENTS

PREFACE

Wireless/mobile communications network technologies have been dramatically advanced in recent years, inculding the third generation (3G) wireless networks, wireless LANs, Ultra-wideband (UWB), ad hoc and sensor networks. However, wireless network security is still a major impediment to further deployments of the wireless/mobile networks. Security mechanisms in such networks are essential to protect data integrity and confidentiality, access control, authentication, quality of service, user privacy, and continuity of service. They are also critical to protect basic wireless network functionality.

This edited book covers the comprehensive research topics in wireless/mobile network security, which include cryptographic co-processor, encryption, authentication, key management, attacks and countermeasures, secure routing, secure medium access control, intrusion detection, epidemics, security performance analysis, security issues in applications, etc. It can serve as a useful reference for researchers, educators, graduate students, and practitioners in the field of wireless/network network security.

The book contains 15 refereed chapters from prominent researchers working in this area around the world. It is organized along five themes (parts) in security issues for different wireless/mobile networks.

- **Part I: Security in General Wireless/Mobile Networks**: Chapter 1 by Lutz and Hasan describes a high performance and optimal elliptic curve processor as well as an optimal co-processor using Lopez and Dahab's projective coordinate system. Chapter 2 by Lufei and Shi proposes an adaptive encryption protocol to dynamically choose a proper encryption algorithm based on application-specific requirements and device configurations.

- **Part II: Security in Ad Hoc Networks**: The next five chapters focus on security in ad hoc networks. Chapter 3 by Hoeper and Gong introduces a security framework for pre-authentication and authenticated models in ad hoc networks. Chapter 4 by Pan, Cai, and Shen promotes identity-based key management in ad hoc networks. Chapter 5 by Wu *et al.* provides a survey of attacks and countermeasures in ad hoc networks. Chapter 6 by Giruka and Singhal presents several routing protocols for ad-hoc networks, the security issues related to routing, and securing routing protocols in ad hoc networks. Chapter 7 by Anantvalee and Wu classifies the architectures for intrusion detection systems in ad hoc networks.

- **Part III: Security in Mobile Cellular Networks**: The next two chapters discuss security in mobile cellular networks. Chapter 8 by Sun, Xiao, and Wu introduces intrusion detection systems in mobile cellular networks. Chapter 9 by Zheng *et al.* proposes an epidemics spread model for smartphones.

- **Part IV: Security in Wireless LANs**: The next three chapters study the security in wireless LANs. Chapter 10 by Kim and Shin focuses on cross-domain authentication over wireless local area networks, and proposes an enhanced protocol called the *Mobility-adjusted Authentication Protocol* that performs mutual authentication and hierarchical key derivation. Chapter 11 by Shi *et al.* proposes Authentication, Authorization and Accounting (AAA) architecture and authentication for wireless LAN roaming. Chapter 12 by Agarwal and Wang studies the cross-layer interactions of security protocols in wireless LANs, and presents an experimental study.

- **Part V: Security in Sensor Networks**: The last three chapters focus on security in sensor networks. Chapter 13 by Mišić and Mišić reviews confidentiality and integrity polices for clinical information systems and compares candidate technologies IEEE 802.15.1 and IEEE 802.15.4 from the aspect of resilience of MAC and PHY layers to jamming and denial-of-service attacks. Chapter 14 by Rayi *et al.* provides a survey of key management schemes in sensor networks. The last chapter by Su, Xiao, and Boppana introduces security attacks, and reviews the recent approaches of secure network routing protocols in both mobile ad hoc and sensor networks.

Although the covered topics may not be an exhaustive representation of all the security issues in wireless/mobile networks, they do represent a rich and useful sample of the strategies and contents.

This book has been made possible by the great efforts and contributions of many people. First of all, we would like to thank all the contributors for putting together excellent chapters that are very comprehensive and informative. Second, we would like to thank all the reviewers for their valuable suggestions and comments which have greatly enhanced the quality of this book. Third, we would like to thank the staff members from Springer, for putting this book together. Finally, We would like to dedicate this book to our families.

YANG XIAO
Tuscaloosa, Alabama, USA

XUEMIN (SHERMAN) SHEN
Waterloo, Ontario, CANADA

DING-ZHU DU
Richardson, Texas, USA

Part I

SECURITY IN GENERAL WIRELESS/MOBILE NETWORKS

1

HIGH PERFORMANCE ELLIPTIC CURVE CRYPTOGRAPHIC CO-PROCESSOR

Jonathan Lutz
General Dynamics - C4 Systems
Scottsdale, Arizona
E-mail: Jonathan.Lutz@gdc4s.com

M. Anwarul Hasan
Department of Electrical and Computer Engineering
University of Waterloo, Waterloo, ON, Canada
E-mail: ahasan@ece.uwaterloo.ca

For an equivalent level of security, elliptic curve cryptography uses shorter key sizes and is considered to be an excellent candidate for constrained environments like wireless/mobile communications. In FIPS 186-2, NIST recommends several finite fields to be used in the elliptic curve digital signature algorithm (ECDSA). Of the ten recommended finite fields, five are binary extension fields with degrees ranging from 163 to 571. The fundamental building block of the ECDSA, like any ECC based protocol, is elliptic curve scalar multiplication. This operation is also the most computationally intensive. In many situations it may be desirable to accelerate the elliptic curve scalar multiplication with specialized hardware.

In this chapter a high performance elliptic curve processor is described which is optimized for the NIST binary fields. The architecture is built from the bottom up starting with the field arithmetic units. The architecture uses a field multiplier capable of performing a field multiplication over the extension field with degree 163 in 0.060 microseconds. Architectures for squaring and inversion are also presented. The co-processor uses Lopez and Dahab's projective coordinate system and is optimized specifically for Koblitz curves. A prototype of the processor has been implemented for the binary extension field with degree 163 on a Xilinx XCV2000E FPGA. The prototype runs at 66 MHz and performs an elliptic curve scalar multiplication in 0.233 msec on a generic curve and 0.075 msec on a Koblitz curve.

1. INTRODUCTION

The use of elliptic curves in cryptographic applications was first proposed independently in [15] and [23]. Since then several algorithms have been developed whose strength relies on the difficulty of the discrete logarithm problem over a group of elliptic curve points. Prominent examples include the Elliptic Curve Digital Signature Algorithm (ECDSA) [24], EC El-Gammal and EC Diffie Hellman [12]. In each case the underlying cryptographic primitive is elliptic curve *scalar* multiplication. This operation is by far the most computationally intensive step in each algorithm. In applications where many clients authenticate to a single server (such as a server supporting SSL [7, 26] or WTLS [1]), the computation of the scalar multiplication becomes the bottle neck which limits throughput. In a scenario such as this it may be desirable to accelerate the elliptic curve scalar multiplication with specialized hardware. In doing so, the scalar multiplications are completed more quickly and the computational burden on the server's main processor is reduced.

The selection of the ECC parameters is not a trivial process and, if chosen incorrectly, may lead to an insecure system [12, 24, 22]. In response to this issue NIST recommends ten finite fields, five of which are binary fields, for use in the ECDSA [24]. The binary fields include GF(2^{163}), GF(2^{233}), GF(2^{283}), GF(2^{409}) and GF(2^{571}) defined by the reduction polynomials in Table 1. For each field a specific curve, along with

Table 1. NIST Recommended Finite Fields

Field	Reduction Polynomial
GF(2^{163})	$F(x) = x^{163} + x^7 + x^6 + x^3 + 1$
GF(2^{233})	$F(x) = x^{233} + x^{74} + 1$
GF(2^{283})	$F(x) = x^{283} + x^{12} + x^7 + x^5 + 1$
GF(2^{409})	$F(x) = x^{409} + x^{87} + 1$
GF(2^{571})	$F(x) = x^{571} + x^{10} + x^5 + x^2 + 1$

a method for generating a pseudo-random curve, are supplied. These curves have been intentionally selected for both cryptographic strength and efficient implementation.

Such a recommendation has significant implications on design choices made while implementing elliptic curve cryptographic functions. In standardizing specific fields for use in elliptic curve cryptography (ECC), NIST allows ECC implementations to be heavily optimized for curves over a single finite field. As a result, performance of the algorithm can be maximized and resource utilization, whether it be in code size for software or logic gates for hardware, can be minimized.

Described in this chapter are hardware architectures for multiplication, squaring and inversion over binary finite fields. Each of these architectures is optimized for a specific finite field with the intent that it might be implemented for any of the five NIST recommended binary curves. These finite field arithmetic units are then integrated together along with control logic to create an elliptic curve cryptographic co-processor capable of computing the scalar multiple of an elliptic curve point. While the co-processor supports all curves over a single binary field, it is optimized for the special Koblitz curves [16].

To demonstrate the feasibility and efficiency of both the finite field arithmetic units and the elliptic curve cryptographic co-processor, the latter has been implemented in hardware using a field programmable gate array (FPGA). The design was synthesized, timed and then demonstrated on a physical board holding an FPGA.

This chapter is organized as follows. Section 2 gives an overview of the basic mathematical concepts used in elliptic curve cryptography. This section also provides an introduction to the hardware/software system used to implement the elliptic curve scalar multiplier. Section 3 presents efficient hardware architectures for finite field multiplication and squaring. A method for high speed inversion is also discussed. In Section 4 and Section 5 a hardware architecture of an elliptic curve scalar multiplier is presented. This architecture uses the multiplication, squaring and inversion methods discussed in Section 3. Finally Section 6 provides concluding remarks and a summary of the research contributions documented in this report.

2. BACKGROUND

The fundamental building block for any elliptic curve-based cryptosystem is elliptic curve scalar multiplication. It is this operation that is to be performed by the co-processor. Provided in this section is an overview of the mathematics behind elliptic curve scalar multiplication, including both field arithmetic and curve arithmetic.

2.1. Arithmetic over Binary Finite Fields

The elements of the binary field GF(2^m) are interrelated through the operations of addition and multiplication. Since the additive and multiplicative inverses exist for all fields, the subtraction and division operations are also defined. Discussed in this section are basic methods for computing the sum, difference and product of two elements. Also presented is a method for computing the inverse of an element. The inverse, along with a multiplication, is used to implement division.

Addition and Subtraction: If two field elements $a, b \in$GF(2^m) are represented as polynomials $A(x) = a_{m-1}x^{m-1} + \cdots + a_1x + a_0$ and $B(x) = b_{m-1}x^{m-1} + \cdots + b_1x + b_0$ respectively, then their sum is written

$$S(x) = A(x) + B(x) = \sum_{i=0}^{m-1} (a_i + b_i)x^i. \tag{1}$$

A field of characteristic two provides two distinct advantages. First, the bit additions $a_i + b_i$ in (1) are performed modulo 2 and translate to an exclusive-OR (XOR) operation. The entire addition is computed by a component-wise XOR operation and does not require a carry chain. The second advantage is that in GF(2) the element 1 is its own additive inverse (i.e. $1 + 1 = 0$ or $1 = -1$). Hence, addition and subtraction are equivalent.

Multiplication: The product of field elements a and b is written as

$$P(x) = A(x) \times B(x) \mod F(x) = \sum_{i=0}^{m-1} \sum_{j=0}^{m-1} a_i b_j x^{i+j} \mod F(x)$$

where $F(x)$ is the field reduction polynomial. By expanding $B(x)$ and distributing $A(x)$ through its terms we get

$$P(x) = b_{m-1} x^{m-1} A(x) + \cdots + b_1 x A(x) + b_0 A(x) \mod F(x).$$

By repeatedly grouping multiples of x and factoring out x we get

$$\begin{aligned} P(x) = (\cdots(((A(x)b_{m-1})x + A(x)b_{m-2})x + \cdots + A(x)b_1)x \\ + A(x)b_0) \mod F(x). \end{aligned} \tag{2}$$

A bit level algorithm can be derived from (2). However, many of the faster multiplication algorithms rely on the concept of group-level multiplication. Let g be an integer less than m and let $s = \lceil m/g \rceil$. If we define the polynomials

$$B_i(x) = \begin{cases} \sum_{j=0}^{g-1} b_{ig+j} x^j & 0 \le i \le s-2, \\ \sum_{j=0}^{(m \mod g)-1} b_{ig+j} x^j & i = s-1, \end{cases}$$

then the product of a and b is written

$$P(x) = A(x) \left(x^{(s-1)g} B_{s-1}(x) + \cdots + x^g B_1(x) + B_0(x) \right) \mod F(x).$$

In the derivation of equation (2) multiples of x were repeatedly grouped then factored out. This same grouping and factoring procedure will now be implemented for multiples of x^g arriving at

$$\begin{aligned} P(x) = (\cdots((A(x)B_{s-1}(x))x^g + A(x)B_{s-2}(x))x^g + \cdots)x^g \\ + A(x)B_0(x) \mod F(x) \end{aligned}$$

which can be computed using Algorithm 1.

Algorithm 1. Group-Level Multiplication

Input: $A(x)$, $B(x)$, and $F(x)$

Output: $P(x) = A(x)B(x) \mod F(x)$

$P(x) \leftarrow B_{s-1}(x)A(x) \mod F(x)$;

for $k = s - 2$ **downto** 0 **do**

  $P(x) \leftarrow x^g P(x)$;

  $P(x) \leftarrow B_k(x)A(x) + P(x) \mod F(x)$;

Inversion: For any element $a \in$ GF(2^m) the equality $a^{2^m-1} \equiv 1$ holds. When $a \neq 0$, dividing both sides by a results in $a^{2^m-2} \equiv a^{-1}$. Using this equality the inverse, a^{-1}, can be computed through successive field squarings and multiplications. In Algorithm 2 the inverse of an element is computed using this method.

Algorithm 2. Inversion by Square and Multiply

Input: Field element a

Output: $b \equiv a^{(-1)}$

$b \leftarrow a$;

for $i = 1$ **to** $m - 2$ **do**

  $b \leftarrow b^2 * a$;

$b \leftarrow b^2$;

The primary advantage to this inversion method is the fact that it does not require hardware dedicated specifically to inversion. The field multiplier can be used to perform all required field operations.

2.2. Arithmetic over the Elliptic Curve Group

The field operations discussed in the previous section are used to perform arithmetic over an elliptic curve. This chapter is aimed at the elliptic curve defined by the non-supersingular Weierstrass equation for binary fields. This curve is defined by the equation

$$y^2 + xy = x^3 + \alpha x^2 + \beta \tag{3}$$

where the variables x and y are elements of the field GF(2^m) as are the curve parameters α and β. The points on the curve, defined by the solutions, (x, y), to (3) form an additive group when combined with the "point at infinity". This extra point is the group identity and is denoted by the symbol $\mathcal{O}$. By definition, the addition of two elements in a group results in another element of the group. As a result any point on the curve, say P, can be added to itself an arbitrary number of times and the result will also be a point on the curve. So for any integer k and point P adding P to itself $k - 1$ times results in the point

$$kP = \underbrace{P + P + \cdots + P}_{k \text{ times}}.$$

Given the binary expansion $k = 2^{l-1}k_{l-1} + 2^{l-2}k_{l-2} + \cdots + 2k_1 + k_0$ the scalar multiple kP can be computed by

$$Q = kP = 2^{l-1}k_{l-1}P + 2^{l-2}k_{l-2}P + \cdots + 2k_1P + k_0P.$$

By factoring out 2, the result is

$$Q = (2^{l-2}k_{l-1}P + 2^{l-3}k_{l-2}P + \cdots + k_1P)2 + k_0P.$$

By repeating this operation it is seen that

$$Q = (\cdots((k_{l-1}P)2 + k_{l-2}P)2 + \cdots + k_1P)2 + k_0P$$

which can be computed by the well known (left-to-right) double and add method for scalar multiplication shown in Algorithm 3.

Two basic operations required for elliptic curve scalar multiplication are point ADD and point DOUBLE. The mathematical definitions for these operations are derived from the curve equation in (3). Consider the points P_1 and P_2 represented by the coordinate pairs (x_1, y_1) and (x_2, y_2) respectively. Then the coordinates, (x_a, y_a), of point $P_a = P_1 + P_2$ (or ADD(P_1, P_2)) are computed using the equations

$$\begin{aligned} x_a &= \left(\frac{y_1 + y_2}{x_1 + x_2}\right)^2 + \frac{y_1 + y_2}{x_1 + x_2} + x_1 + x_2 + \alpha \\ y_a &= \left(\frac{y_1 + y_2}{x_1 + x_2}\right)(x_1 + x_a) + x_a + y_1. \end{aligned}$$

Similarly the coordinates (x_d, y_d) of point $P_d = 2P_1$ (or DOUBLE(P_1)) are computed using the equations

$$\begin{aligned} x_d &= x_1^2 + \left(\frac{\beta}{x_1^2}\right) \\ y_d &= x_1^2 + \left(x_1 + \frac{y_1}{x_1}\right)x_d + x_d. \end{aligned}$$

Algorithm 3. Scalar Multiplication by Double and Add Method

Input: Integer $k = (k_{l-1}, k_{l-2}, \ldots, k_1, k_0)_2$, Point P

Output: Point $Q = kP$

$Q \leftarrow \mathcal{O}$;

if $(k_{l-1} == 1)$ **then**

$Q \leftarrow P$;

for $i = l - 2$ **downto** 0 **do**

$Q \leftarrow \texttt{DOUBLE}(Q)$;

if $(k_i == 1)$ **then**

$Q \leftarrow \text{ADD}(Q, P)$;

So the addition of two points can be computed using two field multiplications, one field squaring, eight field additions and one field inversion. The double of a point can be computed using two field multiplications, one field squaring, six field additions and one field inversion.

3. HIGH PERFORMANCE FINITE FIELD ARITHMETIC

In order to optimize the curve arithmetic discussed in Section 2.2 the underlying field operations must be implemented in a fast and efficient way. The required field arithmetic operations are addition, multiplication, squaring and inversion. Each of these operations have been implemented in hardware for use in the prototype discussed in Section 5. Generally speaking, field multiplication has the greatest effect on the performance of the entire elliptic curve scalar multiplication.[1] For this reason, focus will be primarily on the field multiplier when discussing hardware architectures for field arithmetic.

This section is organized as follows. Section 3.1 presents a hardware architecture designed to perform finite field multiplication. In Section 3.2 the ideas presented for multiplication are extended to create a hardware architecture optimized for squaring. Section 3.3 gives a method for inversion due to Itoh and Tsujii. This method does not require any additional hardware but instead uses the multiplication and squaring units described in Sections 3.1 and 3.2. Section 3.4 gives a description of a comparator/adder

[1] Inversion takes much longer than multiplication, but its effect on performance can be greatly reduced through use of projective coordinates. This is discussed in greater detail in Section 4.1.

which both compares and adds finite field elements. Finally, Section 3.5 summarizes results gleaned from a hardware prototype of each arithmetic unit/routine.

3.1. Multiplication

In [11] a digit serial multiplier is proposed which is based on look-up tables. This method was implemented in software for the field GF(2^{163}) and reported in [14]. To the best of our knowledge this performance of 0.540 μ-seconds for a single field multiplication is the fastest reported result for a software implementation. In this section the possibilities of using this look-up table-based algorithm in hardware will be explored.

First to be described in this section is the algorithm used for multiplication. Then we present a hardware structure designed to compute $R(x)W(x) \mod F(x)$ where $R(x)$ and $W(x)$ are polynomials with degrees $g-1$ and $m-1$ respectively and $g << m$. A description of the multiplier's data path follows. In conclusion there will be a discussion behind the reasons for the choice of digit sizes.

Multiplication Algorithm: The computations of

$$\begin{aligned} P(x) &\leftarrow x^g P(x) \mod F(x) \text{ and} \\ P(x) &\leftarrow B_k(x)A(x) + P(x) \mod F(x) \end{aligned}$$

from the **for** loop of Algorithm 1 on page 7 can be broken up into the following steps.

$$\begin{aligned} V_1 &= x^g \sum_{i=0}^{m-g-1} p_i x^i, \\ V_2 &= x^g \sum_{i=m-g}^{m-1} p_i x^i \mod F(x) \\ V_3 &= B_k(x)A(x) \mod F(x) \text{ and} \\ P(x) &= V_1 + V_2 + V_3 \end{aligned}$$

Note that V_1 is a g-bit shift of the lower $m-g$ bits of $P(x)$. V_2 is a g-bit shift of the upper g bits of $P(x)$ followed by a modular reduction. V_3 requires a polynomial multiplication and reduction where the operand polynomials have degree $g-1$ and $m-1$. Algorithm 1 can be modified to create Algorithm 4.

In [11] polynomials V_2 and V_3 are computed with the assistance of look-up tables mainly for software implementation. The look-up tables used to compute V_2 and V_3 are referred to as the M-Table and T-Table respectively. The M-Table is addressed by the bit string $(p_{m-1}, p_{m-2}, \ldots, p_{m-g})$ interpreted as the integer $2^{g-1}p_{m-1} + 2^{g-2}p_{m-2} + \cdots + p_{m-g}$. Similarly the T-Table is addressed by the coefficients of $B_k(x)$, or the integer $B_k(x=2)$. The elements of the M-Table are a function of the reduction polynomial $F(x)$ and can be precomputed. The elements of the T-Table are a function

Algorithm 4. Efficient Group Level Multiplication

Input: $A(x)$, $B(x)$, and $F(x)$

Output: $P(x) = A(x)B(x) \mod F(x)$

$P(x) \leftarrow B_{s-1}(x)A(x) \mod F(x)$;

for $k = s - 2$ **downto** 0 **do**

$V_1 \leftarrow x^g \sum_{i=0}^{m-g-1} p_i x^i$;

$V_2 \leftarrow x^g \sum_{i=m-g}^{m-1} p_i x^i \mod F(x)$;

$V_3 \leftarrow B_k(x)A(x) \mod F(x)$;

$P(x) \leftarrow V_1 + V_2 + V_3$;

of $A(x)$ and hence are dynamic. These values need to be computed each time a new $A(x)$ is used.

Computation of $R(x)W(x) \mod F(x)$: Instead of using tables, below the polynomials V_2 and V_3 are computed on the fly. The computation of V_2 and V_3 are similar in that they both require a multiplication of two polynomials followed by a reduction, where the first polynomial has degree $g - 1$ and the other has degree less than m. This is obvious for V_3 and can be shown easily for V_2. Note that

$$V_2 = p_{m-1}x^{m+g-1} + \cdots + p_{m-g+1}x^{m+1} + p_{m-g}x^m \mod F(x)$$
$$= x^m \left(p_{m-1}x^{g-1} + \cdots + p_{m-g+1}x + p_{m-g}\right) \mod F(x).$$

The field reduction polynomial $F(x) = x^m + x^d + \cdots + 1$ provides us the equality $x^m \equiv x^d + \cdots + 1$. Substituting for x^m we see that

$$V_2 = \left(x^d + \cdots + 1\right)\left(p_{m-1}x^{g-1} + \cdots + p_{m-g+1}x + p_{m-g}\right) \mod F(x).$$

Provided $d + g < m$, V_2 results in a polynomial of degree less than m which does not need to be reduced. Since d is relatively small for all five NIST polynomials, it is reasonable to assume that $d+g < m$. For the remainder of this chapter, this assumption is used.

With this said, the following method can be used to compute both V_2 and V_3. Consider the polynomial multiplication and reduction $R(x)W(x) \mod F(x)$ where

$R(x) = \sum_{i=0}^{g-1} r_i x^i$ and $W(x)$ is a polynomial with degree less than m. Then

$$\begin{aligned} R(x)W(x) \mod F(x) = & r_{g-1}(x^{g-1}W(x) \mod F(x)) \\ & + r_{g-2}(x^{g-2}W(x) \mod F(x)) \\ & \vdots \\ & + r_1(xW(x) \mod F(x)) \\ & + r_0(W(x) \mod F(x)) \end{aligned}$$

The value $x^iW(x) \mod F(x)$ is just a shifted and reduced version of $x^{i-1}W(x) \mod F(x)$. So each value $x^iW(x) \mod F(x)$ can be generated sequentially starting with $x^0W(x)$ as shown in Figure 1. When using a reduction polynomial with a low Hamming weight, such as a trinomial or pentanomial, these terms can be computed quickly at very little cost. Once these values are determined, the final result is computed using a g-input modulo 2 adder. The inputs to the adder are enabled by their corresponding coefficient r_i. This is shown in Figure 2. Note that the polynomial $x^iW(x)$ affects the output of the adder only if the coefficient bit r_i is a one. Otherwise the input associated with $x^iW(x)$ is driven with zeros.

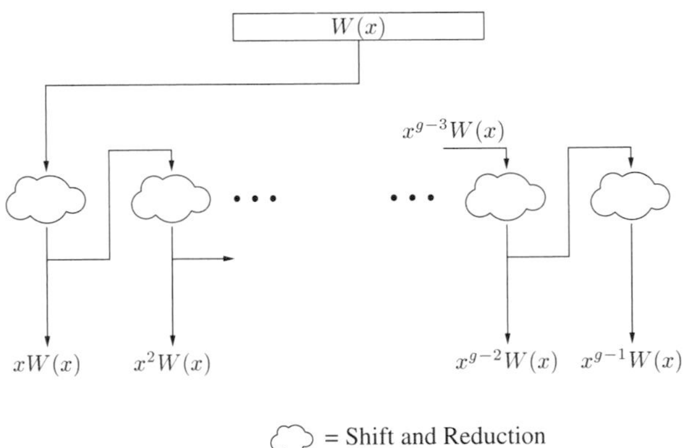

Figure 1. Generating $x^iW(x) \mod F(x)$

Each individual output bit of the g-operand mod 2 adder is computed using $g-1$ XOR gates and g AND gates. The AND gates are used to enable each input bit and the XOR gates compute the mod 2 addition. Figure 3 demonstrates how this is done. The depth of the logic in the figure is linearly related to g.

This method for multiplication is implemented for computation of both V_2 and V_3. In the case of V_3, the polynomial $W(x)$ has degree $m-1$ and will change for every

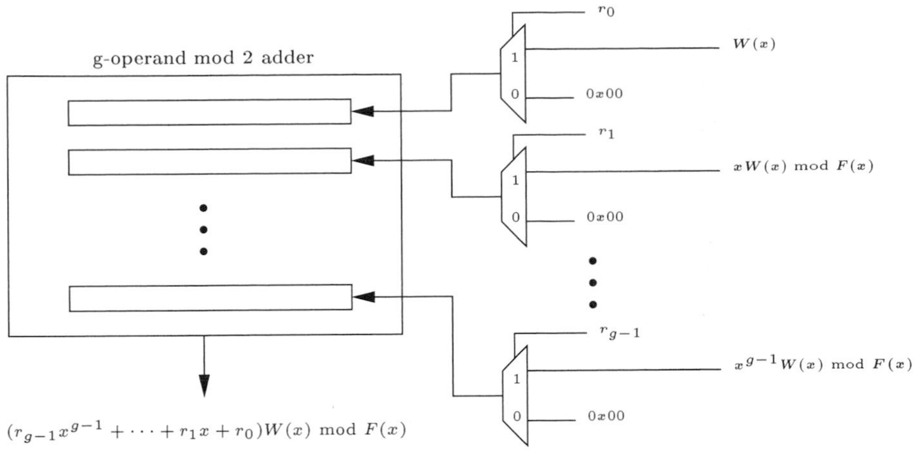

Figure 2. Computing $R(x)W(x) \mod F(x)$

field multiplication. For V_2 the polynomial $W(x)$ has degree d and is fixed. The value d is the degree of the second leading non-zero coefficient of $F(x)$. For reasonable digit sizes this computation can be performed in a single clock cycle.

Multiplier Data Path: The multiplier's data path connecting the V_2 and V_3 generators along with the adder used to compute $P(x) = V_1 + V_2 + V_3$ is shown in Figure 4. A buffer is inserted at the output of the V_3 generator to separate its delay from the delay of the adder for $V_1 + V_2 + V_3$. This, in effect, increases the maximum possible value for the digit size g. If added by itself, this buffer would add a cycle of latency to the multiplier's performance time. This extra cycle is compensated for by bypassing the $P(x)$ register and driving the multiplier's output with the output of the 3-operand mod2 adder. It is important to note that the delay of the 3-operand mod2 adder is being merged with the delay of the bus which connects the multiplier to the rest of the design. In this case the relatively relaxed bus timing has room to accommodate the delay.

Choice of Digit Size: The multiplier will complete a multiplication in $\lceil m/g \rceil$ clock cycles. Since this is a discrete value, the performance may not change for every value of g. To minimize cost of the multiplier (which increases with g) the smallest digit size g should be chosen for a given performance $\lceil m/g \rceil$. For example, the digit sizes $g = 21$ and $g = 22$ for field size $m = 163$ result in the same performance, $\lceil \frac{163}{21} \rceil = \lceil \frac{163}{22} \rceil = 8$, but $g = 22$ requires a larger multiplier.

Implementation results of a prototype of this multiplier for the field GF(2^{163}) and NIST polynomial for various digit sizes are shown in Table 2. For each digit size, the table lists the corresponding cycle performance and resource cost. A maximum digit

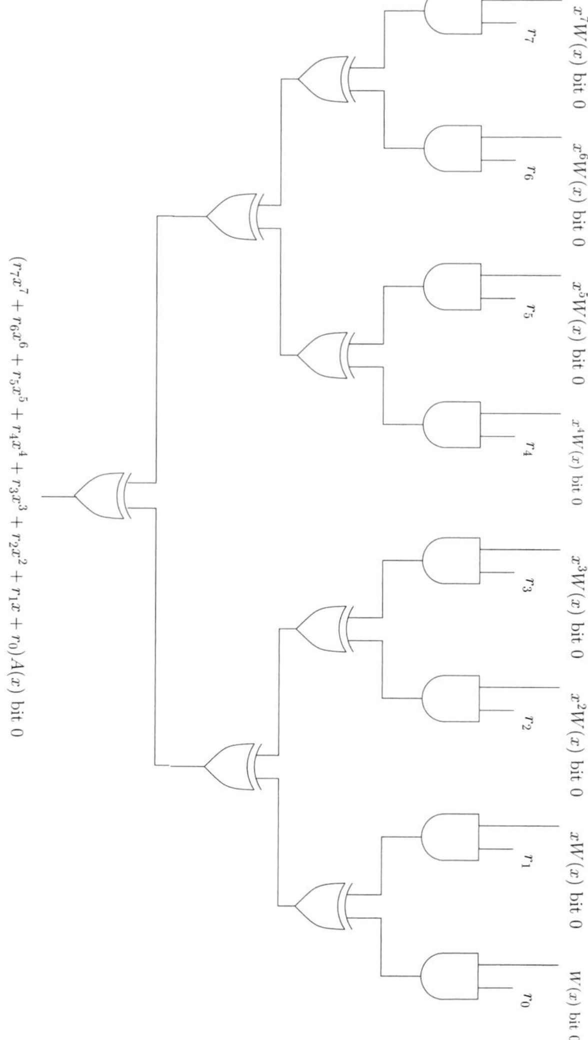

Figure 3. Computation of a Single Bit in $R(x)W(x) \mod F(x)$

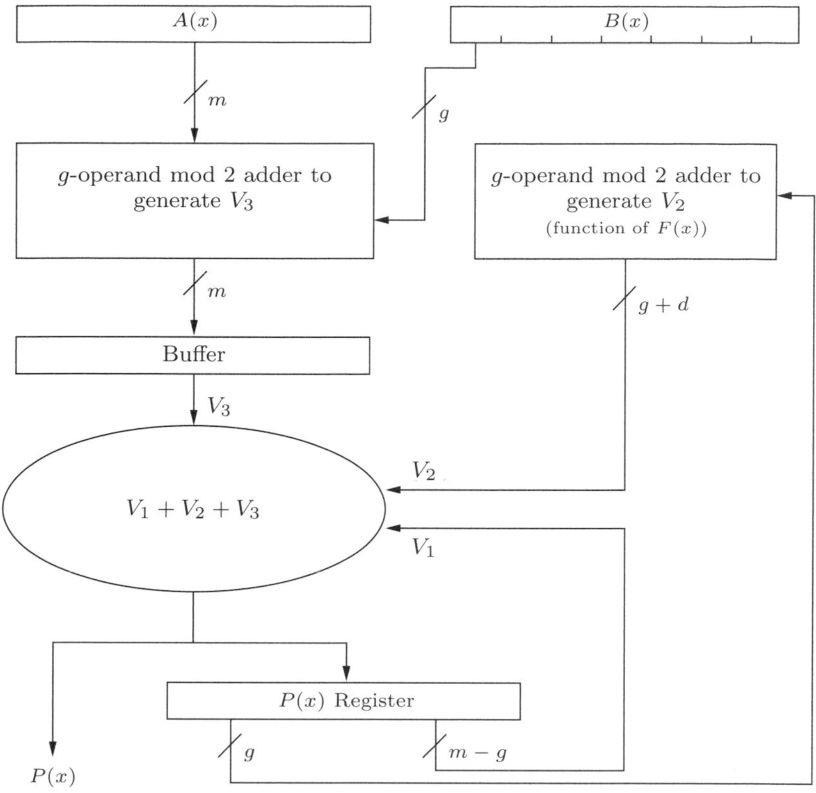

Figure 4. Multiplier Data-Path

size of $g = 41$ is a good choice for several reasons. First, as the performance cost of the actual field multiplication decreases, the relative cost of loading and unloading the multiplier increases. So as the digit size increases, its affect on the total performance (including time to load and unload the multiplier) decreases. Second, results showed that $g > 41$ had difficulty meeting timing at the target operating frequency of 66 MHz.

3.2. Squaring

While squaring is a specific case of general multiplication and can be performed by the multiplier, performance can be improved significantly by optimizing the architecture specifically for the case of squaring. The square of an element a represented by $A(x)$ involves two mathematical steps. The first is the polynomial multiplication of $A(x)$ resulting in

$$A^2(x) = a_{m-1}x^{2m-2} + \cdots + a_2x^4 + a_1x^2 + a_0.$$

Table 2. Performance/Cost Trade-off for Multiplication over GF(2^{163})

Digit Size	Performance in clock cycles	# LUTs	# Flip Flops
$g = 1$	163	677	670
$g = 4$	41	854	670
$g = 28$	6	3,548	670
$g = 33$	5	4,040	670
$g = 41$	4	4,728	670

The second is the reduction of this polynomial modulo $F(x)$. Assuming that m is an odd integer, which is the case for all five NIST recommended binary fields, if the terms with degree greater than $m - 1$ are separated and x^{m+1} is factored out where possible the result will be $A^2(x) = A_h(x)x^{m+1} + A_l(x)$ where

$$A_h(x) = a_{m-1}x^{m-3} + \cdots + a_{\left(\frac{m+3}{2}\right)}x^2 + a_{\left(\frac{m+1}{2}\right)}$$
$$A_l(x) = a_{\left(\frac{m-1}{2}\right)}x^{m-1} + \cdots + a_1x^2 + a_0,$$

The polynomial $A_l(x)$ has degree less than m and does not need to be reduced. The product $A_h(x)x^{m+1}$ may have degree as large as $2m - 2$. The reduction polynomial gives us the equality $x^m = x^d + \cdots + 1$. Multiplying both sides by x, we get $x^{m+1} = x^{d+1} + \cdots + x$. So

$$A_h(x)x^{m+1} = A_h(x)\left(x^{d+1} + \cdots + x\right).$$

This multiplication can be performed using a method similar to the one described in Section 3.1. The same architecture used to compute $R(x)W(x) \mod F(x)$ in the multiplier is used here to compute $x^{m+1}A_h(x)$. The digit size is set to $g = d + 2$ and the elements of g-operand mod 2 adder are generated from $A_h(x)$. $A_h(x)$ is in turn generated by expanding $A(x)$ (i.e., inserting zeros between the coefficient bits of $A(x)$). Since the digit size is set to $d + 2$, the multiplication is completed in a single cycle. This method only works if $d + 2 < m$ which is the case for each of the NIST polynomials. Figure 5 shows the data flow for the squaring operation. Note that the flow does not include any buffers and so is implemented in pure combinational logic.

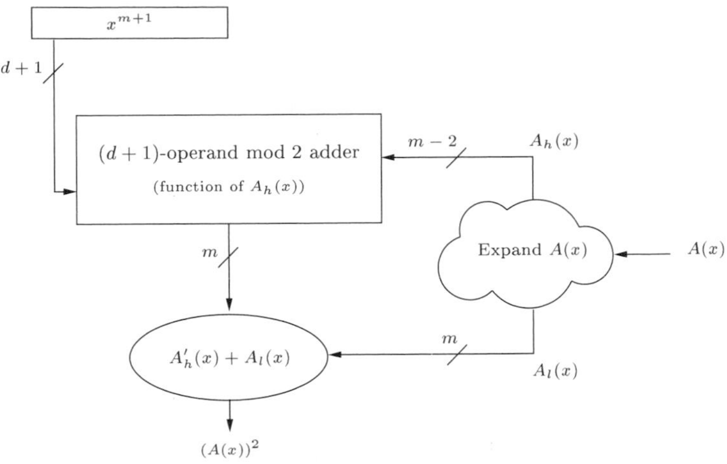

Figure 5. Data-Path of the Squaring Unit

The prototype of this squaring unit for field GF(2^{163}) using the NIST reduction polynomial runs at 66 MHz and is capable of performing a squaring operation in a single clock cycle. This implementation requires 330 LUTs and 328 Flip Flops.

3.3. Inversion

The inversion method described in Algorithm 2 on page 7 requires $m-1$ squarings and $m-2$ multiplications. In order to accurately estimate the cycle performance of the inversion, consideration must be given to the performance of the multiplication and squaring units as well as the time required to load and unload these units. The architecture of the elliptic curve scalar multiplier will be discussed in detail in Section 5. For now, it is sufficient to know that the arithmetic units are loaded using two independent m bit data buses and unloaded using a single m bit data bus. The operands are stored in a dual port memory which takes two clock cycles to read from and one cycle to write to. These combined makes three cycles that are required to both load and unload any arithmetic unit. Further analysis assumes that these three cycles remain constant for all m. If C_s and C_m denote the number of clock cycles required to complete a squaring and multiplication respectively, then an inversion can be completed in

$$(C_s + 3)(m-1) + (C_m + 3)(m-2)$$

clock cycles. For the field GF(2^{163}) where $C_s = 1$ and $C_m = 4$, this translates to 1775 clock cycles.

Performance can be improved by using Algorithm 5 due to Itoh and Tsujii [13]. This algorithm is derived from the equation $a^{(-1)} \equiv a^{2^m - 2} \equiv \left(2^{2^{m-1}}{}_{-1}\right)^2$

which is true for any non-zero element $a \in$GF(2^m). From

$$a^{2^t-1} \equiv \begin{cases} \left(a^{2^{t/2}-1}\right)^{2^{t/2}} \left(a^{2^{t/2}-1}\right) & \text{for } t \text{ even,} \\ a\left(a^{2^{t-1}-1}\right)^2 & \text{for } t \text{ odd,} \end{cases} \tag{4}$$

the computation required for the exponentiation $2^{2^{m-1}-1}$ can be iteratively broken down. Algorithm 5 requires $\lfloor \log_2(m-1) \rfloor + H(m-1) - 1$ multiplications and $m-1$ squarings. Using the notation defined earlier, this translates to

$$(C_s + 3)(m-1) + (C_m + 3)(\lfloor \log_2(m-1) \rfloor + H(m-1) - 1)$$

clock cycles. For GF(2^{163}) this translates to 711 clock cycles.

Algorithm 5. Optimized Inversion by Square and Multiply

Inputs: Field element $a \neq 0$,

Binary representation of $m - 1 = (m_{l-1}, \ldots, m_2, m_0)_2$

Output: $b \equiv a^{(-1)}$

$b \leftarrow a^{m_{l-1}}$;

$e \leftarrow 1$;

for $i = l - 2$ **downto** 0 **do**

$b \leftarrow b^{2^e} b$;

$e \leftarrow 2e$;

if $(m_i == 1)$ **then**

$b \leftarrow b^2 a$;

$e = e + 1$;

$b \leftarrow b^2$;

Now, the majority of the time spent for each squaring operation is used to load and unload the squaring unit (three out of the four cycles). Algorithm 5 requires several sequences of repetitive squaring (i.e. computations of the form x^{2^t}). These repeated squarings do not require intermediate values to be stored outside the squaring unit. By modifying the squaring unit to support the *re-square* of an element, most of the memory accesses otherwise required to load and unload the squaring unit are eliminated. In fact, the squaring unit only needs to be loaded and unloaded once for each multiplication. Hence the number of clock cycles is reduced to

$$(C_s(m-1) + 3(\lfloor \log_2(m-1) \rfloor + H(m-1) - 1)) \\ + (C_m + 3)(\lfloor \log_2(m-1) \rfloor + H(m-1) - 1)$$

clock cycles. For the field GF(2^{163}) with $C_s = 1$ and $C_m = 4$, this results in 252 clock cycles.

This is a competitive value since a typical hardware implementation of the Extended Euclidean Algorithm (EEA) is expected to complete an inversion in approximately $2m$ clock cycles or 326 cycles for GF(2^{163}). This corresponds to a 60 clock cycle reduction or 20% performance improvement without requiring hardware dedicated specifically for inversion. Table 3 lists the performance numbers of the previously mentioned inversion methods when implemented over the field GF(2^{163}).

Table 3. Comparison of Various Inversion Methods for GF(2^{163})

Method	# Squarings	# Multiplications	# Cycles
Square & Multiply	$m-1$	$m-2$	1127
Itoh & Tsujii	$m-1$	$\lfloor \log_2(m-1) \rfloor + H(m) - 1$	711
Itoh & Tsujii w/ *re-square*	$m-1$	$\lfloor \log_2(m-1) \rfloor + H(m) - 1$	252
EEA	-	-	326

The actual time to complete an inversion using the ECC co-processor architecture discussed in Section 5 is 259 clock cycles. The 7 extra cycles are due to control related instructions executed in the micro-sequencer.

3.4. Comparator/Adder

The primary purpose of the Comparator/Adder is to compute the sum of two field elements. This is done with an array of m exclusive OR gates. To minimize register usage as well as time to complete the addition, the sum of the two operands is the only value stored in a register. In this way, the sum is available immediately after the operands are loaded into the Comparator/Adder. In other words, it takes no extra clock cycles to complete a finite field addition.

In addition to computing the sum of two finite field elements, the Comparator/Adder also acts as a comparator. The comparison is performed by taking the logical NOR of all the bits in the sum register. If the result is a one, then the sum is zero and the two operands are equal. If operand a is set to zero, then operand b can be tested for zero.

The logic depth for the zero detect circuitry (the m-bit NOR gate) is $\log_2(m)$ and is registered before being sent out of the module. Figure 6 provides a functional diagram of the Comparator/Adder.

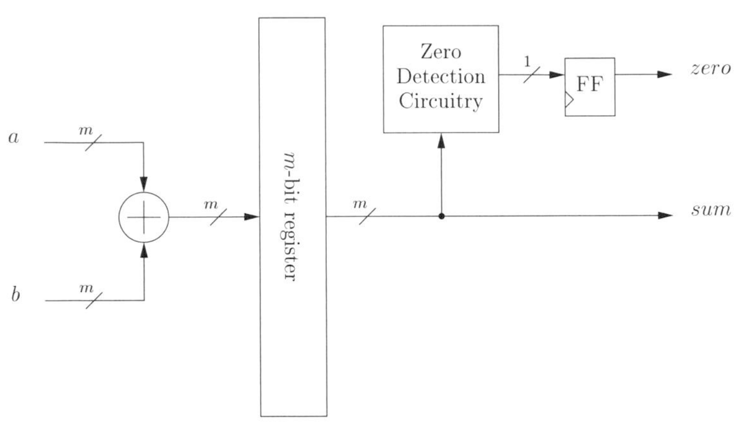

Figure 6. Data-Path of the Comparator/Adder

3.5. Remarks

In this section, we have discussed hardware architectures designed to perform finite field addition, multiplication and squaring. Also discussed was an efficient method for inversion which uses the squaring and multiplication units. The performance results associated with these arithmetic units are summarized in Table 4.

4. ECC SCALAR MULTIPLICATION

The section is organized as follows. Section 4.1 introduces projective coordinates and discusses some of the reasons for using a projective system. Section 4.2 presents two methods for recoding the scalar. They are non-adjacent form (NAF) and τ-adic non-adjacent form (τ-NAF).

4.1. Choice of Coordinate Systems

Projective coordinates allow the inversion required by each `DOUBLE` and `ADD` to be eliminated at the expense of a few extra field multiplications. The benefit is measured by the ratio of the time to complete an inversion to the time to complete a multiplication. The inversion algorithm proposed by Itoh and Tsujii [13] will be used

Table 4. Performance of Finite Field Operations

Operation ($g = 41$)	# Cycles	# Cycles Including Initial and Final Data Movement
Multiplication	4	7
Squaring	1	4
Addition	0	3
Inversion	256	259

and therefore, the above ratio is guaranteed to be larger than $\lfloor \log_2(m-1) \rfloor$ and could be larger depending on the efficiency of the squaring operations. Therefore, projective coordinates will provide us the best performance for NIST curves. Several flavors of projective coordinates have been proposed over the last few years. The prominent ones are *Standard* [21], *Jacobian* [4, 12] and López & Dahab [18] projective coordinates.

If the affine representation of P be denoted as (x, y) and the projective representation of P be denoted as (X, Y, Z), then the relation between affine and projective coordinates for the Standard system is

$$x = \tfrac{X}{Z} \quad \text{and} \quad y = \tfrac{Y}{Z}.$$

For Jacobian projective coordinates the relation is

$$x = \tfrac{X}{Z^2} \quad \text{and} \quad y = \tfrac{Y}{Z^3}.$$

Finally for López & Dahab's, the relation between affine and projective coordinates is

$$x = \tfrac{X}{Z} \quad \text{and} \quad y = \tfrac{Y}{Z^2}.$$

For López & Dahab's system the projective equation of the elliptic curve in (3) then becomes

$$Y^2 + XYZ = X^3Z + \alpha X^2 Z^2 + \beta Z^4.$$

It is important to note that when using the left-to-right double and add method for scalar multiplication all point additions are of the form ADD(P, Q). The base point P is never modified and as a result will maintain its affine representation (i.e. $P = (x, y, 1)$). The constant Z coordinate significantly reduces the cost of point addition (from 14 field multiplications down to 10). The addition of two distinct points $(X_1, Y_1, Z_1) + (X_2, Y_2, 1) = (X_a, Y_a, Z_a)$ using *mixed* coordinates (one projective point and one affine point) is then computed by

$$\begin{array}{ll} A = Y_2 \cdot Z_1^2 + Y_1 & E = A \cdot C \\ B = X_2 \cdot Z_1 + X_1 & X_a = A^2 + D + E \\ C = Z_1 \cdot B & F = X_a + X_2 \cdot Z_a \\ D = B^2 \cdot (C + \alpha \cdot Z_1^2) & G = X_a + Y_2 \cdot Z_a \\ Z_a = C^2 & Y_a = E \cdot F + Z_a \cdot G \end{array} \tag{5}$$

Similarly, the double of a point (X_1, Y_1, Z_1) is $(X_d, Y_d, Z_d) = 2(X_1, Y_1, Z_1)$ is computed by

$$\begin{aligned} Z_d &= Z_1^2 \cdot X_1^2 \\ X_d &= X_1^4 + \beta \cdot Z_1^4 \\ Y_d &= \beta \cdot Z_1^4 \cdot Z_d + X_d \cdot (\alpha \cdot Z_d + Y_1^2 + \beta \cdot Z_1^4) \end{aligned} \tag{6}$$

In Table 5, the number of field operations required for the affine, Standard, Jacobean and López & Dahab coordinate systems are provided. In the table the symbols $\mathcal{M}$, $\mathcal{S}$, $\mathcal{A}$ and $\mathcal{I}$ denote field multiplication, squaring, addition and inversion respectively.

Table 5. Comparison of Projective Point Systems

System	Point Addition	Point Doubling
Affine	$2\mathcal{M} + 1\mathcal{S} + 8\mathcal{A} + 1\mathcal{I}$	$3\mathcal{M} + 2\mathcal{S} + 4\mathcal{A} + 1\mathcal{I}$
Standard	$13\mathcal{M} + 1\mathcal{S} + 7\mathcal{A}$	$7\mathcal{M} + 5\mathcal{S} + 4\mathcal{A}$
Jacobian	$11\mathcal{M} + 4\mathcal{S} + 7\mathcal{A}$	$5\mathcal{M} + 5\mathcal{S} + 4\mathcal{A}$
López & Dahab	$10\mathcal{M} + 4\mathcal{S} + 8\mathcal{A}$	$5\mathcal{M} + 5\mathcal{S} + 4\mathcal{A}$

The projective coordinate system defined by López and Dahab will be used since it offers the best performance for both point addition and point doubling.

4.2. Scalar Multiplication using Recoded Integers

The binary expansion of an integer k is written as $k = \sum_{i=0}^{l-1} k_i 2^i$ where $k_i \in \{0, 1\}$. For the case of elliptic curve scalar multiplication the length l is approximately equal to m, the degree of the extension field. Assuming an average Hamming weight, a scalar multiplication will require approximately $l/2$ point additions and $l - 1$ point doubles. Several recoding methods have been proposed which in effect reduce the number of additions. In this section two methods are discussed, namely NAF [9, 29] and τ-adic NAF [16, 29].

Scalar Multiplication using Binary NAF: The symbols in the binary expansion are selected from the set $\{0, 1\}$. If this set is increased to $\{0, 1, -1\}$ the expansion is referred to as *signed binary* (SB) representation. When using this representation, the double and add scalar multiplication method must be slightly modified to handle the -1 symbol (often denoted as $\bar{1}$). If the expansion $k'_{l-1}2^{l-1} + \cdots + k'_1 2 + k'_0$ where $k'_i \in \{0, 1, \bar{1}\}$ is denoted by $(k'_{l-1}, \ldots, k'_1, k'_0)_{SB}$, then Algorithm 6 computes the scalar multiple of point P. The negative of the point (x, y) is $(x, x + y)$ and can be computed

Algorithm 6. Scalar Multiplication for Signed Binary Representation

Input: Integer $k = (k'_{l-1}, k'_{l-2}, \ldots, k'_1, k'_0)_{\text{SB}}$, Point P

Output: Point $Q = kP$

$Q \leftarrow \mathcal{O}$;

if $(k'_{l-1} \neq 0)$ **then**

$Q \leftarrow k'_{l-1}P$;

for $i = l - 2$ **downto** 0 **do**

$Q \leftarrow \text{DOUBLE}(Q)$;

if $(k'_i \neq 0)$ **then**

$Q \leftarrow \text{ADD}(Q, k'_i P)$;

with a single field addition. The signed binary representation is redundant in the sense that any given integer has more than one possible representation. For example, 17 can be represented by $(1001)_{\text{SB}}$ as well as $(101\bar{1})_{\text{SB}}$.

Interest here is in a particular form of this signed binary representation called NAF or non-adjacent form. A signed binary integer is said to be in NAF if there are no adjacent non-zero symbols. The NAF of an integer is unique and it is guaranteed to be no more than one symbol longer than the corresponding binary expansion. The primary advantage gained from NAF is its reduced number of non-zero symbols. The average Hamming weight of a NAF is approximately $l/3$ [29] compared to that of the binary expansion which is $l/2$. As a result, the running time of elliptic curve scalar multiplication when using binary NAF is reduced to $(l + 1)/3$ point additions and l point doubles. This represents a significant reduction in run time.

In [29], Solinas provides a straightforward method for computing the NAF of an integer. This method is given here in Algorithm 7.

Algorithm 7. Generation of Binary NAF

Input: Positive integer k

Output: $k' = \text{NAF}(k)$

```
i ← 0;
while (k > 0) do
    if (k ≡ 1 (mod 2)) then
        k'_i ← 2 − (k mod 4);
        k ← k − k'_i;
    else
        k'_i ← 0;
    k ← k/2;
    i ← i + 1;
```

Scalar Multiplication using τ-NAF: Anomalous Binary Curves (ABC's), first proposed for cryptographic use in [16], provide an efficient implementation when the scalar is represented as a complex algebraic number. ABC's, often referred to as the Koblitz curves, are defined by

$$y^2 + xy = x^3 + \alpha x^2 + 1 \tag{7}$$

with $\alpha = 0$ or $\alpha = 1$. The advantage provided by the Koblitz curves is that the DOUBLE operation in Algorithm 6 can be replaced with a second operation, namely Frobenius mapping, which is easier to perform.

If point (x, y) is on a Koblitz curve then it can be easily checked that (x^2, y^2) is also on the same curve. Moreover, these two points are related by the following Frobenius mapping

$$\tau(x, y) = (x^2, y^2)$$

where τ satisfies the quadratic equation

$$\tau^2 + 2 = \mu\tau. \tag{8}$$

In (8), $\mu = (-1)^{1-\alpha}$ and α is the curve parameter in (7) and is 0 or 1 for the Koblitz curves.

The integer k can be represented with radix τ using signed representation. In this case, the expansion is written

$$k = \kappa_{l-1}\tau^{l-1} + \cdots \kappa_1\tau + \kappa_0,$$

where $\kappa_i \in \{0, 1, \bar{1}\}$. Using this representation, Algorithm 6 can be rewritten, replacing the DOUBLE(Q) operation with τQ or a Frobenius mapping of Q. The modified algorithm is shown in Algorithm 8. Since τQ is computed by squaring the coordinates of Q, this suggests a possible speed up over the DOUBLE and ADD method.

Algorithm 8. Scalar Multiplication for τ-adic Integers

Input: Integer $k = (\kappa_{l-1}, \kappa_{l-2}, \ldots, \kappa_1, \kappa_0)_\tau$, Point P

Output: Point $Q = kP$

$Q \leftarrow \mathcal{O}$;

if $(\kappa_{l-1} \neq 0)$ **then**

  $Q \leftarrow \kappa_{l-1}P$;

for $i = l - 2$ **downto** 0 **do**

  $Q \leftarrow \tau Q$;

  if $(\kappa_i \neq 0)$ **then**

    $Q \leftarrow$ ADD$(Q, \kappa_i P)$;

This complex representation of the integer can be improved further by computing its non-adjacent form. Solinas proved the existence of such a representation in [29] by providing an algorithm which computes the τ-adic non-adjacent form or τ-NAF of an integer. This algorithm is provided here in Algorithm 9. In most cases, the input to Algorithm 9 will be a binary integer, say k (i.e. $r_0 = k$ and $r_1 = 0$). If k has length l then TNAF(k) will have length $2l$, roughly twice the length of NAF(k).

The length of the representation generated by Algorithm 9 can be reduced by either preprocessing the integer k, as is done in [29], or by post processing the result. A method for post processing the output of Algorithm 9 is presented here.

Remember that $\tau(x, y) = (x^2, y^2)$. Since $z^{2^m} = z$ for all $z \in$GF(2^m), it follows that

$$\tau^m(x, y) = (x^{2^m}, y^{2^m}) = (x, y).$$

This relation gives us the general equality

$$(\tau^m - 1)P \equiv 0$$

Algorithm 9. Generation of τ-adic NAF

Input: $r_0 + r_1\tau$ where $r_0, r_1 \in \mathbb{Z}$

Output: $u =$TNAF$(r_0 + r_1\tau)$

```
i ← 0;
while (r0 ≠ 0 or r1 ≠ 0) do
    if (r0 ≡ 1 (mod 2)) then
        ui ← 2 − (r0 − 2r1 mod 4);
        r0 ← r0 − ui;
    else
        ui ← 0;
    t ← r0;
    r0 ← r1 + μr0/2;
    r1 ← −t/2;
    i ← i + 1;
```

where P is a point on a Koblitz curve. As a result, any integer k expressed with radix τ can be reduced modulo $\tau^m - 1$ without changing the scalar multiple kP. This reduction is performed easily with a few polynomial additions. Consider the τ-adic integer

$$u = u_{2m-1}\tau^{2m-1} + \cdots + u_{m+1}\tau^{m+1} + u_m\tau^m + u_{m-1}\tau^{m-1} + \cdots + u_1\tau + u_0.$$

Factoring out τ^m wherever possible, the result is

$$\begin{aligned} u = {} & (u_{2m-1}\tau^{m-1} + \cdots + u_{m+1}\tau + u_m)\tau^m \\ & +(u_{m-1}\tau^{m-1} + \cdots + u_1\tau + u_0) \end{aligned}$$

Substituting τ^m with 1 and combining terms results in

$$u = ((u_{2m-1} + u_{m-1})\tau^{m-1} + \cdots + (u_{m+1} + u_1)\tau + (u_m + u_0).$$

The output of Algorithm 9 is approximately twice the length of the input but may be slightly larger. Assuming the length of the input to be approximately m symbols, the reduction method must be capable of reducing τ-adic integers with length slightly greater $2m$. Algorithm 10 describes this method for reduction.

Algorithm 10. Reduction mod τ^m

Input: $u = u_{l-1}\tau^{l-1} + \cdots + u_1\tau + u_0$ with $m \leq l < 3m$

Output: v =REDUCE_TM(u)

$v \leftarrow 0$;

if $(l > 2m)$ **then**

$v \leftarrow (u_{l-1}\tau^{l-2m-1} + \cdots + u_{2m+1}\tau + u_{2m})$;

if $(l > m)$ **then**

$v \leftarrow v + (u_{2m-1}\tau^{m-1} + \cdots + u_{m+1}\tau + u_m)$;

$v \leftarrow v + (u_{m-1}\tau^{m-1} + \cdots + u_1\tau + u_0)$;

Now the result of Algorithm 10 has length m but is no longer in τ-adic NAF form. There may be adjacent non-zero symbols and the symbols are not restricted to the set $\{0, 1, \bar{1}\}$.

The input of Algorithm 9 is of the form $r_0 + r_1\tau$ where $r_0, r_1 \in \mathbb{Z}$. The output is the τ-adic representation of the input. For $v \in \mathbb{Z}[\tau]$ we can write

$$\begin{aligned} v &= v_{m-1}\tau^{m-1} + \cdots + v_2\tau^2 + v_1\tau + v_0 \\ &= v_{m-1}\tau^{m-1} + \cdots + v_2\tau^2 + \text{TNAF}(v_1\tau + v_0) \end{aligned}$$

Now the two least significant symbols of v are in τ-adic NAF. Repeating this procedure for every bit in v the entire string can be converted to τ-adic NAF. This process is described in Algorithm 11.

The output of Algorithm 11 is in τ-adic NAF and has a length of approximately m symbols. If the result is larger than m symbols, it is possible to repeat Algorithms 10 and 11 to further reduce the length. Algorithms 9, 10 and 11 have been implemented in C and were used to generate test vectors for the prototype discussed later in this section. During testing, it was found that a single pass of these algorithms generates a τ-adic representation with average length of m and a maximum length of $m + 5$.

Like radix 2 NAF the τ-adic NAF uses the symbol set $\{1, 0, \bar{1}\}$ and has an average Hamming weight of approximately $l/3$ for an l-bit integer [29]. So Algorithm 8 has a running time of $l/3$ point additions and $l - 1$ Frobenius mappings.

Summary and Analysis: A point addition using López & Dahab's projective coordinates requires ten field multiplications, four field squarings and eight field additions. A point double requires five field multiplications, five field squarings and four field additions. Using this information, the run time for scalar multiplication can be written in terms of field operations. Typically scalar multiplication is measured in terms of field

Algorithm 11. Regeneration of τ-adic NAF

Input: $v = v_{m-1}\tau^{m-1} + \cdots + v_1\tau + v_0$

Output: w =REGEN_TNAF(v)

$w \leftarrow v$;

$i \leftarrow 0$;

while ($w_j \neq 0$ for some $j \geq i$) **do**

if ($w_i == 0$) **then**

$i \leftarrow i + 1$;

else

$t_0 \leftarrow w_i$;

$t_1 \leftarrow w_{i+1}$;

$w_i \leftarrow 0$;

$w_{i+1} \leftarrow 0$;

$w \leftarrow w$+TNAF($t_1\tau + t_0$);

$i \leftarrow i + 1$;

multiplications, inversions and squarings, ignoring the cost of addition. In the case of this architecture, field multiplication and squaring are completed quickly enough that the cost of field addition becomes significant. The run times using binary, binary NAF and τ-adic NAF representations are shown in Table 6. These values are based on the curve addition and doubling equations defined in (5) and (6) assuming arbitrary curve parameters α and β and the average Hamming weights discussed in the previous sections. For the case of τ-NAF, a Frobenius mapping is assumed to require three squaring operations. The symbols $\mathcal{M}$, $\mathcal{S}$, $\mathcal{A}$ and $\mathcal{I}$ correspond to field multiplication, squaring, addition and inversion respectively. In each case it is assumed that the length of the integer is approximately equal to m.

5. A CO-PROCESSOR ARCHITECTURE FOR ECC SCALAR MULTIPLICATION

In the recent past, several articles have proposed various hardware architectures/ accelerators for ECC. These elliptic curve cryptographic accelerators can be categorized into three functional groups. They are

Table 6. Cost of Scalar Multiplication in terms of Field Operations

	Generic m	$m = 163$
Binary	$(10\mathcal{M} + 7\mathcal{S} + 8\mathcal{A})m + \mathcal{I}$	$1630\mathcal{M} + 1141\mathcal{S} + 1304\mathcal{A} + \mathcal{I}$
NAF	$(\frac{25}{3}\mathcal{M} + \frac{19}{3}\mathcal{S} + \frac{20}{3}\mathcal{A})m + \mathcal{I}$	$1359\mathcal{M} + 1033\mathcal{S} + 1087\mathcal{A} + \mathcal{I}$
τ-NAF	$(\frac{10}{3}\mathcal{M} + \frac{13}{3}\mathcal{S} + \frac{8}{3}\mathcal{A})m + \mathcal{I}$	$544\mathcal{M} + 706\mathcal{S} + 435\mathcal{A} + \mathcal{I}$

1. Accelerators which use general purpose processors to implement curve operations but implement the finite field operations using hardware. References [2] and [30] are examples of this. Both of these implementations support the composite field GF(2^{155}).

2. Accelerators which perform both the curve and field operations in hardware but use a small field size such as GF(2^{53}). Architectures of this type include those proposed in [28] and [8]. In [28], a processor for the field GF(2^{168}) is synthesized, but not implemented. Both works discuss methods to extend their implementation to a larger field size but do not actually do so.

3. Accelerators which perform both curve and field operations in hardware and use fields of cryptographic strength such as GF(2^{163}). Processors in this category include [3, 10, 17, 25, 27].

The work discussed in this section falls into category three. The architectures proposed in [25] and [27] were the first reported cryptographic strength elliptic curve co-processors. Montgomery scalar multiplication with an LSD multiplier was used in [27]. In [25] a new field multiplier is developed and demonstrated in an elliptic curve scalar multiplier. In both [17] and [3] parameterized module generation is discussed. To the best of our knowledge the architecture proposed in [10] offers the fastest scalar multiplication using FPGA technology at 0.144 milliseconds. This architecture uses Montgomery scalar multiplication with López and Dahab's projective coordinates. They use a shift and add field multiplier but also compare LSD and Karatsuba multipliers.

This section describes a hardware architecture for elliptic curve scalar multiplication. The architecture uses projective coordinates and is optimized for scalar multiplication over the Koblitz curves using the arithmetic routines discussed in Section 3 to perform the field arithmetic.

5.1. Co-processor Architecture

The architecture, which is detailed in this section, consists of several finite field arithmetic units, field element storage and control logic. All logic related to finite field arithmetic is optimized for specific field size and reduction polynomial. Internal curve computations are performed using López & Dahab's projective coordinate system.

While generic curves are supported, the architecture is optimized specifically for the special Koblitz curves.

The processor's architecture consists of the data path and two levels of control. The lower level of control is composed of a micro-sequencer which holds the routines required for curve arithmetic such as DOUBLE and ADD. The top level control is implemented using a state machine which parses the scalar and invokes the appropriate routines in the lower level control. This hierarchical control is shown in Figure 7.

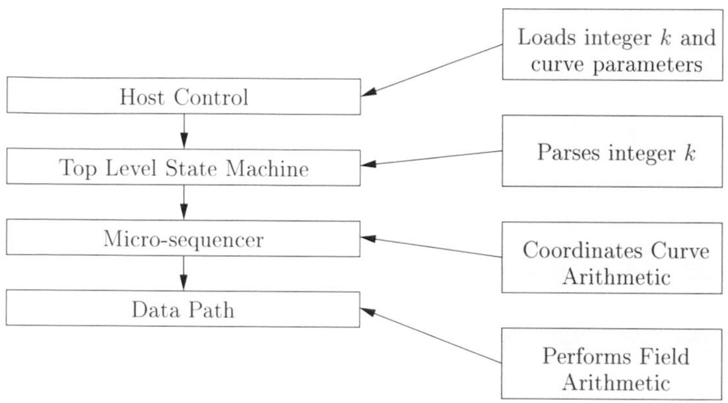

Figure 7. Co-Processor's Hierarchical Control Path

Co-processor Data Path

The data path of the co-processor consists of three finite field arithmetic units as well as space for operand storage. The arithmetic units include a multiplier, adder, and squaring unit. Each of these are optimized for a specific field and corresponding field polynomial. In an attempt to minimize time lost to data movement, the adder and multiplier are equipped with dual input ports which allow both operands to be loaded at the same time (the squaring unit requires a single operand and cannot benefit from an extra input bus). Similarly, the field element storage has two output ports used to supply data to the finite field units. In addition to providing field element storage, the storage unit provides the connection between the internal m-bit data path and the 32-bit external world. Figure 8 shows how the arithmetic units are connected to the storage unit.

The internal m-bit busses connecting the storage and arithmetic units are controlled to perform sequences of field operations. In this way the underlying curve operations DOUBLE and ADD as well as field inversion are performed.

Field Element Storage: The field element storage unit provides storage for curve points and parameters as well as temporary values. Parameters required to perform

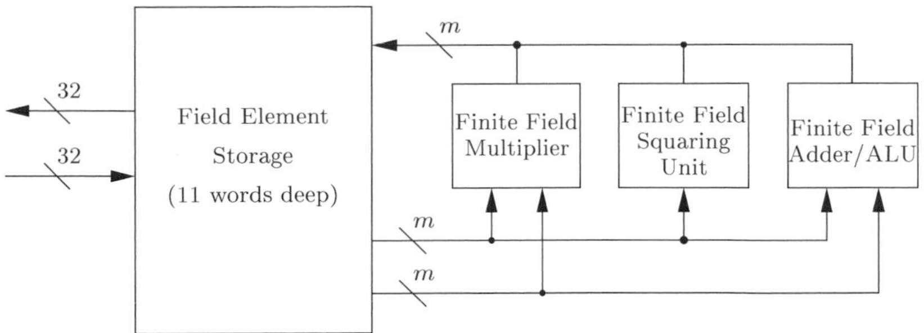

Figure 8. Co-Processor Data-Path

elliptic curve scalar multiplication include the field elements α and β and coordinates of the base point P. Storage will also be required for the coordinates of the scalar multiple Q. The point addition routine developed for this design also requires four temporary storage locations for intermediate values. Figure 9 shows how the storage space is organized.

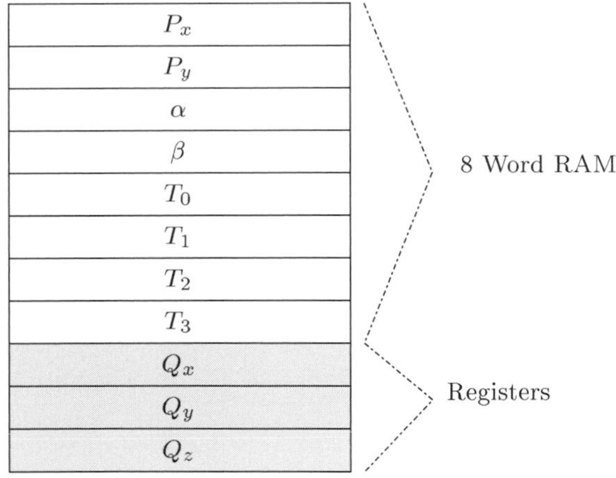

Figure 9. Field Element Storage

The top eight field element storage locations are implemented using 32-bit dual-port RAMs generated by the Xilinx Coregen tool and the bottom three storage locations[2]

[2] These locations are shaded gray in Figures 9 and 10.

are made of register files with 32-bit register widths. The dual 32-bit/m-bit interface support is achieved by instantiating $\lceil \frac{m}{32} \rceil$ dual-port storage blocks (either memories or register files) with 32-bit word widths as shown in Figure 10. The figure assumes $m = 163$. If the 32-bit storage locations in Figure 10 are viewed as a matrix then the rows of the matrix hold the m-bit field words. Each 32-bit location is accessible by the 32-bit interface and each m-bit location is accessible by the m-bit interface. For simplicity sake the field elements are aligned at 32 byte boundaries.

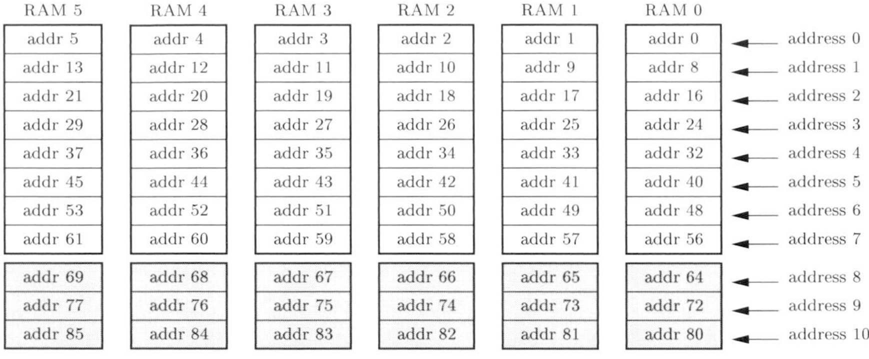

Figure 10. 32-bit/163-bit Address Map

Computation of τQ: In addition to providing storage, the registers in the bottom three m-bit locations are capable of squaring the resident field element. This is accomplished by connecting the logic required for squaring directly to the output of the storage register. The squared result is then muxed in to the input of the storage register and is activated with an enable signal. Figure 11 provides a diagram of this connection. This allows the squaring operations required to compute τQ to be performed in parallel. Furthermore, it eliminates the data movement otherwise required if the squaring unit were to be loaded and unloaded for each coordinate of Q. This provides significant performance improvement when using Koblitz curves.

The Micro-sequencer

The micro-sequencer controls the data movement between the field element storage and the finite field arithmetic units. In addition to the fundamental load and store operations, it supports control instructions such as jump and branch. The following list briefly summarizes the instruction set supported by the micro-sequencer.

- ld: Load operand(s) from storage location into specified field arithmetic unit.
- st: Store result from field arithmetic unit into specified storage location.
- j: Jump to specified address in the micro-sequencer.

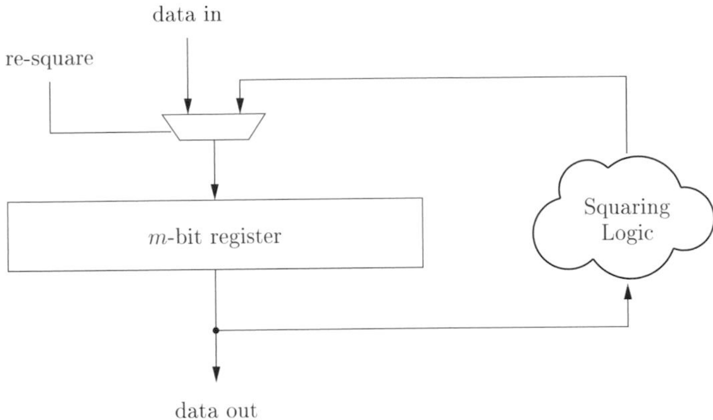

Figure 11. Efficient Frobenius Mapping

- jr: Jump to specified micro-sequencer address and push current address onto the program counter stack.
- ret: Return to micro-sequencer address. The address is supplied by the program counter stack.
- bne: Branch if the last field elements loaded into the ALU are NOT equal.
- nop: Increment program counter but do nothing.
- set: Set internal counter to specified value.
- rsq: Resquares the contents of the squaring unit.
- dbnz: Decrement internal counter and branch if the new value of the counter is zero. This opcode also causes the contents of the squaring unit to be resquared.

A two-pass perl assembler was developed to generate the micro-sequencer bit code. The assembler accepts multiple input files with linked addresses and merges them into one file. This file is then used to generate the bit code. The multiple input file support allows different versions of the ROM code to be efficiently managed. Different implementations of the same micro-sequencer routine can be stored in different files allowing them to be easily selected at compile time.

Micro-sequencer Routines: The micro-sequencer supports the curve arithmetic primitives, field inversion as well as a few other miscellaneous routines. The list below provides a summary of routines developed for use in the design.

- `POINT_ADD` (P, Q): Adds the elliptic curve points P and Q where P is represented in affine coordinates and Q is represented using projective coordinates. The result is given in projective coordinates.

- POINT_SUB (P, Q): Computes the difference $Q - P$. P is represented using affine coordinates and Q is represented using projective coordinates. The result is given in projective coordinates. This routine calls the POINT_ADD routine.
- POINT_DBL (Q): Doubles the elliptic curve point Q. Both Q and the result are in projective coordinates.
- INVERT (X): Computes the inverse of the finite field element X.
- CONVERT (Q): Computes the affine coordinates of an elliptic curve point Q given the point's projective coordinates. This routine calls the INVERT routine.
- COPY_P2Q (P, Q): Copies the x and y coordinates of point P to the x and y coordinates of point Q. The z coordinate of point Q is set to 1.
- COPY_MP2Q (P, Q): Computes the x and y coordinates of point $-P$ and copies them to the x and y coordinates of point Q. The z coordinate of point Q is set to 1.

Several versions of the POINT_ADD routine have been developed. The most generic one supports any curve over the field GF(2^m). In this version, the values of α and β are used when computing the sum of two points. This curve also checks if $Q \neq P$, $Q \neq -P$ and $Q \neq \mathcal{O}$. The second version of the point addition routine is optimized for a Koblitz curve by assuming α and β are equal to the NIST recommended values. The number of field multiplications required to compute the addition of two points is reduced from 10 to 9. The third version of the routine is optimized for a Koblitz curve and also forgoes the checks of point Q. If the base point P has a large prime order and the integer k is less than this order[3], it will never be the case that $Q = \pm P$ or $Q = \mathcal{O}$. This final version of the routine is the fastest of the three routines and is the one used to achieve the results reported at the end of the section.

Top Level Control

The routines listed above along with the POINT_FRB(Q) operation are invoked by the top level state machine. The POINT_FRB(Q) routine computes the Frobenius map of the point Q. This operation is not as complex as the other operations and is not implemented in the micro-sequencer. It is invoked by the top level state machine all the same.

The state machine parses the scalar k and calls the routines as needed. Since integers in NAF and τ-NAF require use of the symbol -1 (denoted $\bar{1}$), the scalar requires more than just an m-bit register for storage. In the implementation given here, each symbol in the scalar is represented using two bits; one for the magnitude and one for the sign. Table 7 provides the corresponding representation. For each bit k_i in the scalar k the magnitude is stored in the register $k_i^{(m)}$ and the sign is stored in register

[3] These are fair assumptions since the security of the ECC implementation relies on these properties.

Table 7. Representation of the Scalar k

Symbol	Magnitude	Sign
0	0	-
1	1	0
$\bar{1}$	1	1

$k_i^{(s)}$. Table 8 provides example representations for integers in binary form, NAF, and τ-adic NAF using $m = 8$.

Table 8. Example Representations of the Scalar

k	$k^{(m)}$	$k^{(s)}$
$(01001100)_2$	$(01001100)_2$	$(00000000)_2$
$(0100\bar{1}010)_{NAF}$	$(01001010)_2$	$(00001000)_2$
$(0100\bar{1}010)_{\tau-NAF}$	$(01001010)_2$	$(00001000)_2$

The top level state machine is designed to support binary, NAF and τ-adic NAF representations of the scalar. This effectively requires the state machine to perform Algorithms 3, 6 and 8. By taking advantage of the similarities between these algorithms, the top level state machine can perform this task with the addition of a single mode. This is shown in Algorithm 12. The algorithm is written in terms of the underlying curve and field primitives provided by the micro-sequencer (listed in Section 5.1).

The first step of Algorithm 12 is to search for the first non-zero bit in $k^{(m)}$. Once found, either P or $-P$ is copied to Q depending on the sign of the non-zero bit. The **while** loop then iterates over all the remaining bits in the scalar performing "doubles and adds" or "Frobenius mappings and adds" depending on the mode. Since the curve arithmetic is performed using projective coordinates, the result must be converted to affine coordinates at the end of computation.

Choice of Field Arithmetic Units

The use of redundant arithmetic units, specifically field multipliers, has been suggested in [3] and should be considered when designing an elliptic curve scalar multiplier.

Algorithm 12. State Machine Algorithm

Inputs: $k^{(m)} = (k_{l-1}^{(m)}, k_{l-2}^{(m)}, \ldots, k_1^{(m)}, k_0^{(m)})_2$,

$k^{(s)} = (k_{l-1}^{(s)}, k_{l-2}^{(s)}, \ldots, k_1^{(s)}, k_0^{(s)})_2$,

Point P and *mode* (NAF or τ-NAF)

Output: Point $Q = kP$

```
i ← l − 1;
while (k_i^(m) == 0) do
    k ← i − 1;
if (k_i^(s) == 1) then
    COPY_MP2Q(P, Q);
else
    COPY_P2Q(P, Q);
i ← i − 1;
while (i ≥ 0) do
    if (mode == τ-NAF) then
        Q ← POINT_FRB(Q);
    else
        Q ← POINT_DBL(Q);
    if (k_i^(m) == 1) then
        if (k_i^(s) == 1) then
            Q ← POINT_SUB(Q, P);
        else
            Q ← POINT_ADD(Q, P);
    i ← i − 1
Q ← CONVERT(Q);
```

It seems the advantage provided remains purely theoretical. This can be seen by examining the top performing ECC multipliers in [10] and [27], both of which use a single field multiplier. Reasons for doing the same for this ECC accelerator are twofold. (1) One of the limiting factors for the performance of the design is data movement. As shown in Figures 12 and 13 the bus usage for point addition and point doubling is very high (83% and 80% respectively). If another multiplier is added to the design there may not be enough free bus cycles to capitalize on the extra computational power. For the field GF(2^{163}), the multiplier computes a product in four clock cycles and requires three cycles to load and unload the unit. If a second multiplier is added, then two multiplications can be completed in four cycles but six cycles are required to unload the multiplier. (2) Many of the multiplications in point addition and point doubling are dependent on each other and must be performed in sequence. For this reason, the second multiplier may sit idle much of the time. The combination of these observations seems to argue against the use of multiple field multiplication units in the design.

5.2. FPGA Prototype

A prototype of the architecture has been implemented for the field GF(2^{163}) using the NIST recommended field polynomial. The design was coded using Verilog HDL and synthesized using Synopsys FPGA Compiler II. Xilinx' Foundation software was used to place, route and time the netlist. The prototype was designed to run at 66 MHz on a Xilinx' Virtex 2000E FPGA.

The resulting design was verified on the Rapid Prototyping Platform (RPP) provided by Canadian Microelectronics Corporation (CMC) [5, 6]. The hardware/software system includes an ARM Integrator/LM-XCV600E+ (board with a Virtex 2000E FPGA) and an ARM Integrator/ARM7TDMI (board with an ARM7 core) connected by the ARM Integrator/AP board. The design was connected to an AHB slave interface which made it directly accessible by the ARM7 core. Stimulated by compiled C-code, the core read from and wrote to the prototype. The Integrator/AP's system clock had a maximum frequency of 50 MHz. In order to run our design at 66 MHz it was necessary to use the oscillator generated clock provided with the Integrator/LM-SCV600E+. The data headed to and coming from the design was passed across the two clock domains.

5.3. Results

Table 9 shows the performance in clock cycles of the prototypes field and curve operations. These values were gathered using a field multiplier digit size of $g = 41$.

Note that the multiple instantiations of the squaring logic allow for the Frobenius mapping of a projective point to be completed in a single cycle. This significantly improves the performance of scalar multiplication when using the Koblitz curves.

The prototype of the scalar multiplier has been implemented using several digit sizes in the field multiplier. Table 10 reports the area consumption and resulting performance of the architecture given the different digit sizes. Table 11 provides a comparison of published performance results for scalar multiplication.

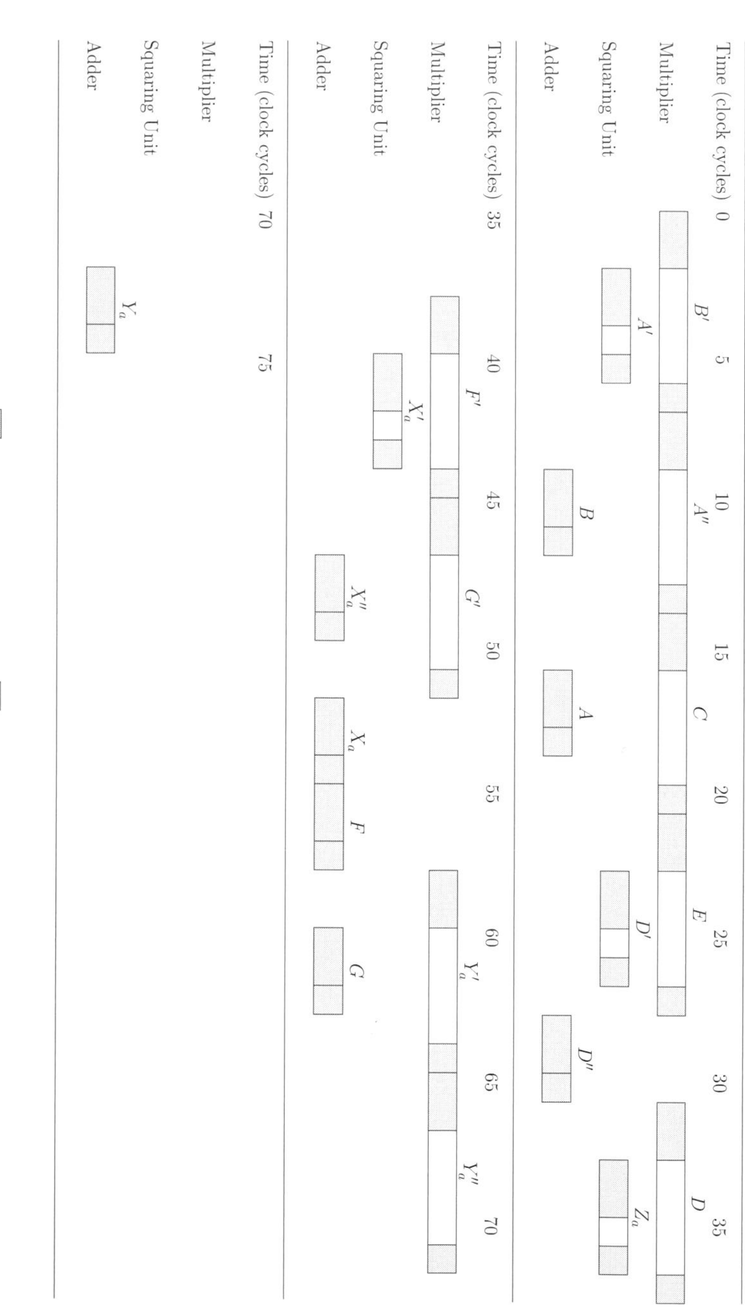

Figure 12. Utilization of Finite Field Units for Point Addition

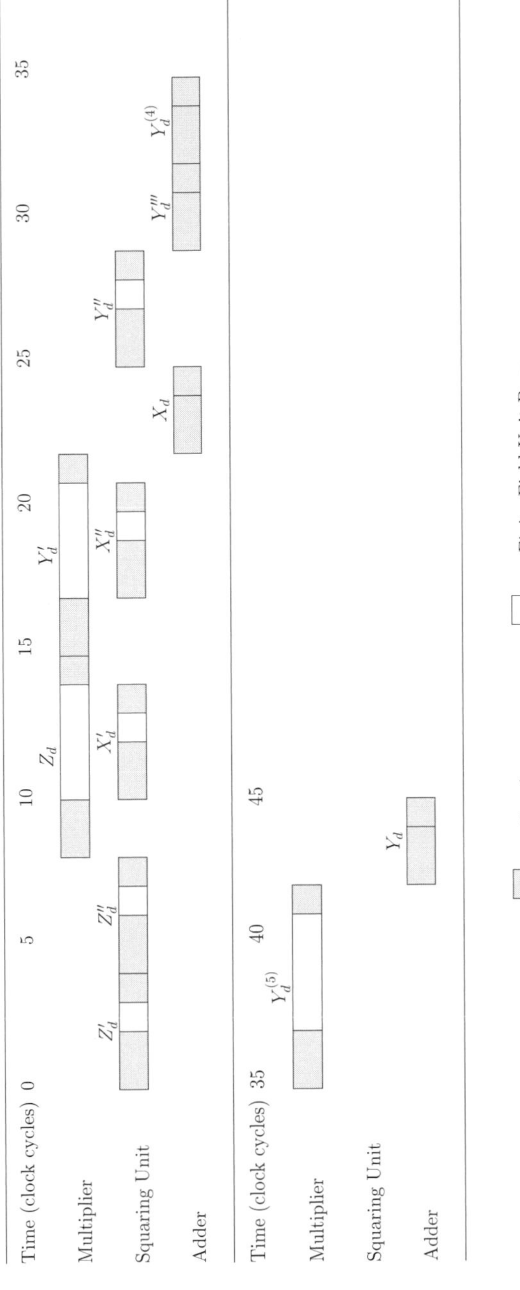

Figure 13. Utilization of Finite Field Units for Point Doubling

Table 9. Performance of Field and Curve Operations

Operation ($g = 41$)	# Cycles
Point Addition	79
Point Subtraction	87
Point Double	68
Frobenius Mapping	1

Table 10. Performance and Cost Results for Scalar Multiplication

Multiplier Digit Size	# LUTs	# FFs	Binary (ms)	NAF (ms)	τ-NAF (ms)
$g = 4$	6,144	1,930	1.107	0.939	0.351
$g = 14$	7,362	1,930	0.446	0.386	0.135
$g = 19$	7,872	1,930	0.378	0.329	0.113
$g = 28$	8,838	1,930	0.309	0.272	0.090
$g = 33$	9,329	1,930	0.286	0.252	0.083
$g = 41$	10,017	1,930	0.264	0.233	0.075

Table 11. Comparison of Published Results

Implementation	Field	FPGA	Scalar Mult. (ms)
S. Okada et. al. [25]	$GF(2^{163})$	Altera EPF10K250	45
Leong & Leung [17]	$GF(2^{155})$	Xilinx XCV1000	8.3
M. Bednara et. al. [3]	$GF(2^{191})$	Xilinx XCV1000	0.27
Orlando & Paar [27]	$GF(2^{167})$	Xilinx XCV400E	0.210
N. Gura et. al. [10]	$GF(2^{163})$	Xilinx XCV2000E	0.144
Our design ($g = 14$)	$GF(2^{163})$	Xilinx XCV2000E	0.135
Our design ($g = 41$)	$GF(2^{163})$	Xilinx XCV2000E	0.075

The performance of 0.144 ms reported in [10] is the fastest reported scalar multiplication using FPGA technology. The design presented in this report provides almost double (0.075 ms) the performance for the specific case of Koblitz curves.

The co-processor discussed in this chapter requires approximately half the CLBs used in the co-processor of [10] using the same FPGA. It must be noted that the co-processor presented in [10] is robust in that it supports all fields up to GF(2^{256}). In applications where support for a only single field size is required it is overkill to support elliptic curves over many fields. In scenarios such as this, this new elliptic curve co-processor offers an improved cost effective solution.

6. CONCLUDING REMARKS

In this chapter, the development of an elliptic curve cryptographic co-processor has been discussed. The co-processor takes advantage of multiplication and squaring arithmetic units which are based on the look-up table-based multiplication algorithm proposed in [11]. Field elements are represented with respect to the polynomial basis. While the base point and resulting scalar are given in affine coordinates, internal arithmetic is performed using projective coordinates. This choice of coordinate system allows the scalar multiple of a point to be computed with a single field inversion alleviating the need for a highly efficient inversion method. The processor was designed to support signed, unsigned and τ-NAF integer representation. All curves over a specific field are supported, but the architecture is optimized specifically for the Koblitz curves.

7. ACKNOWLEGEMENTS

This work was supported in part by the Security Technology Center in the Semiconductor Products Sector of Motorola. Dr. Hasan's work was supported in part by NSERC. Pieces of the work were presented at SPIE 2003 [19] and ITCC 2004 [20].

8. REFERENCES

1. *Wireless Application Protocol - Version 1.0*, 1998.

2. G. B. Agnew, R.C. Mullin, and S. A. Vanstone. An implementation of elliptic curve cryptosystems over $F_{2^{155}}$. *IEEE Journal on Slected Areas in Communications*, 11:804–813, June 1993.

3. Marcus Bednara, Michael Daldrup, Joachim von zur Gathen, Jamshid Shokrollahi, and Jurgen Teich. Implementation of elliptic curve cryptographic coprocessor over GF(2^m) on an FPGA. In *International Parallel and Distributed Processing Symposium: IPDPS Workshops*, April 2002.

4. D. Chudnovsky and G. Chudnovsky. Sequences of numbers generated by addition in formal groups and new primality and factoring tests. *Advances in Applied Mathematics*, 1987.

5. Canadian Microelectronics Corporation. *CMC Rapic-Prototyping Platform: Design Flow Guide*, 2002.

6. Canadian Microelectronics Corporation. *CMC Rapic-Prototyping Platform: Installation Guide*, 2002.

7. T. Dierks and C. Allen. *The TLS Protocol - Version 1.0 IETF RFC 2246*, 1999.

8. Lijun Gao, Sarvesh Shrivastava, and Gerald E. Sobelman. Elliptic curve scalar multiplier design using FPGAs. In *Cryptographic Hardware and Embedded Systems (CHES)*, 1999.

9. Daniel M. Gordon. A survey of fast exponentiation methods. *J. Algorithms*, 27(1):129–146, 1998.

10. Nils Gura, Sheueling Chang Shantz, Hans Eberle, Summit Gupta, Vipul Gupta, Daniel Finchelstein, Edouard Goupy, and Douglas Stebila. An end-to-end systems approach to elliptic curve cryptography. In *Cryptographic Hardware and Embedded Systems (CHES)*, 2002.

11. M. Anwarul Hasan. Look-up table-based large finite field multiplication in memory constrained cryptosystems. *IEEE Transactions on Computers*, 49(7), July 2000.

12. IEEE. *P1363: Editorial Contribution to Standard for Public Key Cryptography*, February 1998.

13. T. Itoh and S. Tsujii. A fast algorithm for computing multiplicative inverses in GF(2^m) using normal bases. *Information and Computing*, 78(3):171–177, 1988.

14. Brian King. An improved implementation of elliptic curves over GF(2^n) when using projective point arithmetic. In *Selected Areas in Cryptography*, 2001.

15. Neal Koblitz. Elliptic curve cryptosystems. *Mathematics of Computation*, 1987.

16. Neal Koblitz. CM curves with good cryptographic properties. In *Advances in Cryptography, Crypto '91*, pages 279–287. Springer-Verilag, 1991.

17. Philip H. W. Leong and Ivan K. H. Leung. A microcoded elliptic curve processor using FPGA technology. *IEEE Transactions on VLSI Systems*, 10(5), October 2002.

18. Julio Lopez and Ricardo Dahab. Improved algorithms for elliptic curve arithmetic in GF(2^n). In *Selected Areas in Cryptography*, pages 201–212, 1998.

19. Jonathan Lutz and Anwarul Hasan. High performance finite field multiplier for cryptographic applications. In *SPIE's Advanced Signal Processing Algorithms, Architectures, and Implemenations*, Volume 5205, pages 541-551, 2003.

20. Jonathan Lutz and Anwarul Hasan. High performance fpga based elliptic curve cryptographic co-processor. In *IEEE International Conference on Information Technology (ITCC)*, Volume II, pages 486-492, 2004.

21. Alfred Menezes. Elliptic curve public key cryptosystems. *Kluwer Academic Publishers*, 1993.

22. A. Menezes, E. Teske, A. Weng. Weak Fields for ECC. Technical Report CORR 2003-15, Centre for Applied Cryptographic Research, University of Waterloo, 2003. *See http://www.cacr.math.uwaterloo.ca*

23. Victor Miller. Uses of elliptic curves in cryptography. In *Advances in Cryptography, Crypto '85*, 1985.

24. NIST. *FIPS 186-2 draft, Digital Signature Standard (DSS)*, 2000.

25. Souichi Okada, Naoya Torii, Kouichi Itoh, and Masahiko Takenaka. Implementation of elliptic curve cryptographic coprocessor over GF(2^m) on an FPGA. In *Cryptographic Hardware and Embedded Systems (CHES)*, pages 25–40. Springer-Verlag, 2000.

26. OpenSSL. *See http://www.openssl.org*.

27. Gerardo Orlando and Christof Paar. A high-performance reconfigurable elliptic curve processor for GF (2^m). In *Cryptographic Hardware and Embedded Systems (CHES)*, 2000.

28. Martin Christopher Rosner. Elliptic curve cryptosystems on reconfigurable hardware. Master's thesis, Worcester Polytechnic Institute, 1998.

29. Jerome A. Solinas. Improved algorithms for arithmetic on anomalous binary curves. In *Advances in Cryptography, Crypto '97*, 1997.

30. S. Sutikno, R. Effendi, and A. Surya. Design and implemntation of arithmetic processor $F_{2^{155}}$ for elliptic curve cryptosystems. In *IEEE Asia-Pacific Conference on Circuits adn Systems*, pages 647–650, November 1998.

2

AN ADAPTIVE ENCRYPTION PROTOCOL IN MOBILE COMPUTING

Hanping Lufei and Weisong Shi
Department of Computer Science
Wayne State University
E-mails: hlufei@wayne.edu, weisong@wayne.edu

Use of encryption for secure communication plays an important role in building applications in mobile computing environments. With the emergence of more and more heterogeneous devices and diverse networks, it is difficult, if not impossible, to use a one-size-fits-all encryption algorithm that always has the best performance in such a dynamic environment. We envision that the only way to accelerate the deployment of encryption algorithms is providing a flexible adaptation of choosing an appropriate encryption algorithm from multiple diverse algorithms according to the characteristics of heterogeneous mobile computing environments.

Based on the Fractal framework [1], we propose and implement an adaptive encryption protocol, which can dynamically choose a proper encryption algorithm based on application-specific requirements and device configurations. Performance evaluation results show that in the divergent environment with different devices and applications, the adaptive encryption protocol successfully selects the best encryption algorithm from the candidate algorithms, and minimizes the total time overhead and insures the security as well.

1. INTRODUCTION

Use of encryption for secure communication is important for building distributed applications. With the development of computer and communication technologies, more and more heterogeneous devices, like desktops, laptops, PocketPCs, and cellular phones are connected to the Internet using diverse networks, like Ethernet, Wi-Fi, Bluetooth, 3G/4G wireless technology. On one hand, different technologies have different characteristics. On the other hand, a heterogeneous environment makes it possible to dynamically change between different devices and network environments. For instance, a person uses a laptop with a cable modem at home, a cell phone with 3G/4G or Bluetooth on the way to the office, a desktop with Ethernet LAN in the office and a PDA with Wi-Fi in the meeting room. Diverse network connections and heterogeneous devices demand the adaptation functionality in a distributed fashion because no one-size-fits-all single function or protocol can perform well over all these networks and devices. Although many symmetric or asymmetric encryption algorithms have been proposed, none of them takes the diversities of device and network into the design. It is difficult, if not impossible, to build a one-size-fits-all encryption protocol which can run well in the dynamic environment. The only way to accelerate the deployment of encryption algorithms is to provide a flexible adaptation of choosing multiple diverse algorithms.

Adaptation has been considered as a general approach to address the mismatch problem between clients and servers [2, 3, 4, 5]. From the perspective of adaptation locations, some of them propose the in-network adaptation, such as CANS [2], Rover [3], Odyssey [4], and Active Names [5], which focus on how to do the adaptation step by step across an overlay path. From the network OSI model's point of view, some of them work in the network layer [6], which adapts the TCP/IP protocol dynamically according to the changing situations on both ends. The Fractal framework [1], a dynamic application level protocol adaptation approach, utilizes the mobile code technology for protocol adaptation and leverages existing content distribution networks (CDN) for protocol adaptors (mobile codes) deployment. The protocol adaptation in Fractal is based on the assumption that an application protocol is composed of a series of components, also called protocol adaptors (PAD). When a protocol needs to be adapted, the application simply needs to add or remove some PADs into or from it. We will give a brief introduction about the Fractal framework in Section 3.

Based on the Fractal framework, we propose and implement an adaptive encryption protocol, which dynamically chooses a proper encryption algorithm based on application-specific requirements and device configurations. Evaluation results show that the adaptive encryption protocol can choose the best encryption algorithm from the candidates to minimize the total time overhead and ensure the security as well.

The rest of the chapter is organized as follows. After a brief introduction of background in Section 2, the Fractal framework and platform of the adaptive encryption protocol are depicted in Section 3. Section 4 describes the adaptation model for the adaptive encryption protocol. Performance evaluation and related work are described in Section 5 and Section 6 respectively. We summarize the chapter in Section 7.

2. BACKGROUND

In the design and implementation of the adaptive encryption protocol, several background topics are involved, such as: mobile code [7, 8], content distribution network [9, 10], protocol adaptation [11, 6, 12], and encryption algorithms. In this section, we explain the general background of each related research field.

2.1. Mobile Code

Mobile code [8] is defined as the data that can be executed as a program. The code can be pre-compiled for immediate execution on the recipient's processor, compiled upon receipt for subsequent execution or interpreted. The mobile code system has been used to build a distributed processing environment that is flexible in the communication abstractions it provides to applications and to enhance existing distributed applications. For the benefit of mobile code [7], a major asset provided by code mobility is that it enables service customization. The ability to request the remote execution of code helps increase application server flexibility without permanently affecting the size or complexity of the server. In Fractal we implement each protocol adaptor as a mobile code module, which is sent and executed remotely on the client side to build a new protocol allowing the client to talk with the application server.

2.2. Content Distribution Network

Content Distribution Networks (CDN) [10] is an intermediate layer of infrastructure between origin servers and clients. CDN can achieve scalable content delivery by distributing load among its edgeservers, by serving client requests from edgeservers that are close to requests, and by bypassing congested network paths. Currently CDNs are only used to deliver Web-based content. In Fractal framework, CDN is used to deliver protocol adaptor (PAD). If we consider the PAD as a Web-based object, most of the current techniques in CDN can be leveraged to the delivery of PAD. Fractal framework extends the utilization of CDNs from traditional Web-based content to Web-based objects like mobile code and mobile agent.

2.3. Protocol Adaptation

Changing protocols to adapt link condition and network environment is not the new idea, e.g., Reno and Vegas congestion control in TCP/IP protocol [13] is a kind of adaptation. More sophisticated protocol adaptation approaches, such as STP proposed in [6], but most of them are in the network layer which makes them hard to have a general view of the whole system status. The problem of adapting to a changing network environment is further complicated because changes in network conditions are usually transparent to higher layers of the protocol stack. When higher layers, e.g., application layer, are aware of network variation, protocol adaptation can be done more adaptively and intelligently. Based on these observations, Fractal works entirely in the application layer to adapt the application protocol according to heterogeneous client environments.

2.4. Three Symmetric-Key Encryption Algorithms

Many symmetric key encryption algorithms have been proposed. DES, AES, and RC4 are three of the most popular shared-key encryption algorithms.

1. *DES/Triple DES* [14] Data Encryption Standard is addressed in FIPS PUB 46. Data are encrypted in 64-bit blocks using a 56-bit key. DES transforms 64-bit input in a series of steps into a 64-bit output. The same steps and the same key are used to decrypt the data. With the development of hardware technology,

DES shows potential vulnerability to a brute-force attack. Triple DES (3DES) is an alternative of traditional DES algorithm. Triple DES provides a security level of 2^{112}, independent of the key size. National Institutes of Standards and Technology (NIST) requires all new applications should use triple DES or more advanced encryption algorithms, while DES is still supported for legacy applications. DES can be broken by brute force attack because of the limited key length. Triple DES is secure but with the computation time as three times slower than DES. The poor performance of triple DES triggered the call for an advanced encryption standard (AES).

2. *AES* [15] AES is a relatively new algorithm compared with DES. Observing that DES is more and more out of date and 3DES is not a long term replacement candidate for the widely used DES algorithm. NIST called a new Advanced Encryption Standard (AES). AES is more secure than DES. It can has key length as long as 256 bits. It also have high computation efficiency and flexibility to be practical in a wide range of applications. The security level of AES is $2^{128,192,256}$ depending on the used key size, where the AES block sizes are 128, 192, and 256.

3. *RC4 Stream Cipher* [16] RC4 is a contemporary variable key-size stream cipher with byte-oriented operations. It is based on the use of a random permutation. Key length is in a range from 1 to 256 bytes. RC4 is easy to be implemented even on resource-constraint devices, such as Berkeley Motes and smart cards. Adjustment of key length can achieve a tradeoff between running speed and security level.

There are several other symmetric algorithms have been proposed; however, we believe these three algorithms are diverse enough to show the basic idea of adaptive encryption in this case study.

3. PLATFORM OF THE ADAPTIVE ENCRYPTION PROTOCOL

The adaptive encryption protocol is utilized between two communication parties: *application server* and *client*. We assume that some clients use legacy applications, which support only old encryption algorithms, while some clients have more flexibility to choose different algorithms. Three encryption algorithms, namely DES [14], AES [15], RC4 [16] are the candidates of encryption algorithms. The sender side adopts the Fractal framework [1] to choose proper encryption algorithms based on their diverse characteristics and different client applications configurations. Note that we focus on how to choose different algorithms in the context of symmetric encryption. The procedure to set up the symmetric key(s) is beyond the scope of this chapter. It is very easy to set up the symmetric keys using the Diffie-Hellman [17] key exchange mechanism.

Figure 1 shows the platform of the adaptive encryption protocol including five components: *Application server*, *Adaptation proxy*, *CDN edgeservers*, *Protocol adap-*

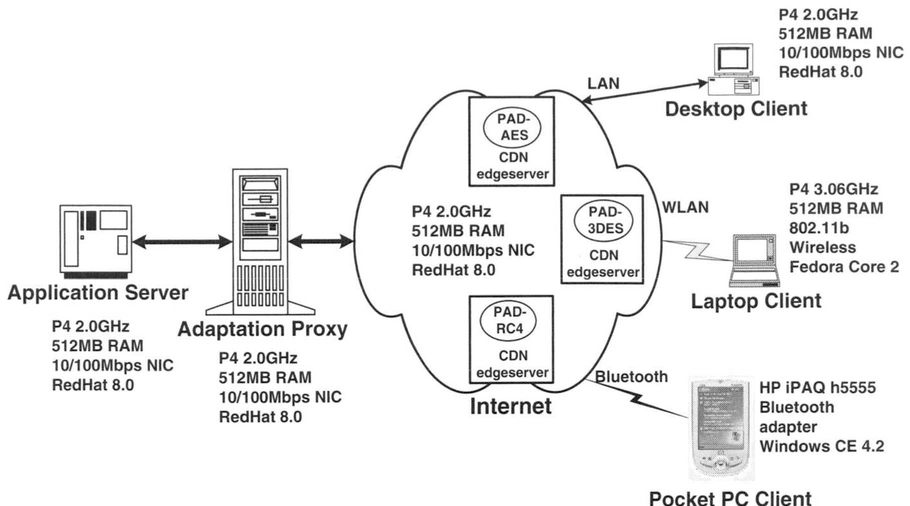

Figure 1. Platform for the adaptive encryption protocol.

tors (PADs), and *Clients* (e.g., desktop, laptop, PocketPC). The application server is the application service provider. In order to provide the functionality to heterogeneous clients in diverse environments, the application server usually communicates with clients through different encryption protocols. Although the application server can talk in many different encryption protocols, the client may not have the necessary protocol to talk with the sender. To help the client talk with the application server, the PAD, which is a protocol adaptor, encapsulates the encryption protocol candidates into a mobile code module and deploys them across the CDN edgeservers that locates on the edge of the Internet. By downloading and deploying one or more PADs, the client is then capable of starting communication with the application server using required encryption protocols. On the sender side, we assume the application server has already deployed all PADs in advance. An important issue for the sender is which PADs should be used and where to find them. Close to the application server, an adaptation proxy is set up to handle the issues about PAD negotiations. Before the initialization of communication between the sender and the client, the client has to negotiate with the adaptation proxy to find proper PADs. The client will be asked to provide some metadata about his environments, such as computing ability, memory space, and network configurations to the adaptation proxy. Having these metadata, the adaptation proxy will generate the metadata of the proper PADs for the client and send the metadata of PADs back to the client. Inside these metadata is enough information for the client to download the PADs from the closest edgeserver of CDNs with which the application server is associated. Next, we will give more details about the adaptation proxy.

3.1. Adaptation Proxy

Adaptation proxy plays an important role in the adaptive encryption protocol. Usually it is deployed in the same administration domain as the application server and is responsible for negotiation with the client. A general structure of the adaptation proxy is shown in Figure 2, which includes a *negotiation manager* module and a *distribution manager* module. Each module is running as a daemon on the adaptation proxy. Next we will explain the structure and functionality of each module respectively.

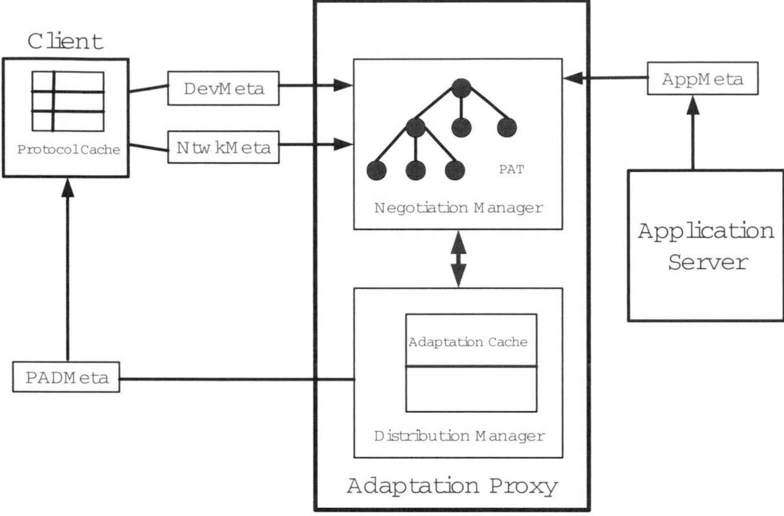

Figure 2. Structure of the adaptation proxy.

Negotiation Manager As shown in Figure 2, the negotiation manager is the key in the adaptation proxy which negotiates with the client. Some application level metadata is needed to be transmitted between the adaptation proxy and the application server, and between the adaptation proxy and the client to support the negotiation function. We define these metadata formats in Figure 3. In the rest of the chapter, we will use the acronyms in the parentheses to refer to them.

*Device Metadata (**DevMeta**)* = { Operating system type, CPU type, CPU speed, memory size }

*Network Metadata (**NtwkMeta**)* = { Network type, Network bandwidth }

*PAD Metadata (**PADMeta**)* = { PAD ID, PAD size, PAD overhead, Message digest, URL, Parent link, Child link, ... , Child link }

*Application Metadata (**AppMeta**)* = { Application ID, *PADMeta* 1, ... , *PADMeta* n}

Figure 3. Definitions of metadata.

DevMeta and *NtwkMeta*, provided by clients, contain the hardware information and the network environment of the client. The application server supplies *PADMeta* to the negotiation manager, who holds the general information of each PAD. *PAD ID* is a unique identification generated by the application server. *PAD overhead* consists of the computing overhead at both the client side and server side, and corresponding traffic overhead in the network. *Message digest* is computed using the SHA-1 [18] function and used by clients to verify the integrity of the PAD. *URL* is the link to download the PAD. Note that it is the CDN's responsibility to find the closest edgeserver which holds the PAD, and to redirect the request to that edgeserver. *Parent link* and *Child link* are used to build the protocol adaptation topology in the negotiation manager. *AppMeta* is comprised of *Application ID*, which marks different applications, and some *PADMeta*, which forms a protocol adaptation topology. The application server pushes new *AppMeta* to the negotiation manager when the protocol adaptation topology is first created or changed later. Usually the protocol adaptation topology is represented by a protocol adaptation tree (PAT) structure as shown in Figure 2 in the upper box located in the negotiation manager. We will give more details about the PAT tree in Section 4.1.

When the negotiation manager receives a request from a client, it first checks its adaptation cache, located in the distribution manager. The cache has entries mapping client side information to an array of *PADMeta* that the client needs. Each mapping entry is structured as follows:

$$\{\ \mathit{DevMeta, Application\ ID, NtwkMeta}\ \} \Rightarrow \{\ \mathit{PADMeta\_1}, \ldots\ ,\mathit{PADMeta\_n}\ \}$$

If the adaptation cache does not have the entry corresponding to the client side metadata, the negotiation manager then will use a path search algorithm described in Section 4.2 to form a new entry and transfer it to the distribution manager.

Distribution Manager The distribution manager is in charge of further processing of these *PADMeta* received from the negotiation manager, updating the adaptation cache, and finally sending *PADMeta* back to the client. When the distribution manager receives the *PADMeta* generated by the negotiation manager, it inserts message digest and URL data into the *PADMeta* and hides the parent and child links since the exposure to the client is unnecessary. After the negotiation procedure, which will be discussed in the following section, the distribution manager will update the adaptation cache so that the negotiation result can be directly retrieved from the cache if the same client configuration occurs later. Finally the distribution manager will handle the network communication details and send these *PADMeta* back to the client. Next we will explain the interactive negotiation protocol.

3.2. Interactive Negotiation Protocol

An interactive negotiation protocol(INP) is proposed for the interactions among these components, as shown in Figure 4. We assume both the client side and server side understand the protocol definitions. The application server has pre-deployed PADs in the application context and already pushed the *AppMeta* to the adaptation proxy, which has built a PAT inside the negotiation manager. The PADs have been distributed across the CDNs edgeservers.

At the beginning of the negotiation, a client first checks its own protocol cache, which contains some *PADMeta* saved for previous requests. If there is an entry of the protocol cache which matches the current request, the client will directly start the application communication with the application server. If not, the client sends `INIT_REQ`, which contains application request in payload, to the adaptation proxy [1] to initialize the protocol negotiation. Each packet has an *INP header* segment, which is used to maintain the interactive negotiation protocol integrity, and we will omit the details in the *INP header*. The adaptation proxy then sends `INIT_REP` as well as `Cli_META_REQ`, having empty *DevMeta* and *NtwkMeta* to be filled by the client, to acknowledge the request and ask some information about the client. After getting the reply, the client gets the content of *DevMeta* and *NtwkMeta* locally by probing the system using system calls and sends out the `Cli_META_REP`. Based on the `Cli_META_REP`, *PADMeta* is computed and sent back to the client in `PAD_META_REP` by the adaptation proxy. Next, the client updates his protocol cache and sends `PAD_DOWNLOAD_REQ` containing PAD ID to the URL of the PAD. The CDN will automatically choose a close CDN edgeserver and send back the PAD code in `PAD_DOWNLOAD_REP`. If multiple PADs are required, it is not necessary that those PADs downloaded from the same edgeserver. It is up to the CDN to manage the delivery of PADs. After the security check and PAD(s) deployment, the client sends out the `APP_REQ` to the application server. The `APP_REQ` contains the application request as well as the negotiated protocol identifications, which notify the application server to choose the proper PADs to talk with the client. From now on the client and the application server continue the application session using the negotiated protocol. The formats of all message types used in INP are listed on the bottom of Figure 4.

4. ADAPTATION MODEL OF THE ADAPTIVE ENCRYPTION PROTOCOL

Adaptation is the major function of the adaptive encryption protocol. In this section, we will show how the adaptation model works. First, we will explain the protocol adaptation topology, the protocol adaptation tree (PAT), which is the main data structure in the procedure of adaptation. Then we will clarify the adaptation path search algorithm.

4.1. Protocol Adaptation Tree and Protocol Adapters

Figure 5 shows a general example of the protocol adaptation tree (PAT), which is built by the negotiation manager based on *AppMeta* received from the application server. Each node of PAT is a protocol adaptor. The child PAD is an auxiliary component of the parent PAD. In order to run the parent PAD, one and only one of the children PADs

[1] Note that the client does not have to realize the existence of the adaptation proxy. The application server will automatically redirect the request to its corresponding adaptation proxy.

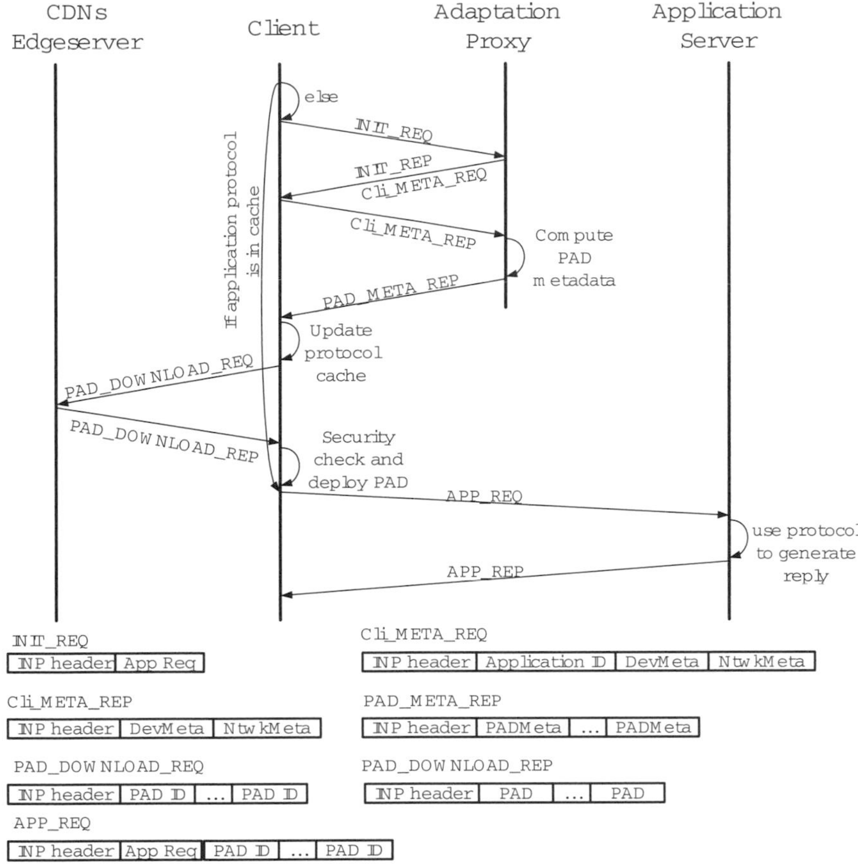

Figure 4. The Interactive Negotiation Protocol.

must work together with the parent PAD. For example, in Figure 5, if PAD2 is the FTP protocol, PAD7 is the TCP protocol, and PAD8 is the UDP protocol, the PAD2 can choose either PAD7 or PAD8, but not both. It is possible that one PAD is needed by multiple PADs, like TCP protocol is needed by both FTP and HTTP protocols. For the purpose of maintaining the tree structure, we use a symbolic copy of the child PAD if it is required by more than one parent PAD. For instance, in Figure 5, PAD6 is a symbolic link of PAD7, which is needed by both PAD1 and PAD2. So in order to satisfy an application protocol, a path should be found from the root application to one leaf, e.g., the path composed of PAD2 and PAD7 in the dotted line in Figure 5. The PADs along the path forms the adaptive protocol. The PAT in the adaptive encryption protocol is a one-level tree as shown in Figure 6 with each leaf is an encryption PAD. Key length and size of each PAD is shown in Table 1. Next, we will explain how to select the PADs in the adaptation path to build the adaptive encryption protocol.

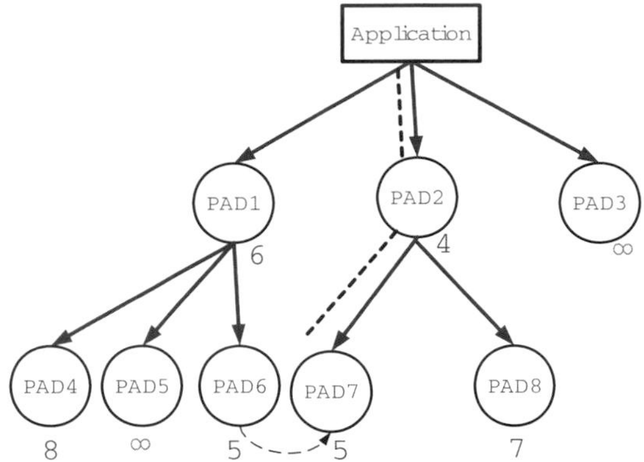

Figure 5. A general protocol adaptation tree.

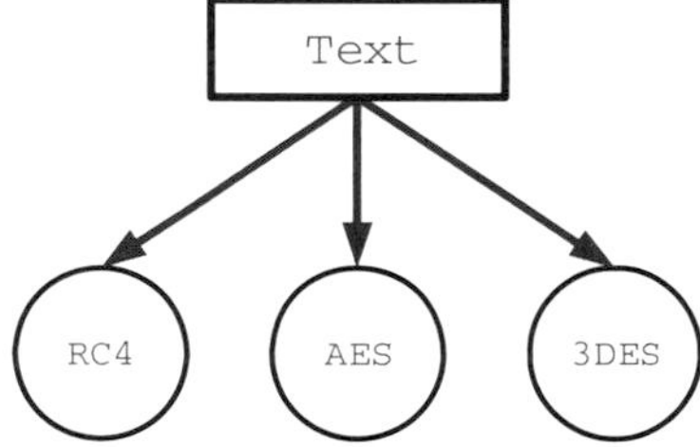

Figure 6. Protocol adaptation tree of the adaptive encryption protocol.

Table 1. The key length and size of each PAD.

PAD name	3DES-64	3DES-128	3DES-192	AES-128	AES-192	AES-256	RC4
Key Length	64 bits	128 bits	192 bits	128 bits	192 bits	256 bits	64 bits
Size	24KB	24KB	24KB	21KB	21KB	21KB	10KB

4.2. The Adaptation Path Search Algorithm

The goal of the adaptation path search algorithm is to find certain PADs from PAT to form an adaptation path for a client so that the overhead occurred by using the PADs along the path, is reduced as much as possible. For more complicated PAT such as the one in Figure 5, the Fractal framework [1] proposed an adaptation path search algorithm to find the path efficiently. For the one-level PAT in the adaptive encryption protocol the adaptation path search algorithm reduces to evaluate the overhead of each encryption PAD algorithm one by one and choose the one with the least overhead.

$$\mathbf{PAD_{total}} = PAD_{\text{download time}} + PAD^{server}_{comp} + PAD^{client}_{comp} + PAD_{\text{traffic overhead}} \quad (1)$$

We define the total overhead of each PAD as the sum of PAD download time, server side and client side computing time for unit data, i.e. RC4 encryption time for 1KB data on server side and decryption time for 1KB data on client side, finally the traffic overhead incurred by this PAD as shown in Equation 1. Since the PAD size is very small as we can see in Table 1, large amount PAD download experiment from the three PlanetLab nodes shows that the average download time are as close as 1 millisecond difference. Furthermore, each PAD is at most downloaded once in the whole application procedure. We consider the PAD download time as a constant and eliminate it from the PAD total overhead evaluation. On the other hand, since the three PADs, 3DES, AES, and RC4 do not change the size of the input data even with different key length, the traffic overhead of each PAD can also be excluded. Eventually, the PAD total overhead is simplified as Equation 2.

$$\mathbf{PAD_{total}} = PAD^{server}_{comp} + PAD^{client}_{comp} \quad (2)$$

Server side computing time of each PAD can be obtained proactively by testing each PAD on the application server. In order to evaluate the client side computing time of each PAD, running each PAD on each client configuration to get the overhead is not a wise solution because there are so many different client configurations. Instead, we use a linear model to estimate the overhead, which inspired by the observation that the computing overhead of each PAD is roughly proportional to the processor speed. As shown in the second part of the new total overhead equation 3, if the computing overhead of a PAD on a standard processor speed, Std_{cpu}, i.e. 500MHz Pentium IV in the platform, is known as the PAD^{Std}_{comp}, the computing overhead on the client side can be deducted from the linear ratio of the speed of the standard processor and client processor. However, this linear model is not so accurate because other parameters of the system introduces error into the linear model, i.e. the operating system. We abstract normalized ratio parameters about two key properties: *processor types* as $\mathcal{A}$ and *application types* as $\mathcal{B}$ in the equation. Note that it is easy to introduce more parameters if necessary, e.g., the operating system types and the network types defined in Fractal framework [1].

$$\mathbf{PAD_{total}} = PAD_{comp}^{server} + \mathcal{A} \times \mathcal{B} \times \frac{Std_{cpu}}{Cli_{cpu}} \times PAD_{comp}^{Std} \tag{3}$$

Usually, the normalized ratios such as $\mathcal{A}$ and $\mathcal{B}$ are in the form of matrix, as shown in Equation 4, to measure the performance ratios of 7 PADs on 3 kinds of processor types and on 2 kinds of application types, legacy and new system since they have different encryption requirements. P, D, and L represent the Intel PXA 255 processor in Pocket PC, Pentium IV 2.0GHz processor in Desktop, and Pentium IV 3.06GHz processor in Laptop respectively. We use the following simple example to explain the normalized matrix.

$$\begin{array}{cc} & \begin{array}{cc} WinCE & PalmOS \end{array} \\ \begin{array}{c} WinMedia \\ Kinoma \end{array} & \begin{pmatrix} 1 & \infty \\ \infty & 1 \end{pmatrix} \end{array}$$

The above matrix shows the impacts of two operating systems (the top line) on two multimedia players (the left most column). The values in the matrix mean the Windows Media works fine in the WinCE operating system (WinCE) [19] but not in PalmOS, while Kinoma player [20] runs well in PalmOS instead of WinCE. The value of ratios does not have to be an integer. Suppose now we are about to find the better one in terms of the computing time from these two players on WinCE platform. We get the time value using the linear method as, for instance, 5 sec for WinMedia and 2 sec for Kinoma. Without the normalized matrix, Kinoma will be chosen as the better player; however, the fact is that Kinoma can not run on WinCE at all. To get the correct result, we can use the first column of this normalized matrix to adjust the linear results by multiplying 2 sec with ratio 1 for WinMedia and multiplying 5 sec with ratio ∞ for Kinoma. Then the computing time of Kinoma becomes ∞, which immediately disqualifies itself. Go back to the normalized matrix $\mathcal{A}$ and $\mathcal{B}$, because most of the operations in these encryption algorithms are bit operations instead of float-point operations, they have almost same running efficiency in these client CPU types. We set all values as 1. Different encryption requirements of applications are reflected in $\mathcal{B}$. For example, the legacy systems only use the DES algorithm while the new applications will utilize the new encryption algorithms. Correspondingly in the normalized ratio matrix, we set the ratio as 1 for 3DES algorithm and ∞ for others in legacy systems. In our experimental platform, we specify the client applications on desktop as a legacy system and that on laptop and PocketPC as a new system. This may not be always true in reality, but just for comparison purpose in this experimental platform.

$$\mathcal{A} = \begin{array}{c} \\ pad_0 \\ \vdots \\ pad_n \end{array} \begin{array}{c} \begin{array}{ccc} cpu_0 & \cdots & cpu_a \end{array} \\ \begin{pmatrix} \alpha_{0(0)} & \cdots & \alpha_{0(a)} \\ \vdots & \ddots & \vdots \\ \alpha_{n(0)} & \cdots & \alpha_{n(a)} \end{pmatrix} \end{array}$$

$$= \begin{array}{c} \\ \mathit{3DES-64} \\ \mathit{3DES-128} \\ \mathit{3DES-192} \\ \mathit{AES-128} \\ \mathit{AES-192} \\ \mathit{AES-256} \\ \mathit{RC4-64} \end{array} \begin{array}{c} \begin{array}{ccc} P & D & L \end{array} \\ \begin{pmatrix} 1 & 1 & 1 \\ 1 & 1 & 1 \\ 1 & 1 & 1 \\ 1 & 1 & 1 \\ 1 & 1 & 1 \\ 1 & 1 & 1 \\ 1 & 1 & 1 \end{pmatrix} \end{array}$$

$$\mathcal{B} = \begin{array}{c} \\ pad_0 \\ \vdots \\ pad_n \end{array} \begin{array}{c} \begin{array}{ccc} app_0 & \cdots & app_b \end{array} \\ \begin{pmatrix} \beta_{0(0)} & \cdots & \beta_{0(b)} \\ \vdots & \ddots & \vdots \\ \beta_{n(0)} & \cdots & \beta_{n(b)} \end{pmatrix} \end{array}$$

$$= \begin{array}{c} \\ \mathit{3DES-64} \\ \mathit{3DES-128} \\ \mathit{3DES-192} \\ \mathit{AES-128} \\ \mathit{AES-192} \\ \mathit{AES-256} \\ \mathit{RC4-64} \end{array} \begin{array}{c} \begin{array}{cc} LegacySystem & NewSystem \end{array} \\ \begin{pmatrix} 1 & \infty \\ 1 & \infty \\ 1 & \infty \\ \infty & 1 \\ \infty & 1 \\ \infty & 1 \\ \infty & 1 \end{pmatrix} \end{array}$$

Now for a specific incoming client with processor type i and application type j available in metadata, the adaptation proxy will find the corresponding ratio vector $\begin{pmatrix} \alpha_{0(i)} & \alpha_{1(i)} & \ldots & \alpha_{n(i)} \end{pmatrix}^T$, and $\begin{pmatrix} \beta_{0(j)} & \beta_{1(j)} & \ldots & \beta_{n(j)} \end{pmatrix}^T$ from $\mathcal{A}$, and $\mathcal{B}$ based on its processor and application types. Given that we have only a limited number of consumer-used processors, the vector will be found with high probability. Otherwise a similar type with close parameters will be chosen instead. After the application session, the normalized matrix will be extended to include the new processor types. Then the normalized ratio matrix can be formed to estimate the total time overhead of each PAD for this new client using Equation 4. After obtaining the total time overhead of each PAD, The adaptive encryption protocol can be decided using the reduced adaptation path search algorithm. For the comprehensive descriptions of total overhead, normalized matrix, and adaptation path search algorithm, please refer to the Fractal framework [1].

$$\begin{pmatrix} pad^{total}_{3DES-64} \\ pad^{total}_{3DES-128} \\ pad^{total}_{3DES-192} \\ pad^{total}_{AES-128} \\ pad^{total}_{AES-192} \\ pad^{total}_{AES-256} \\ pad^{total}_{RC4-64} \end{pmatrix} = \begin{pmatrix} pad^{svr-comp}_{3DES-64} \\ pad^{svr-comp}_{3DES-128} \\ pad^{svr-comp}_{3DES-192} \\ pad^{svr-comp}_{AES-128} \\ pad^{svr-comp}_{AES-192} \\ pad^{svr-comp}_{AES-256} \\ pad^{svr-comp}_{RC4-64} \end{pmatrix} + \frac{cpu}{Cli_{cpu}} * \begin{pmatrix} \alpha_{3DES-64(i)} \\ \alpha_{3DES-128(i)} \\ \alpha_{3DES-192(i)} \\ \alpha_{AES-128(i)} \\ \alpha_{AES-192(i)} \\ \alpha_{AES-256(i)} \\ \alpha_{RC4-64(i)} \end{pmatrix}^T$$

$$*I* \begin{pmatrix} \beta_{3DES-64(j)} \\ \beta_{3DES-128(j)} \\ \beta_{3DES-192(j)} \\ \beta_{AES-128(j)} \\ \beta_{AES-192(j)} \\ \beta_{AES-256(j)} \\ \beta_{RC4-64(j)} \end{pmatrix}^T * I * \begin{pmatrix} pad^{Std-comp}_{3DES-64} \\ pad^{Std-comp}_{3DES-128} \\ pad^{Std-comp}_{3DES-192} \\ pad^{Std-comp}_{AES-128} \\ pad^{Std-comp}_{AES-192} \\ pad^{Std-comp}_{AES-256} \\ pad^{Std-comp}_{RC4-64} \end{pmatrix}$$

5. PERFORMANCE EVALUATION AND ANALYSIS

In our experimental platform, as shown in Figure 1, three kinds of client hosts, *desktop*, *laptop*, and *Pocket PC*, use different message receiver applications, to connect to the message sender and an adaptation proxy. The hardware and software configurations of the servers and clients are also shown in Figure 1. The message sender has 100 messages with size as 100K bytes. We implement three encryption algorithms, 3DES, AES and RC4 in C code as three protocol adaptors. The first two encryption algorithms have three different key length settings. Key length and size of each algorithm is shown in Table 1. We also implement an adaptation proxy connected with the application server in the same LAN domain. Similar to the previous section. To emulate the behavior of the real content distribution network and edgeservers, we utilize three nodes, in Wayne State University, New York University, and University of California at Berkeley respectively, from PlanetLab [21] as the distributed PAD servers.

We test the total time overhead of each algorithm for desktop, laptop, and PocketPC clients, as shown in Figure 7. The x-axis lists different encryption algorithms, the y-axis shows the total time for each algorithm including the sender encryption time and the receiver decryption time. In Figure 7(a), since the receiver application of the desktop is a legacy application in our experimental setup, which accepts only DES algorithms, the output of the adaptive path selection algorithm will set all other encryption algorithms except DES algorithms to infinite, which is denoted as N/A in the figure. However, for comparison purpose, we also show their corresponding computing overhead on the same figure. As a matter of fact, although AES-class algorithms have less computing overhead, they will not be chosen as the proper encryption algorithm for the desktop, which runs legacy applications only. Now only 3DES algorithms are eligible candidates. It is trivial that 3DES with 64 bits key should run faster than 3DES with 128 bits or 192 bits length key. Usually the adaptive encryption protocol will choose 3DES-64 since it has the fastest running speed with reasonable security enforcement. But this does not prevent application from choosing 128 or 192 bits 3DES. By introducing more adaptation parameters, like a normalized matrix for application security requirements, more secure algorithm could be selected. We believe this is a trivial task and decide not to be discussed in this chapter.

For the applications running on the laptop, 3DES is obviously not considered because it is out of date (and replaced by AES algorithms) for new applications. AES-128 which has slightly less total time overhead than other three algorithms have, as shown in Figure 7(b), will be selected by the adaptive encryption protocol. Note that similar to the case for desktop, other AES algorithms could also be selected for more secure purpose by extending the total time overhead evaluation formula. Finally, in Figure 7(c), we can see that the major part of the total time overhead is contributed by the receiver decryption time because the hardware of PocketPC on which receiver application executes is not as powerful as desktop or laptop hardware configurations. Not surprising, RC4-64 is selected as the most appropriate encryption algorithm, which is much faster than other algorithms. This is compatible with the fact that RC4 is almost the default encryption algorithm for small resource-constraint devices. It is worth noting that the choice made by the adaptive encryption protocol is straightforward in this case study. However, our work is the first effort to make the choice making in a formal way. We believe that the adaptive encryption protocol will be more useful in complicated applications in the foreseeable future work, includes investigating more encryption algorithms in heterogeneous environments, and applying this technique to the distributed computer-assistant surgery application [22].

6. RELATED WORK

The adaptive encryption protocol shares its goals with some recent efforts that are aimed at injecting functionality into application for adaptation. We categorize related

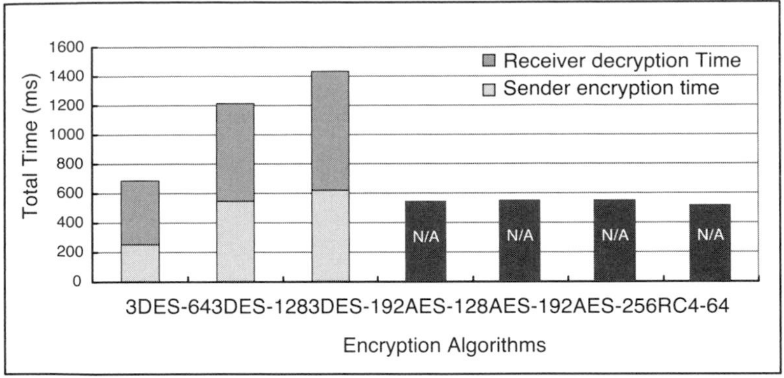

(a) Message receiver on Desktop

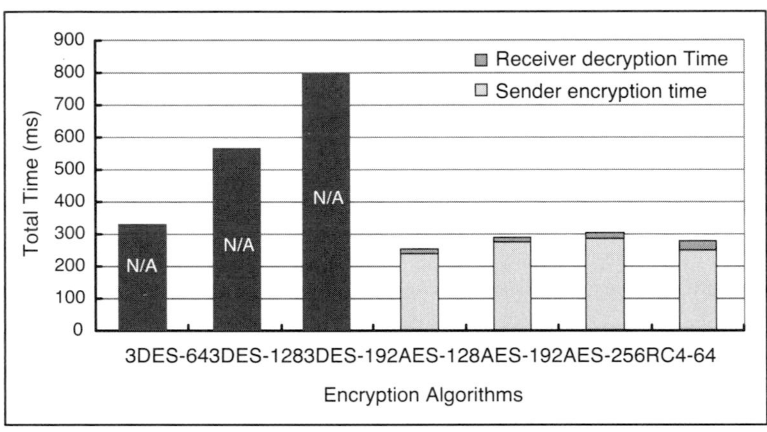

(b) Message receiver on Laptop

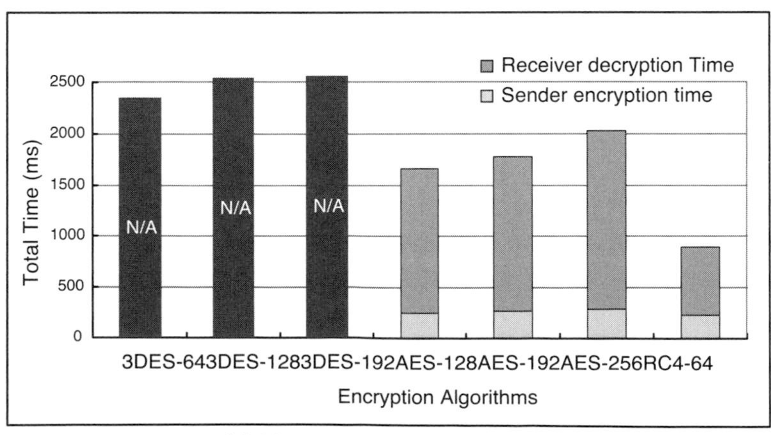

(c) Message receiver on Pocket PC

Figure 7. A comparison of the total time overhead for different message receivers: (a) Desktop, (b) Laptop, and (c) PocketPC.

research into three groups as *distributed adaptation, protocol adaptation*, and *mobile code and mobile agent.*

Distributed adaptation From the Internet topology's point of view, adaptation functionality can be introduced either at the end-points or distributed on intermediate nodes. Odyssey [4], Rover [3] and InfoPyramid [23] are examples of systems that support end point adaptation. Conductor [24] and CANS [2] provide an application transparent adaptation framework that permits the introduction of arbitrary adaptors in the data path between applications and end services. While these approaches provide an extremely general adaptation mechanism, significant change to existing infrastructure is required for their deployment. However, the adaptive encryption protocol does not have the deployment problem for leveraging the existing CDNs technology to distributed protocol adaptors, which are implemented using mobile code.

From the network structure's perspective, there are two issues: whether adaptation functionality is introduced at network layer with application-transparency or at the application level with application-awareness. Systems such as transformer tunnels [12] and protocol boosters [11] are examples of application-transparent adaptation efforts that work at the network level. Such systems can cope with localized changes in network conditions but cannot adapt to behaviors that differ widely from the norm. Moreover, their transparency hinders composability of multiple adaptations. More general are programmable network infrastructures, such as COMET [25], which supports flow-based adaptation, and Active Networks [26, 27], which permit special code to be executed for each packet at each visited network element. While these approaches are very general adaptation mechanisms, significant change to existing infrastructure is required for their deployment. The adaptive encryption protocol overcomes this shortcoming because it works entirely on the application level. Similar efforts also work at the application level. The cluster-based proxies in BARWAN/ Daedalus [28], TACC [29], and MultiSpace [30] are examples of systems where application-transparent adaptation happens in intermediate nodes (typically a small number) in the network. Active Services [31] extend these systems to a distributed setting by permitting a client application to explicitly start one or more services on its behalf that can transform the data it receives from an end service. Our work is different from other application level approaches in the following ways: first, it is not using intermediate nodes which may occur with deployment problems. Second it does not rely on any specific data stream or client conditions. On the contrary, it is designed to cope with any applications and client environments as long as one has the proper protocol adaptor.

Protocol adaptation There are some research work about the protocol adaptation. In network level systems such as [6], in which communicating end hosts use untrusted mobile code to remotely upgrade each other with the transport protocols that they use to communicate. Transformer tunnels [12] and protocol boosters [11] are doing application-transparent adaptation by tuning the network protocol according to the change of network situations. Such systems can deal with localized changes in network conditions but cannot react to changing environments outside the network layer. Since the adaptive encryption protocol works at the application layer, it can maximally adapt application level protocols which have no way to be completed in the network layer. It is also different from the Web browser plugins, e.g., Realplay, Flash, and so on. Plugin is an application component which completes part of the functionality, incapable of doing protocol adaptation. Although today some Web sites provide multiple choices of plugins to do the similar function, they still need the client to manually select one, but maybe not the best. The adaptive encryption protocol adapts the functionality by means of protocol adaptation which has transparency to the client and other characteristics, such as flexibility and extendibility, which plugins do not have.

Mobile code and mobile agent Mobile code is a good candidate for carrying a protocol module since it has long been known as a mechanism for providing a late binding of function to systems [32, 33, 34]. Mobile code and related technologies also have been proposed and studied as effective means of implementing content adaptation, protocol update, and program migration in distributed applications. In [35, 6] they propose a system in which communicating end hosts use untrusted mobile code to remotely upgrade each other with the transport protocols that are used to communicate. Our work is complimentary to their work because our proposal works in the application level. A new lightweight, component-based mobile agent system that can adapt to diverse devices and features resource saving is proposed in [36]. In this system, mobile code is brought in and associated execution states of an application dynamically after migration. NWSLite [37] provides a sophisticated predicting tools for the remote code execution offloaded from mobile client to the close server. To our best knowledge, the adaptive encryption protocol is the first approach to use mobile code to do encryption protocol adaptation that extends the utilization of mobile code technology.

Finally, a lot of encryption algorithms have been proposed [38], e.g., DES [14], AES [15], and RC4 [16], however, the focus of this paper is on selecting an appropriate encryption algorithm for a specific client configuration. Therefore, we envision our work complements to the research of cryptography algorithms very well.

7. SUMMARY

In this chapter, an adaptive encryption protocol is proposed to benefit the application from choosing the appropriate encryption algorithm according to dynamic client devices and application requirements. With the emergence of more and more cryptography algorithms, adaptation becomes a necessity because each algorithm has distinct characteristics from others although they are for the same application purposes. For the whole encryption algorithms family, some of them are very secure but require more computing power, while some of them are less secure but can run on tiny resource-constraint devices. We believe that the proposed adaptive encryption protocol makes an initial step towards using mobile code to support the application-level encryption protocol adaptation. The adaptive encryption protocol shows great flexibility in adapting encryption algorithms among multiple encryption algorithms in various client environments to reduce the total overhead without sacrificing of security, and it is a very promising approach in both conventional distributed applications and ever-increasing resource-constraint information appliances, such as smart phones, PocketPCs, and so on. Furthermore, we have observed that our prediction of the PAD overhead is partially based on the ratio between the client CPU speed and a standard CPU speed. This may not completely reflect the real situation in which client may have one or more of the following: (1) multiple processors, (2) multiple pipelines within in a single processor, (3) varying cache sizes, etc. All these can affect the computation time and hence the total delay time. Thus, it is very interesting to investigate the influence of diverse CPU architectures and adjust the adaptation model to reflect the variation accordingly.

8. REFERENCES

1. H. Lufei, W. Shi, Fractal: A mobile code based framework for dynamic application protocol adaptation in pervasive computing, in: Proc. of IPDPS'05.
2. X. Fu, W. Shi, A. Akkerman, V. Karamcheti, CANS: Composable, Adaptive Network Services Infrastructure, in: Proc. of USITS'01, pp. 135–146.
3. A. D. Joseph, J. A. Tauber, M. F. Kasshoek, Mobile Computing with the Rover Toolkit, IEEE Transaction on Computers 46 (3) (1997) 337–352.
4. B. D. Noble, Mobile Data Access, Ph.D. thesis, School of Computer Science, Carnegie Mellon University (May 1998).
5. A. Vahdat, M. Dahlin, T. Anderson, A. Aggarwal, Active names: Felxible location and transport of wide area resources, in: Proc. of USITS'99.
6. P.Patel, A.Whitaker, D.Wetherall, J.Lepreau, T.Stack, Upgrading transport protocols using untrusted mobile code, in: Proc. of ACM SOSP'03.
7. A. Fuggetta, G. P. Picco, G. Vigna, Understanding code mobility, in: IEEE TRANSACTIONS ON SOFTWARE ENGINEERING, 1998, p. Vol.24 No. 5.
8. D. Halls, Applying mobile code to distributed systems, Ph.D. thesis, Computer Laboratory University of Cambridge (1997).
9. Akamai Technologies Inc., Edgesuite services.
URL `http://www.akamai.com/html/en/sv/edgesuite_over.html`
10. B. Krishnamurthy, J. Rexford, Web Protocols and Practice: HTTP/1.1, Networking Protocols, Caching and Traffic Measurement, Addison-Wesley, Inc, 2001.
11. A. Mallet, J. Chung, J. Smith, Operating System Support for Protocol Boosters, in: Proc. of HIPPARCH Workshop, Uppsala Sweden, 1997.
12. P. Sudame, B. Badrinath, Transformer Tunnels: A Framework for Providing Route-Specific Adaptations, in: Proc. of USENIX'98.
13. N. W. Group, Tcp slow start, congestion avoidance, fast retransmit, and fast recovery algorithms.
URL `http://rfc.net/rfc2001.html`
14. Data Encryption Standard.
URL `http://www.itl.nist.gov/fipspubs/fip46-2.htm/`
15. Advanced Encryption Standard.
URL `http://csrc.nist.gov/CryptoToolkit/aes/`
16. RC4 RFC3268.
URL `http://www.faqs.org/rfcs/rfc3268.html/`
17. W. Diffie, M. Hellman, Multiuser cryptographic techniques, IEEE Transactions on Information Theory 22 (1) (1976) 644–654.

18. FIPS 180-1, Secure Hash Standard, U.S. Department of Commerce/N.I.S.T., National Technical Information Service, Springfield, VA, 1995.

19. Windows CE Operating Systems.
URL `http://www.microsoft.com/windowsce/`

20. Kinoma player.
URL `http://www.kinoma.com/`

21. Planetlab.
URL `http://planet-lab.org/`

22. H. Lufei, W. Shi, L. Zamorano, Communication optimization for image transmission in computer-assisted surgery, in: Proceedings of 2004 Congress of Neurological Surgeons Annual Meeting (abstract)), San Francisco, CA, 2004.

23. R. Mohan, J. R. Simth, C. Li, Adapting Multimedia Internet Content for Universal Access, IEEE Transactions on Multimedia 1 (1) (1999) 104–114.

24. M. Yarvis, A. Wang, A. Rudenko, P. Reiher, G. J. Popek, Conductor: Distributed Adaptation for complex Networks, in: Proc. of HotOS, 1999.

25. A. T. Campbell, et al., A Survey of Programmable Networks, ACM SIGCOMM Computer Communication Review 29 (2) (1999) 7–23.

26. D. Tennenhouse, D. Wetherall, Towards an Active Network Architecture, Computer Communications Review 26 (2).

27. D. J. Wethrall, J. V. Guttag, D. L. Tennenhouse, ANTS: A toolkit for building and dynamically deploying network protocols, in: Proc. of 2nd IEEE OPENARCH.

28. A. Fox, S. Gribble, Y. Chawathe, E. A. Brewer, Adapting to Network and Client Variation Using Infrastructural Proxies: Lessons and Prespectives, IEEE Personal Communication 5 (4) (1998) 10–19.

29. A. Fox, S. Gribble, Y. Chawathe, E. A. Brewer, P. Gauthier, Cluster-based Scalable Network Services, in: Proc. of SOSP'97.

30. S. D. Gribble, M. Welsh, E.A.Brewer, D. Culler, The MultiSpace: An Evolutionary Platform for Infrastructual Services, in: Proc. of Usenix'99.

31. E. Amir, S. McCanne, R. Katz, An active service framework and its application to real-time multimedia transcoding, in: Proc. of the SIGCOMM'98.

32. A. Birrell, G. Nelson, S. Owicki, E. Wobber, Network objects, in: Software-Practice and Experience, 1995, pp. 25(S4):87–130.

33. A. D. Joseph, A. F. deLespinasse, J. Tauber, D. Gifford, M. F. Kaashoek, Rover:a toolkit for mobile information access, in: Proc. of ACM SOSP'95.

34. E. Jul, H. Levy, N. Hutchinson, A. Black, Fine-grained mobility in the emerald system, in: ACM Tran. on Computer Systems, 1988, pp. 6(1):109–133.

35. P. Patel, D. Wetherall, J. Lepreau, A. Whitake, Tcp meets mobile code, in: Proc. of HTOS'03.

36. Y. Chow, W. Zhu, C. Wang, F. C. Lau, The state-on-demand execution for adaptive component-based mobile agent systems, in: Proc. of ICPADS'04.

37. S. Gurun, C. Krintz, R. Wolski, Nwslite: A light-weight prediction utility for mobile devices, in: Proc. of MobiSys'04, Boston, MA, 2004.

38. B. Schneier (Ed.), Applied Cryptography, Second Edition, Protocols, Algorthms, and Source Code in C, John Wiley & Sons, Inc., 1996.

Part I

SECURITY IN AD HOC NETWORK

3

PRE-AUTHENTICATION AND AUTHENTICATION MODELS IN AD HOC NETWORKS

Katrin Hoeper
Department of Electrical and Computer Engineering
University of Waterloo,
200 University Avenue West, Waterloo, Ontario, N2L 3G1, Canada
E-mail: khoeper@engmail.uwaterloo.ca

Guang Gong
Department of Electrical and Computer Engineering
University of Waterloo,
200 University Avenue West, Waterloo, Ontario, N2L 3G1, Canada
E-mail: ggong@calliope.uwaterloo.ca

Providing entity authentication and authenticated key exchange among nodes are both target objectives in securing ad hoc networks. In this chapter, a security framework for authentication and authenticated key exchange in ad hoc networks is introduced. The framework is applicable to general ad hoc networks and formalizes network phases, protocol stages, and design goals. To cope with the diversity of ad hoc networks, many configuration parameters that are crucial to the security of ad hoc networks are discussed. Special attention is paid to the initial exchange of keys between pairs of nodes (pre-authentication) and the availability of a trusted third party in the network. Next, several pre-authentication and authentication models for ad hoc networks are discussed. The models can be implemented as a part of the proposed security framework and correspond to the wide range of ad hoc network applications. Advantages and disadvantages of the models are analyzed and suitable existing authentication and key exchange protocols are identified for each model.

1. INTRODUCTION

The number of applications that involve wireless communications among mobile devices is rapidly growing. Many of these applications require the wireless network to be spontaneously formed by the participating mobile devices themselves. We call such networks *ad hoc networks*. The idea of ad hoc networks is to enable connectivity among any arbitrary group of mobile devices everywhere, at any time. We distinguish two categories of ad hoc networks, *mobile ad hoc networks (MANETs)* and *smart sensor networks*. Typical devices of MANETs are PDAs, laptops, cell phones, etc., and the devices of smart sensor networks are sensors. MANETs are used at business meetings and conferences to confidentially exchange data, at the library to access the Internet with a laptop, and at hospitals to transfer confidential data from a medical device to a doctor's PDA. Sensor networks can be used for data collection, rescue missions, law enforcement and emergency scenarios. Many more applications exist already or are imaginable in the near future. Caused by the widespread applications, a general security model and protocol framework for authentication and authenticated key establishment in ad hoc networks have not been defined yet.

1.1. Ad Hoc Network Properties

To achieve the ambitious goal of providing ubiquitous connectivity, ad hoc networks have special properties that distinguish them from other networks. We briefly discuss those properties in the following.

Ad hoc networks are *temporary* networks because they are formed to fulfill a special purpose and cease to exist after fulfilling this purpose. Mobile devices might arbitrarily join or leave the network at any time, thus ad hoc networks have a *dynamic infrastructure*. Most mobile devices use radio or infrared frequencies for their communications which leads to a very *limited transmission range*. Usually the transmission range is increased by using *multi-hop* routing paths. In that case a device sends its packets to its neighbor devices, i.e. devices that are in transmission range. Those neighbor nodes then forward the packets to their neighbors until the packets reach their destination. The most distinguishing property of ad hoc networks is that the networks are *self-organized*. All network interactions have to be executable in absence of a trusted third party (TTP), such as the establishment of a secure channel between nodes and the initialization of newly joining nodes. Hence, in contrast to wireless networks, ad hoc networks do not rely on a fixed infrastructure and the accessibility of a TTP. The self-organizing property is unique to ad hoc networks and makes implementing security protocols a very challenging task. Another characteristics of ad hoc networks are the *constrained network devices*. The constraints of ad hoc network devices are a small CPU, small memory, small bandwidth, weak physical protection and limited battery power, as first summarized in [23]. In most ad hoc networks all *devices have similar constraints*. This property distinguishes the architecture of an ad hoc network from a client-server structure.

1.2. Security Challenges

The special properties of ad hoc networks enable all the neat features such networks have to offer, but at the same time, those properties make implementing security protocols a very challenging task. There are four main security problems that need to be dealt with in ad hoc networks: (1) the *authentication* of devices that wish to talk to each other; (2) the *secure key establishment* of a session key among authenticated devices; (3) the *secure routing* in multi-hop networks; and (4) the *secure storage of (key) data* in the devices. Note, that once (1) and (2) are achieved, providing confidentiality is easy. In the remainder of this article, we will focus on entity authentication and authenticated key establishment (AAKE) protocols and their implementation issues in ad hoc networks. Note that most security problems related to such protocols occur in the bootstrapping phase, i.e. at the time nodes wish to securely communicate for the very first time. We refer to this phase as the *pre-authentication phase*, and we define and discuss this stage in great detail later in this chapter.

1.3. Outline

As said earlier, due to the wide range of ad hoc network applications, no general security framework has been introduced yet. In this chapter, we introduce a security framework for authentication and authenticated key exchange in ad hoc networks. The framework is applicable to general ad hoc networks and formalizes network phases, protocol stages, and design goals. To cope with the diversity of ad hoc networks, we discuss many configuration parameters that are crucial to the security of ad hoc networks. We pay special attention to the initial key exchange between pairs of nodes (pre-authentication) and the availability of a TTP in the network. We then categorize several pre-authentication and authentication models that can be implemented as a part of the proposed security framework. The models correspond to the wide range of ad hoc network applications and we analyze their advantages and disadvantages and identify suitable existing authentication and key exchange protocols for each model.

The rest of this chapter is organized as follows. In Section 2, we introduce a security framework for ad hoc networks, including network and authentication phases, protocol stages and design goals. In Section 3, we identify some security related configuration problems that are crucial for protocol implementations in many ad hoc network applications. Taking all previous results into account, we categorize and analyze a number of pre-authentication and authentication models in Section 4 and 5, respectively. Finally, in Section 6, conclusions are drawn.

2. SECURITY FRAMEWORK

In this section, we first discuss the different network phases that occur in the lifecycle of an ad hoc network. Then, we introduce the two authentication phases of communicating nodes in such networks. Next, we define the protocol stages of general AAKE protocols in ad hoc networks. At the end of this section, we summarize the design goals all protocols that are designed for ad hoc networks should meet. All these definitions combined form a security framework for general ad hoc networks. The framework helps designing security solutions for ad hoc networks. In particular, when proposing protocols for ad hoc networks, all network and authentication phases, protocol stages and design goals as defined in this security framework need to be addressed.

2.1. Network Phases

We distinguish two network phases in ad hoc networks, namely the *network initialization phase* and the *running system phase*. In the first phase, the network is set up. All nodes that are present at the network initialization phase, i.e. during the time the network is formed, are initialized. The self-organization property of the network is sometimes not required in this phase. For instance, a TTP might be available in the initialization phase in order to initialize all present nodes with required data, such as system parameters and cryptographic keys. After the initialization phase, nodes can freely join or leave the network at any time. We refer to this as running system phase. Ad hoc networks are generally self-organized in this phase. This follows that no TTP or other fixed infrastructure is longer available. Consequently, current network nodes are responsible to initialize newly joining nodes with required key material, cope with leaving nodes and execute all other necessary administrative tasks in a self-organized manner.

2.2. Authentication Phases

We distinguish two authentication phases for authentications among network nodes. The first phase consist of the initial exchange of data and cryptographic key material among a group of two or more nodes. The data can include secret or public keys, for example. The same data is used to identify each other in all later authentications among the same nodes. The described initial authentication phase is called *imprinting* in the resurrecting duckling model [23], and *initialization* in the Bluetooth protocol [4]. Henceforth, we will adopt the term *pre-authentication* from [2]. The data that is exchanged in the pre-authentication phase needs to be sent over a secure channel, where secure refers to an authentic and confidential channel for exchanging symmetric key data, and to an authentic channel for exchanging public keys in asymmetric schemes. Pre-authentication is not limited to the devices present at the time of the network initialization phase, it also needs to be provided to subsequently joining nodes in the running system phase. All nodes that subsequently join the ad hoc network need to be able to securely obtain shared data and required key material from all potential communication partners. The main challenge is to provide pre-authentication in the running system phase, even though the network environment might have changed and a TTP is not accessible any longer. During the second phase, the *authentication phase*, the nodes identify each other by using the authentic data that was exchanged in the pre-authentication phase. These authentications are executed over an insecure channel and need to be secured by the key material exchanged during pre-authentication.

2.3. Protocol Stages

We now consider the protocol stages of a two party AAKE protocol. The desired AAKE protocol should first provide pairwise pre-authentication, then mutual authentication between the same two nodes, and lastly, a secure establishment of a session key shared between the nodes. All AAKE protocols can be executed in the *running system* in an ad hoc network, i.e. after the network initialization phase. A suitable AAKE protocol should take all ad hoc network properties and constraints into account. Note, that the protocol design goals are defined in the next section.

A typical AAKE protocol in our security framework for ad hoc networks consists of the following three stages:

1. Pre-Authentication
The first stage is the pre-authentication between two devices that wish to communicate with each other at this or a later time. In this phase either a secret key or an authentic copy of a public key are securely shared between the devices. Keys can be shared during pre-authentication using one of the pre-authentication models that we will introduce in Section 4. The best suited model needs to be chosen according to the particular application.

The key data that has been exchanged or established during pre-authentication is used in all subsequent authentications between the same nodes. Hence, the next time the same nodes wish to securely communicate, i.e. to execute an AAKE protocol, the nodes can skip the pre-authentication stage and directly start with the authentication. Pre-authentication needs only to be repeated if keys are revoked or expired.

2. Authentication
In the second stage, the authentication stage, the participants mutually authenticate each other using the key data from the pre-authentication phase. A suited authentication protocol can be chosen out of the authentication models introduced in Section 5. The best suited protocol needs to be chosen according to the respective application. If the authentication of one node fails, the protocol stops and further countermeasures might be taken, for example revoking the key of the rejected node.

3. Session Key Establishment
Upon successful mutual authentication, the nodes start establishing a session key in the third protocol stage. Note that all session keys need to be established over an authentic channel. Otherwise, Oscar could take over Alice's role after her successful authentication to Bob. To overcome this attack, the session key establishment stage can be combined with the previous authentication stage. Again, for suitable AAKE protocols please refer to Section 5.

2.4. Design Goals

After discussing the special properties and needs of ad hoc networks and several of the issues that occur when implementing protocols in such networks, we now derive the design goals that all ad hoc network protocols should meet in order to be suitable.

All protocols should only require *few computational steps* due to the limited battery power of all ad hoc devices. Too many computational steps would drain the battery. For the same reason protocols should only require *few message flows*. Caused by the nature of wireless networks, the communication bandwidth is very small. If messages are too large, they will be split into several packets. Sending many packets contradicts with the previous design goal, therefore *small data packages* are desirable. Due to the limited computational power of ad hoc devices, preferable protocols should mainly require *cheap computations.* As a general trend, the processors of most ad hoc devices, such as PDAs, are becoming more and more powerful, and therefore heavy computations, such as modular exponentiations, are becoming feasible. However, heavy computations require more battery power, and thus, it is important to restrict the number of heavier computations. Based on the assumption that all ad hoc network devices have similar constraints, suited protocol should be *balanced*, i.e. all devices need to perform approximately the same number of equally heavy computations. Considering the very limited memory space of all devices, protocols should neither require much memory space for the protocol code itself nor for the storage of parameters and key material. As a consequence, *short code, short keys and short system parameters* are desirable. When designing protocols the *consequences of data disclosure should be very restricted* because ad hoc network devices and especially sensors provide only a low level of physical protection. Once an attacker gains access to the device, he/she is usually able to obtain the stored data, including the key material. Note that this attack is quite reasonable since such devices cannot be protected as some servers that are locked away in secure rooms, for instance. The protocol should be designed in a way that the disclosure of the stored data does not compromise the entire system. Also the delectability of such disclosures within the system needs to be examined when designing a protocol.

In addition to the previous design objectives, protocol designed for sensor networks should be *scalable* to cope with the large number of sensors in the network and be *fault tolerant* because sensors are very prone to failures.

3. SYSTEM CONFIGURATIONS

In this section, we identify the problems that one might encounter when implementing AAKE protocols in ad hoc networks. Therefore, we consider several system settings that occur in different applications, such as the availability of a TTP, the security of the communication channels, the constraints of the devices, the number of participating domains, etc..

3.1. Availability of Trusted Third Party (TTP)

The availability of a TTP is crucial for a protocol implementation and one of the new challenges of ad hoc networks. A TTP can play several roles in a network, for instance, the TTP could be responsible to initialize devices with secret keys, issue and distribute public keys and certificates, distribute session keys to devices that wish to securely communicate, or help to verify the validity of certificates by providing certificate revocation lists (CRLs). We distinguish among four different settings for the availability of a TTP, described in the following paragraphs and illustrated in Figure 1. The four rows in the figure correspond to the four settings, where the first column describes the network initialization phase, the second column the event of a joining

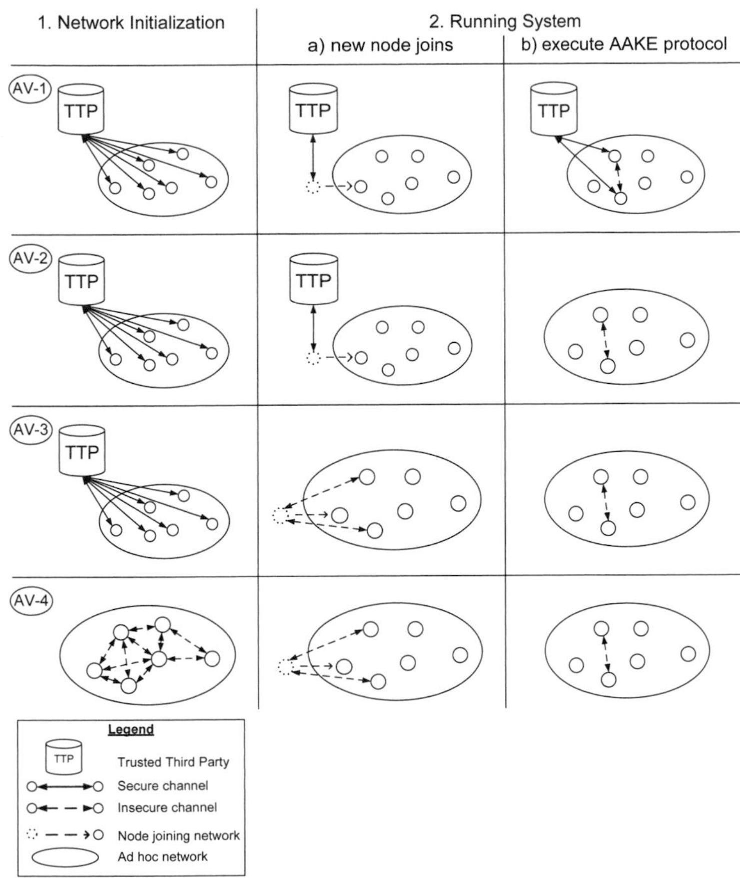

Figure 1. Four scenarios of TTP availabilities *AV-1* – *AV-4*, as described in Section 3.1: (1) during the network initialization; and (2) in the running system when (a) new nodes join or (b) present nodes establish a secure channel, i.e execute an AAKE protocol.

node in the running system phase and the third column the event of present nodes establishing a secure channel, i.e executing an AAKE protocol.

AV-1: TTP is always available

The case that a TTP is accessible by all network nodes at any time is generally not considered as an option in ad hoc networks, because ad hoc networks should be self-organized after their initialization phase. However, in the future it might be reasonable to assume an Internet connection in some as hoc network applications, for example via an access point. In that case, we could adopt WLAN solutions and modify them to cope with the resource constraints and mobility of ad hoc network devices.

AV-2: TTP is available at network initialization phase and every time a node joins
The second option comprises all scenarios where a TTP is available at the network initialization phase and, in addition, the TTP is accessible by all nodes that subsequently join the network. This assumption is not as restrictive as it might seem, because the TTP does not need to be accessible by all network nodes every time a new node joins a network. For instance, there could be applications in which nodes contact a TTP to receive the required system parameters and keys before joining the network. The network itself is still self-organized and the present nodes have no access to a TTP.

AV-3: TTP is available at network initialization phase
In this scenario only the nodes that were present at the time of the network initialization phase are initialized by the TTP. Usually this is called self-organization property of the network. The present network nodes are responsible to take over the tasks of the TTP, such as issuing and distributing keys and/or certificates to subsequently joining nodes.

AV-4: No TTP is available at any network phase
In this scenario network nodes need to take over the tasks of the TTP during the network initialization phase and in the running system phase. If no TTP is available at any time, implementing security protocols such as AAKE protocols is very challenging. If we want to implement a symmetric scheme we would need to develop a security model in which devices can securely exchange their common keys. Whereas implementing a public key encryption schemes would require an authentic channel to exchange public keys without the aid of a TTP that issues keys or key credentials.

3.2. Other Configuration Parameter

There are many other implementation issues that depend on the particular ad hoc network application. We will discuss some of those issues that could affect the implementations of security protocols.

First of all, the *security of the communication channels* is a crucial parameter in ad hoc network applications. We distinguish two communication channels. One channel to exchange the data during the pre-authentication phase and another channel for the authentication and key establishment phases. As discussed earlier, pre-authentication requires a secure channel among the devices to securely exchange authentic public key data or authentic and confidential secret key data. Upon pre-authentication, all communications can be executed over an insecure channel where the communication is secured by the key material that was exchanged during pre-authentication. How a secure pre-authentication or authentication channels can be established is discussed in Section 4 and 5, respectively.

Another implementation issue is the *level of resource constraints*. Depending on the computational constraints of the network devices it might be feasible or infeasible to execute protocols requiring heavy or many computations, as required in most public key schemes. In addition to the computational constraints, we have to consider the communication and power constraints when designing or implementing a protocol. Generally sensors are too constrained for implementing public key protocols.

Hierarchical ad hoc networks haven been proposed as alternative to flat ad hoc topologies to overcome some limitations of the latter, as for instance described in [5]. Hierarchical ad hoc networks have several layers, where each layer consists of a set of similar devices. For instance, the lowest layer consists of the least powerful devices, e.g. sensors, and each higher level consists of some more powerful devices, where the top level could be the Internet. In this way, all heavy computations could be shifted from the very constrained devices to the more powerful ones and thus asymmetric schemes could become feasible. For this reason, the model is attractive for sensor networks. It needs to be analyzed for particular applications if it is reasonable to assume that higher layers can be accessed by all sensor networks at any time.

When Stajano and Anderson [23] were among the first to consider the special properties of ad hoc networks, they assumed a *controller* (mother duck) and several devices that are controlled (ducklings) in all ad hoc networks. In the proposed resurrecting duckling model, the mother duck imprints their ducklings, who, from then on, follow their mother. In another more recent paper Messerges et. al [21] described some applications that require a controller, e.g. sensor networks used for industrial control and building automation. In networks without a controller all nodes have similar roles and are assumed to have similar resource constraints. Whether we have an ad hoc network with or without controller depends on the application.

In some scenarios devices might be *aware of their location* and are able to provide information about their location, such as their geographical coordinates. A simple solution for providing the present location of mobile devices is to embed an additional integrated chip, such as a GPS chip, in all devices. For instance, some high-end PDAs are already equipped with GPS chips. However, there are many different systems that provide location coordinates depending on the network range and location. The most commonly known systems for tracking down devices are: (1) satellite navigation systems, such as GPS, or the European equivalent Galileo; (2) systems for locating devices inside a building using visual, ultra sonic, radio, or infrared channels; and (3) network based positioning system, such as GSM, and WLAN. If a user knows the location of its communication partner the data could be used to build an authentic channel, e.g. for authentication or public key exchange. However, special equipment for tracking devices is unnecessary if the location of devices is predictable. For instance, in some sensor networks, the sensors have an expected location. This knowledge is used in a location-based pairwise key establishment protocol [18], for instance.

The last system property we consider is the *number of domains* in our network. All devices in one domain share the same domain parameters, such as shared keys, that has been distributed during the network initialization, a certificate issued by the domain's certification authority (CA), or system parameters required for some computations. In most sensor networks, it is reasonable to assume one domain. However, in many MANETs, devices are from different domains. Providing authentication in those scenarios is harder to implement. Communicating parties need mechanisms to verify the trustworthiness of devices outside their own domain and to securely agree on some common system parameters. These compatibility issues have to be considered when implementing an AAKE protocol.

4. PRE-AUTHENTICATION MODELS

In this section we discuss several symmetric and asymmetric pre-authentication models (PAMs) for providing pre-authentication in ad hoc networks. We summarize all models for better comparison in Table 1. We reference some papers that introduced protocols in the respective models in the second column and summarize the advantages and disadvantages of each model in the right column.

4.1. Symmetric Solutions

When using symmetric encryption a secret must be shared among the devices that wish to communicate. The secrets are established during the network initialization phase and the pre-authentication phase of the devices. Clearly, an authentic and confidential channel needs to be established to ensure secure pre-authentication. The following models describe how such a secure channel can be established.

PAM-S1. Secure Side Channel Model
In this model the secret information is exchanged over a secure side-channel during the network initialization phase and the pre-authentication phase of the devices. How this secure channel is established is not further specified in the model and left to be done by the users or the administrators that implement the protocol. For instance, the IEEE standard for wireless local area networks (WLAN) IEEE 802.11 [14] does not provide any recommendations and information of how pre-authentication can be achieved and assumes that pre-authentication has taken place before devices start communicating with each other. Hence, IEEE 802.11 is a protocol standard proposed in the discussed model.

PAM-S2. PIN Model
Protocols in this model require that passwords, PINs, or keys are manually entered in all devices that wish to securely communicate. This can be done by an administrator during the network initialization phase or by users as pre-authentication of their devices. Solutions in this model do not scale well because the secret needs to be entered manually in each device. An example for a protocol in this model is the Bluetooth protocol that was introduced by the Bluetooth Special Interest Group (SIG) [4]. The protocol is standardized as IEEE 802.15 [14] for wireless personal area networks (WPANs).

PAM-S3. Physical Contact Model
In this model the symmetric keys are exchanged by physical contact among the devices. Note that the physical contact provides an authentic and confidential channel. The requirement of physical contact among all communicating devices is be too restrictive in some applications. A protocol in this model is introduced in [23].

Table 1. Pre-authentication models for ad hoc networks

Model	Implementation	Comments*
PAM-S1. Secure Side-Channel	Keys exchanged over secure side-channel, e.g. IEEE 802.11 [14]	– secure channel not provided by system itself
PAM-S2. PIN	PIN manually entered in all devices, e.g. Bluetooth [4]	– does not scale well
PAM-S3. Physical Contact	Key exchanged by physical contact, e.g. resurrecting duckling protocol [23]	– requires proximity of the devices
PAM-S4. Pairwise Key Pre-Distribution	Sensors initialized with subset of key pool before deployed, e.g. [8]	– only one domain – requires TTP for every initialization
PAM-A1. Location-Limited	Public key directly exchanged, e.g. [2, 6]	– requires proximity of devices
PAM-A2. ID-Based	Identity used as self-authenticated public key, e.g. [15, 12]	+ implicit pre-authentication – KGC is key escrow
PAM-A3. Self-Certified Public Key	Certificate embedded in public key, e.g. [9]	+ implicit pre-authentication – no AAKE protocols
PAM-A4. Distributed CA	CA represented by n nodes using threshold scheme [27, 16, 19]	+ self-organized – not efficient† – requires many nodes
PAM-A5. Trusted Path	PGP-like; find trusted path between two nodes, e.g. [11, 7]	+ self-organized – not efficient – requires many nodes

* "+"/"–" denote advantages and disadvantages of the model, respectively.
† Efficiency with respect to computation and communication cost.

PAM-S4. Pairwise Key Pre-Distribution Model
Public key cryptography is not feasible in sensor networks and therefore only symmetric schemes are applicable. The approach that all sensors share the same secret key is not suited because once a single key is compromised the entire sensor network would be compromised as well. Due to the weak physical protection of sensors, compromising a single sensor and thus its stored key material is very likely. For this reason, sharing keys in a pair-wise fashion seems to be a more reasonable approach. Since sensors have very constrained memory, they cannot store symmetric keys of every other sensor in the network. To overcome this constraint, the *pairwise key pre-distribution model* is introduced, in which each sensor is initialized with a subset of all network keys. Note that all sensors need to belong to the same domain. However, in most sensor network applications, it can be assumed that a trusted authority can set-up all sensors before they are deployed. An example of a protocol in this model is in [8].

4.2. Asymmetric Solutions

We describe several asymmetric pre-authentication models (PAM-As) in this section. Each model provides a method to obtain an authentic copy of the public key of a communication partner. The lack of a central CA is the main problem when implementing asymmetric protocols in ad hoc networks. We distinguish four categories of PAM-As: (1) with CA and use of certificates; (2) with CA and no use of certificates; (3) without CA and use of certificates; and (4) without CA and no use of certificates. The first category includes the distributed CA model; the second one includes the identity-based model and the self-certified public key; the third category contains the trusted path model; and the fourth contains the location-limited model.

PAM-A1. Location-Limited Model
If proximity of the ad hoc network devices is given, a secure pre-authentication channel can be established by visual or physical contact among the communicating devices. This secure pre-authentication channel enables the devices to directly exchange their public keys, i.e. without the necessity of a CA and public key certificates. This model is based on two assumptions, (1) all participants are located in the same room; and (2) all participants trust each other a priori. The model is well suited in all scenarios that meet those two assumptions and not applicable in any other scenario. Protocols in this model are introduced in [2, 6]. Note that in most cases where devices can perform physical contact implementing the physical contact model (PAM-S3) seems to be more reasonable.

PAM-A2. Identity-Based Model
Identity (ID)-based schemes, introduced in [22], do not require any key exchange prior to the actual authentication, because common information, such as names and email addresses, is used as public key. Since public keys are self-authenticating, certificates are redundant in this model. Pre-authentication is implicitly provided by the system because the (authentic) public keys of all network devices are known prior communicating. As a consequence, protocols in the *identity-based model* do not require any

secure pre-authentication channel. This feature makes the ID-based model attractive for ad hoc networks. ID-based schemes require a TTP that serves as key generation center (KGC) in the network initialization phase in order to generate and distribute the personal secret keys to all users. Drawbacks of this model are: (1) the KGC knows the secret keys of all users; and (2) a confidential and authentic channel between the CA and each network device is required for the securely distribution of the secret keys. The latter problem can be eliminated by using a blinding technique as shown in [17] and the first drawback is shown to have low impact in ad hoc networks in [13]. The first protocol in this model is in [15], but the authors do not provide an actual AAKE protocol and many open questions for a protocol implementation remain. Some AAKE protocols in this model are in [12].

PAM-A3. Self-Certified Public Key Model

In this model the certificates are embedded in the public keys themselves. So-called self-certified public keys are introduced in [9] and, other than in ID-based schemes, the identity itself is not directly used as a key. In fact the identity is a part of the user's public key and signed by a CA and the users themselves. Hence, the public keys are unpredictable and need to be exchanged prior to the communication. The authenticity of the public keys is provided by the keys themselves, and thus we do not need a secure pre-authentication channel. In addition, this approach helps to save some bandwidth and memory space, because certificates do not need to be transmitted and stored. A CA is required to generate the self-certified public keys using the devices' public keys, identifiers, and the CA's master secret key as input. Protocols in this model are an authentication protocol without key agreement and a static DH-like key agreement [9].

PAM-A4. Distributed CA Model

In the distributed CA model the power and the tasks of a CA are distributed to t network nodes by implementing a (t, n)-threshold scheme. The idea is based on the fact that a CA should not be represented by a single node, because nodes can be relatively easily compromised by an adversary. In this model a group of t nodes can jointly issue and distribute certificates. Protocols in this model might support certificate renewing and revocation. We distinguish two cases, (1) a *distributed CA with special nodes* and (2) a *distributed CA without special nodes*. In the first case t special nodes, that have more computational power and that were present at the network initialization phase, represent the CA. The special role of the server nodes contradicts with the property of similar constrained devices as stated in our design goals. An example of a protocol that has been proposed in this model is [27]. In the distributed CA model without special nodes, any t network node represent the CA and can thus issue certificates. Protocols that have been proposed in this model are [16, 19].

PAM-A5. Trusted Path Model

The trusted path model emphasizes the self-organization property which is a unique and challenging feature of ad hoc networks. Network nodes issue and distribute their own certificates and sign other certificates. The model assumes the existence of trust between some nodes and generates trust between nodes in a PGP manner, i.e. by finding a so-called trusted path consisting of certificates between the communicating nodes. The performance of pre-authentication highly depends on the length of the trusted path, which is generally hard to predict. This approach is very efficient in the set-up phase and does not require any heavy computation steps from any parties other than the communicating ones. However, a node probably needs to verify more than one certificate for pre-authentication. An example of a proposed protocol in this model is [11]. This model is also applied to a group case, in which trusted subgroups search for intersections to create a trusted path [10].

5. AUTHENTICATION MODELS

After the pre-authentication phase, the exchanged key material can be used to enable authentication and key establishment in any of the authentication models (AMs) described in this section. We briefly discuss some symmetric, hybrid, and asymmetric authentication models (AM-S, AM-H, AM-A) and summarize the models in Table 2, where we reference proposed protocols in the second column and summarize advantages and disadvantages of the models in the right column. Please note that many more models exist and we only present a small subset, where we limit our focus to models suitable to ad hoc networks.

5.1. Symmetric Solutions

After successful pre-authentication has taken place in any of the previously described symmetric pre-authentication models, we can run any symmetric AAKE protocol.

AM-S1. Challenge-response Using Symmetric Schemes
The devices can use their shared key in a challenge-response type protocol [20], in which devices authenticate each other by demonstrating knowledge of the shared key by encrypting a challenge.

5.2. Hybrid Solutions

Some ad hoc network solutions combine symmetric and asymmetric crypto schemes to provide entity authentication and optionally key establishment after pre-authentication phase has taken place in any of the presented pre-authentication models.

AM-H1. Password Model
Depending on the available memory size and the way the secret is exchanged, it might be desirable to share a short password instead of a long secret key. Note that such passwords are weak secret keys. Due to their shortness, passwords are prone to brute-force attacks, where user-friendly passwords are also prone to off-line dictionary-attacks. The password needs to be securely exchanged by one of the PAM-S discussed in Section 4.1. AAKE protocols in the *password model* combine a weak password and an asymmetric scheme to obtain a strong shared key and are called password-authenticated

Table 2. Authentication models for ad hoc networks

Model	Implementation	Comments*
AM-S1. Challenge-Response	Devices demonstrate knowledge of shared key by encrypting a challenge [20]	+ efficient† – requires long and secure shared secret
AM-H1. Password	Shared password is used for encrypting public keys, e.g. [3, 1]	+ requires short (memorizable) password – not efficient
AM-A1. Challenge-Response	Devices either decrypt a challenge that is encrypted under their public key or sign a challenge [20]	– not efficient
AM-A2. Key Chain	Anchor x_0 of hash chain is private key, x_n public key [24, 25, 26]	+ very efficient – no key agreement

* "+"/"–" denote advantages and disadvantages of the model, respectively.
† Efficiency with respect to computation cost.

key exchange (PAKE) protocols [3]. Due to the use of asymmetric crypto schemes, PAKE protocols require some heavy computational steps, and are thus only applicable to ad hoc networks consisting of powerful devices that have sufficient computation power. Examples of protocols in this model are [3] for the two-party case and [1] for the multi-party case.

5.3. Asymmetric Solutions

We can implement any asymmetric AAKE protocol after pre-authentication has taken place in one of the PAM-As discussed in Section 4.2.

AM-A1. Challenge-Response Using Asymmetric Schemes
Once the devices share authentic copies of their public keys they can use either these public keys or their own private keys to prove their identities. A common method are challenge-response type protocols [20]. The establishment of an encryption key for the current session may be a part of the protocol as well.

AM-A2. Key Chain Model
Hash chains [20] are an asymmetric approach that is attractive for ad hoc network due to the excellent performance. In hash chain schemes, a hash function $h(\cdot)$ is applied n times to a random value x. The initial value $x_0 = x$ is the so-called anchor which serves as the private key, whereas the last value of the hash chain $x_n = h^n(x)$ serves

as public key. Each device first computes its own hash chain, also called key chain, then authentically exchanges x_n with its communication partners in one of the pre-authentication models described in Section 4.2. The value x_0 is kept secret. A device that is challenged by a value x_i from its key chain can prove its identity by responding with the previous value x_{i-1} of the chain. Only a device that knows the anchor x_0 is able to compute the required response. This scheme requires only the computation of hash values which can be implemented very efficiently. Note that schemes implementing key chains provide only unidirectional authentication and no key is established during the protocol execution. Examples of the protocols in this model are [24, 25]. Note that the protocol in [24] is broken and fixed in [26].

6. CONCLUSIONS

Authentication and authenticated key exchange are both identified as primary security objectives in ad hoc networks. In this chapter, we introduced a security framework for general ad hoc networks to achieve these two goals. As part of the security framework we defined network phases, protocol stages, and design goals. Next, we coped with the diversity of ad hoc network applications. Therefore, we identified crucial configuration parameters of particular applications, that need to be taken into account when implementing AAKE protocols in these scenarios. Here, our special focus was on the availability of a TTP, but many other security related configuration parameters were discussed as well. Considering all special network and device constraints, we derived a set of design objectives for general ad hoc network protocols. Taking all previous results into account, we finally categorized a number of pre-authentication and authentication models based on symmetric, hybrid, and asymmetric cryptographic schemes. The models can be implemented as a part of the security framework and they correspond to the diversity of ad hoc network applications. We analyzed the models, pointed out their advantages and disadvantages, and showed for which application particular models are best suited. Furthermore, we identified several previously proposed AAKE protocols that are suitable for each model. Our results can be used as a toolbox for designing and analyzing AAKE protocols as well as a guideline for choosing the best suited protocol for particular ad hoc network applications.

We conclude from our analysis that some commercial ad hoc network applications can be securely and efficiently implemented by existing symmetric solutions. The PIN model is applicable to all PANs, in which a user can set up all of his/her devices with one PIN or password, or an administrator is able to set up all authorized devices in order to share network resources. The physical contact model is suitable for all applications where people or devices, who already trust each other, are located in a small area. Protocols in the pre-distribution scheme are suited in sensor networks in which all sensors belong to one domain. An asymmetric approach which seems to be suitable for mobile device-terminal connections is the exchange of public keys over a location-limited channel. This approach could be implemented in some civil applications, such as virtual classrooms, internet access points, and all communications between PDAs and laptops of different users. The approach is limited to networks with a small number of devices that provide moderate computational power. All approaches in the distributed CA and trusted path model are only suitable for networks with a large number of nodes. Furthermore, we believe that these two models are not efficient in terms of the computational and communication overhead. The identity-based and the self-certified public key model are both promising pre-authentication models because they do not require a secure channel. However, those models need to be further studied and protocols have to be proposed.

7. REFERENCES

1. N. Asokan and P. Ginzboorg, Key Agreement in Ad Hoc Networks, *Computer Communications*, Vol. 23, No. 17, 2000, pp. 1627-1637.
2. D. Balfanz, D.K. Smetters, P. Stewart, and H. Chi Wong, Talking to Strangers: Authentication in Ad-Hoc Wireless Networks, *Proceedings of Network and Distributed System Security Symposium 2002 (NDSS '02)*, 2002.
3. S.M. Bellovin and M. Merritt, Encrypted Key Exchange: Password-Based Protocols Secure Against Dictionary Attacks, *Proceedings of the 1992 IEEE Symposium on Security and Privacy*, IEEE Computer Society, ISBN: 0-8186-2825-1, 1992, pp. 72-84.
4. Bluetooth SIG, *Specification of the Bluetooth System*, Version 1.1; February 22, 2001, available at `https://www.bluetooth.com`.
5. M. Bohge and W. Trappe, An authentication framework for hierarchical ad hoc sensor networks, *Proceedings of the 2003 ACM workshop on Wireless security*, ISBN:1-58113-769-9, ACM Press, 2003, pp.79-87.
6. M. Cagalj, S. Capkun and J.P. Hubaux, Key agreement in peer-to-peer wireless networks, to appear in *Proceedings of IEEE, Special Issue on Security and Cryptography, 2005.*
7. S. Čapkun, J.-P. Hubaux, and L. Buttyán, Self-Organized Public-Key Management for Mobile Ad Hoc Networks, *IEEE Transactions on Mobile Computing*, Vol. 2, No. 1, 2003, pp. 52-64.
8. L. Eschenauer and V.D. Gligor, A Key-Management Scheme for Distributed Sensor Networks, *9th ACM conference on Computer and Communications Security*, ISBN:1-58113-612-9, ACM Press, 2002, pp. 41-47.
9. M. Girault, Self-Certified Public Keys, *Advances in Cryptology- EUROCRYPT '91*, LNCS 547, Springer, 1991, pp. 490-497.
10. S. Gokhale and P. Dasgupta, Distributed Authentication for Peer-to-Peer Networks, *Symposium on Applications and the Internet Workshops 2003 (SAINT'03 Workshops)*, IEEE Computer Society 2003, ISBN 0-7695-1873-7, 2003, pp. 347-353.
11. J.P. Hubaux, L. Buttyán and S. Čapkun, The Quest for Security in Mobile Ad Hoc Networks, *ACM Symposium on Mobile Networking and Computing –MobiHOC 2001*, 2001.
12. K. Hoeper and G. Gong, Identity-Based Key Exchange Protocol for Ad Hoc Networks, *Canadian Workshop of Information Theory -CWIT '05*, 2005.
13. K. Hoeper and G. Gong, Short Paper: Limitations of Key Escrow in Identity-Based Schemes in Ad Hoc Networks, *Security and Privacy for Emerging Areas in Communication Networks –SecureComm '05*, 2005.
14. IEEE 802.11, Standard Specifications for Wireless Local Area Networks, `http://standards.ieee.org/wireless/`.

15. A. Khalili, J. Katz, and W. Arbaugh, Toward Secure Key Distribution in Truly Ad-Hoc Networks, *2003 Symposium on Applications and the Internet Workshops (SAINT 2003)*, IEEE Computer Society, ISBN 0-7695-1873-7, 2003, pp. 342-346.

16. J. Kong, P. Zerfos, H. Luo, S. Lu, and L. Zhang, Providing Robust and Ubiquitous Security Support for Mobile Ad-Hoc Networks, *International Conference on Network Protocols (ICNP) 2001*, 2001.

17. B. Lee, C. Boyd, E. Dawson, K. Kim, J. Yang, and S. Yoo, Secure key issuing in ID-based cryptography, *CRPIT '04: Proceedings of the second workshop on Australasian information security, Data Mining and Web Intelligence, and Software Internationalisation*, Australian Computer Society, Inc., 2004, pp. 69-74.

18. D. Liu and P. Ning, Location-Based Pairwise Key Establishments for Static Sensor Networks, *1st ACM Workshop Security of Ad Hoc and Sensor Networks (SASN) '03*, ISBN:1-58113-783-4, ACM Press, 2003, pp. 72-82.

19. H. Luo, P. Zerfos, J. Kong, S. Lu, and L. Zhang, Self-Securing Ad Hoc Wireless Networks, *Seventh IEEE Symposium on Computers and Communications (ISCC '02)*, 2002.

20. A.J. Menezes, P.C. von Orschot, and S.A. Vanstone, *Handbook of Applied Cryptography*, 1997 by CRC press LLC.

21. T.S. Messerges, J. Cukier, T.A.M. Kevenaar, L. Puhl, R. Struik, and E. Callaway, A security design for a general purpose, self-organizing, multihop ad hoc wireless network, *1st ACM workshop on Security of ad hoc and sensor networks (SASN) '03*, ISBN:1-58113-783-4, ACM Press, 2003, pp. 1-11.

22. A. Shamir, Identity-based Cryptosystems and Signature Schemes, *Advances in Cryptology- CRYPTO '84*, LNCS 196, Springer, 1984, pp. 47-53.

23. F. Stajano and R. Anderson, The Resurrecting Duckling: Security Issues for Ad-Hoc Wireless Networks, *In Proceedings of the 7th International Workshop on Security Protocols*, LNCS 1796, Springer, pp. 172-194, 1999.

24. A. Weimerskirch and D. Westhoff. Zero Common-Knowledge Authentication for Pervasive Networks, *Tenth Annual International Workshop on Selected Areas in Cryptography (SAC 2003)*, 2003.

25. A. Weimerskirch and D. Westhoff, Identity Certified Authentication for Ad-hoc Networks, *Proceedings of the 1st ACM workshop on Security of ad hoc and sensor networks (SASN)*, 2003, ACM Press, ISBN:1-58113-783-4, 2003, pp. 33-40.

26. S. Lucks, E. Zenner, A. Weimerskirch, and D. Westhoff, How to Recognise a Stranger - Efficient and Secure Entity Recognition for Low-End Devices, *submitted for publication*.

27. L. Zhou and Z.J. Haas, Securing Ad Hoc Networks, *IEEE Network Journal*, Vol. 13, No. 6, 1999, pp. 24-30.

4

PROMOTING IDENTITY-BASED KEY MANAGEMENT IN WIRELESS AD HOC NETWORKS

Jianping Pan
Dept. of Computer Science
University of Victoria, BC, Canada
E-mail: pan@uvic.ca

Lin Cai
Dept. of Electrical & Computer Engineering
University of Victoria, BC, Canada
E-mail: cai@uvic.ca

Xuemin (Sherman) Shen
Dept. of Electrical & Computer Engineering
University of Waterloo, ON, Canada
E-mail: xshen@bbcr.uwaterloo.ca

In wireless ad hoc networks, mobile peers communicate with other peers over wireless links, without the support of preexisting infrastructures, which is an attractive form of peer communications for certain applications. Although many enabling technologies have progressed significantly in recent years, the highly-anticipated deployment of large-scale, heterogeneous wireless ad hoc networks still faces considerable technical challenges, among which achieving secure, trustworthy and dependable peer communications is a major one. In this chapter, we promote identity-based key management, which serves as a prerequisite for various security procedures. We first identify that peer identity plays an irreplaceable role in wireless ad hoc networks, where autonomous peers can join or leave such systems and change their location in these systems at any time. Next, we show that identity-based key management schemes are effective and efficient for bootstrapping any chosen security procedures, especially in wireless ad hoc networks where both over-the-air communication and on-board computing resources can be severely constrained. Finally, we illustrate identity-based secure communication schemes with a security enhancement to the Dynamic Source Routing protocol. We find that identity-based schemes are intrinsically suitable for and practically capable of securing wireless ad hoc networks and may have great impact on dealing with other network security issues.

1. INTRODUCTION

With the rapid advance of miniaturized computers and radio communication technologies, wireless ad hoc networks have attracted a lot of attention from both research communities and the industry in recent years [1, 2, 3, 4]: without relying on any preexisting communication and computing infrastructures, autonomous peers are envisioned to communicate with other peers over wireless links, or to assist communications among others when necessary. Also, mobile peers can join or leave such systems at any time; when peers are in these systems, they can change their location at any time. This self-organizing and adaptive form of peer communications is particularly attractive in certain scenarios, where communication or computing infrastructures are either too expensive to build or too fragile to maintain. Wireless ad hoc networks have found many applications in military, commercial and consumer domains; they also have other variants (e.g., wireless sensor networks) with various similarities.

However, the highly-anticipated deployment of large-scale, heterogeneous wireless ad hoc networks still faces considerable technical challenges. Among them, achieving secure, trustworthy and dependable peer communications is a major one, which can hinder the further development of these systems. Due to the absence of properly-protected media and well-trusted infrastructures, and due to the reliance on unknown third-parties to relay data, peer communications in these systems are intrinsically vulnerable to various passive and active attacks [5], which can compromise the confidentiality, integrity and authenticity of information exchange among peers. Also, in some wireless ad hoc networks, peers can become selfish, greedy and even tampered by adversaries, which brings more challenges to secure the already vulnerable peer communications in these systems.

Many efforts have been devoted to securing peer communications in wireless ad hoc networks, and most of them are based on either symmetric-key (SKC) or public-key cryptography (PKC) systems (see [5, 6] and the references therein). Although these systems have successfully demonstrated their capability in securing information infrastructures in other contexts (e.g., the Internet), many of them are found inadequate for wireless ad hoc networks, either due to severe communication or computing constraints, or due to the lack of infrastructure support in such networks. One issue, key management, is of the greatest interest [7], since it is a prerequisite for any security procedures of publicly-known cryptographic algorithms. For example, in SKC, shared keys or preshared secrets should be arranged for involved peers before they can communicate; in PKC, information senders should obtain the public-key of receivers and verify it with trusted third-parties. Pairwise keying is cumbersome in wireless ad hoc networks of many peers with dynamic membership; public-key verification usually relies on centralized key directories or hierarchical certificate authorities, which may not be always available in wireless ad hoc networks. In addition, voluntary public-key verifications may introduce a risk of denial-of-service (DoS) attacks due to the amount of computing and communication resources involved even before the regular communications among peers can happen.

In this chapter, based on the latest advances in identity-based cryptography (IBC), we prompt identity-based key management in wireless ad hoc networks. IBC is a special form of PKC [8]. In regular PKC, an entity (or a peer in ad hoc networks) of known identity generates a pair of public-key and private-key or obtains it from public-key infrastructures (PKIs). The binding between the peer identity and its public-key should be certified by trusted third-parties; otherwise, a peer can easily impersonate others by forging their public-keys and compromise communications intended for those peers. In IBC, such binding and verifying are unnecessary, since the public-key of a peer is exactly its identity (or a known transformation of the identity). As far as a peer can communicate with others by their identity, the peer can apply any security procedures bootstrapped from identities to secure its communications with those peers. We find that the unique features offered by IBC make identity-based key management a strong candidate for securing peer communications in wireless ad hoc networks.

The contributions of this chapter are twofold. First, we present identity-based key management schemes designed for bootstrapping various security procedures in wireless ad hoc networks. We show that these schemes not only accomplish their goals without the support of communication and security infrastructures, but also accommodate dynamic peer membership for potentially a large number of mobile peers. Also, these schemes are effective and efficient. For example, a sender-only peer has no security overhead in terms of verifying the public-key of others or obtaining its own private-key; a peer can send another peer some information only accessible by the latter in the future; a compromised peer can be easily identified and excluded from such systems. Second, we illustrate identity-based secure communication schemes with a security enhancement to the Dynamic Source Routing (DSR) protocol, in order to demonstrate that these schemes are intrinsically suitable for and practically capable of securing wireless ad hoc networks. We also expect that such schemes have great impact on dealing with other network security issues. An IBC and threshold-based key distribution scheme is independently proposed in [9]; in contrast to a conceptual sketch in [9], here we give a concrete design of all necessary building blocks. Although IBC has been explored in other contexts such as IPsec, personal area networks, IPv6 neighbor discovery and grid computing [10, 11, 12, 13], our goal in this chapter is not only to show that IBC-based schemes can support confidentiality, integrity and authenticity, but also to reveal that these security properties can be achieved more effectively and efficiently with IBC-based schemes due to the irreplaceable role of peer identity in wireless ad hoc networks.

The remainder of this chapter is organized as follows. In Section 2, we present a model of wireless ad hoc networks and their security requirements; we also briefly overview identity-based cryptography and its latest advances. In Section 3, we introduce identity-based key management schemes for bootstrapping and managing any chosen security procedures in wireless ad hoc networks. In Section 4, we illustrate identity-based secure communication schemes to ensure the confidentiality, integrity and authenticity of information exchange among autonomous peers in these systems; we also design a security enhancement to DSR, with focus on its route discovery and maintenance procedures and its resistance against various attacks. Section 5 offers

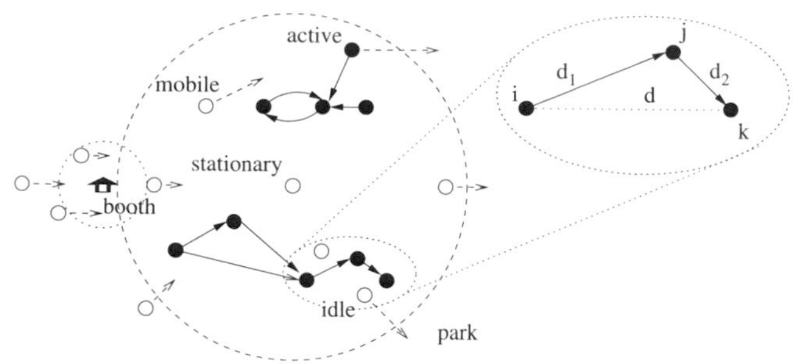

Figure 1. A wireless ad hoc network at a recreation park.

further discussion, and Section 6 reviews related work. Section 7 concludes the chapter with directions of our future work.

2. PRELIMINARIES

2.1. Network Model

Wireless ad hoc networks are fully-distributed systems of self-organizing peers that want to exchange information over wireless links but do not rely on any preexisting infrastructures [1, 2, 3, 4]. Fig. 1 shows such networks in a generic format. Mobile peers (e.g., laptop computers with wireless interfaces as filled or unfilled dots) can join or leave such systems (depicted by a large dashed circle, e.g., a recreation park) at any time. Only peers require keying have to pass by an offline authority regularly (e.g., a ticketing booth within a small dotted circle). However, there are no physical barriers around the vicinity, and peers can join or leave systems at any locations (e.g., a sender-only peer without keying). While peers are in the system, they can remain stationary or change their location, and keep idle or communicate with others. Also, peers can assist communications among others if they choose to do so. Without any centralized online authorities, peers communicate in uni- or bi-direction, single- or multi-hop, single- or multi-path, and single- or multi-point form, or any combinations of these forms.

For a given information exchange between two peers, e.g., transferring a bulk data of b unit amount from peer i to k that is d unit distance away in Fig. 1 (zoomed in a dotted ellipse), i has two strategies. With the first one, i transmits b to k directly, and consumes energy

$$e_i^t(b,d) = (t_1 + t_2 d^n)b, \tag{1}$$

where $2 \leq n \leq 6$ is the path loss exponent, and t_1 and t_2 are the coefficients of distance-independent and distance-related energy consumption, respectively. Some facts may prevent i from adopting this strategy: i) when $d > D$, where D is the maximum transmission range of i; ii) direct wireless communications of i and k may impose strong interference on peers between i and j. With the second strategy, when there is a third peer j that lies in between i and k, i may save energy by requesting j to relay b

to k. Without loss of generality, assume j is d_1 away from i and d_2 from k. If $d_1 < d$, relaying b through j is preferable for i, while j has to volunteer $e_j^r(b) = rb$ to receive b, $e_j^t(b, d_2) = (t_1 + t_2 d_2^m)b$ to transmit b to k, and $e_j^o(b)$ to cover its local expenses. If

$$e_i^t(b, d) - e_i^t(b, d_1) > e_j^r(b) + e_j^t(b, d_2) + e_j^o(b), \tag{2}$$

relaying b through j is also preferable for the entire system, since, overall, it takes less energy to move the same b from i to k. This relaying strategy can be applied recursively to peers in the vicinity of i, j and k.

To enable such relayed communications, peers need to identify other peers of their interest. There are many different naming schemes; e.g., on the Internet, nodes are identified by their IP address or host name. Public IP addresses are location-dependent with regard to the attachment point of addressed nodes to the global Internet routing fabric, which is not available in wireless ad hoc networks. Although host names can be location-invariant, they have to be mapped to IP addresses with the assistance of a hierarchical Domain Name System (DNS), which may not be always available in wireless ad hoc networks. Therefore, mobile peers can only be identified by their own identity of spatial and temporal invariance. For example, peers propose their identity when joining such systems (sender-only peers can have no identity and remain anonymous). To keep collision-free in the identity space, the offline authority can append a timestamp or sequence number to the identity proposed by peers when they request keying. To find a multi-hop path from one peer to another, the source peer initiates a route discovery procedure with the source and destination peer identities. Route requests are forwarded by neighboring peers after their identities have been appended to the request. This process is recursive until the request reaches the destination peer, where a route reply is sent back to the source peer by reversing the forward path identified by the identities of forwarding peers. As we can see, peer identity plays an irreplaceable role in enabling multi-hop communications in wireless ad hoc networks; in the next section, we will see it also plays an important role in key management.

2.2. Security Model

With relaying, peers no longer always communicate with intended peers directly, so they should be assisted with additional security procedures to ensure the confidentiality, integrity and authenticity of their information exchange with intended peers. Without any preexisting communication and security infrastructures, peers may have to deal with unknown relaying peers without the preestablished trustworthiness.

Many security threats appear in ad hoc networks [5]. In addition to poorly-protected communication channels open to various passive and active attacks, pairwise trustworthiness among all involved peers is unpractical to build and difficult to maintain, especially when there is a large number of mobile peers joining or leaving such systems without notice. Selfish peers have the motive and excuse to corrupt relayed data (no matter intentionally or not). Relaying peers and neighboring non-relaying peers have the incentive to eavesdrop relayed data. Malicious or compromised peers can impersonate others to steal genuine information or inject false information into these systems. Besides data plane attacks, there are control plan attacks (e.g., black/gray holes [14, 15], replay attacks [16], network partitions) for specific routing protocols. When collaborative relaying becomes profitable, greedy relaying or non-relaying peers have a strong motive to boost their wealth improperly, by trying to cheat source, destination or other relaying peers. When there is a certain number of malicious peers, they may collude with each other and attempt to beat the entire system (e.g., wormholes [17], rushing attacks [18]). Our focus in this chapter is not on individual or new data plane or control plane attacks, but on the identity-based key management schemes that can be used to bootstrap various security procedures to defend against these and other possible attacks.

Traditional cryptographic techniques have been used to provide certain security properties in networks with trusted infrastructures; similar efforts were attempted in wireless ad hoc networks. For example, source and destination peers should authenticate to each other before information exchange. Also, information should be encrypted by source peers to keep confidentiality, and be verified by destination peers to preserve integrity. These procedures rely on either certified public-keys in PKC-based systems, or pairwise shared-keys in SKC-based systems. If there is a trusted infrastructure (e.g., a generic PKI in corporate networks or a base-station in multi-hop cellular networks), such requirements can be satisfied accordingly.

However, these techniques may not be readily applicable to wireless ad hoc networks. First, there is no generic PKI or central online authority in these systems that can always be involved in communications between any pairs of mobile peers. Second, most end-to-end communications in these systems occur in a hop-by-hop manner, whereby unknown third-parties are required to relay packets; i.e., security proprieties should be achieved not only at the end-to-end level, but also at the per-hop level. Finally, some existing security procedures (e.g., electronic payment) either rely on an online interactive authority (e.g., a bank), or are too heavy (in terms of communication and computing complexity) for wireless ad hoc networks, within which on-board energy constraints are normally the foremost concern.

In summary, security procedures bootstrapped by effective and efficient key management schemes, identity-based ones as we advocate, are highly desirable to ensure the confidentiality, integrity and authenticity of information exchange among autonomous peers in wireless ad hoc networks.

2.3. Identity-Based Cryptography

Motivated by these observations, we approach this challenge from a novel angle and with a new tool — IBC, a special form of PKC. As shown in Fig. 2(a), in regular PKC, the public-key should be certified, since there are no intrinsic bindings between the public-key and the identity of an entity. Otherwise, any entities can impersonate others with a forged public-key. To facilitate public-key certification, hierarchical certificate authorities (CAs) are introduced, and the root CA should be trusted by everyone. This

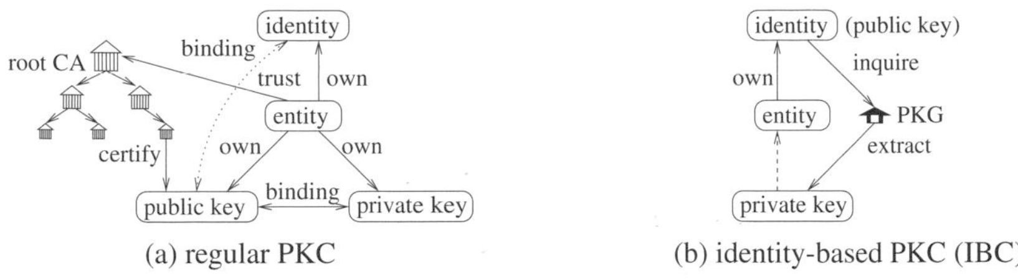

Figure 2. Two forms of public-key cryptography systems.

model may not be applied to wireless ad hoc networks, where neither a PKI nor a CA hierarchy is easy to build or maintain in practice.

Unlike regular PKC, in which an entity generates its public-key and private-key (or obtains them from PKI) and has the public-key certified by CA, in IBC, the entity proposes a unique identity (e.g., $a@b.com$), which is also its public-key. A private-key generator (PKG) extracts a corresponding private-key from the public system parameters and the master-key that is only known to the PKG. The procedure is shown in Fig. 2(b). For example, when a peer i wants to send a message m to another peer k (see Fig. 1), m is encrypted with k's identity id_k and the system parameters; only k can decrypt the encrypted message with its private-key pk_k and the system parameters. When k signs the receipt of m, the receipt is manipulated with pk_k, and is verifiable by everyone knowing id_k. i has to know id_k when communicating with k, and no one else can compromise these procedures without knowing pk_k. Also, IBC can bootstrap symmetric cryptographic procedures by establishing a shared-key $sk_{i,k}$ for i and k.

The concept of IBC was first introduced by Shamir in 1984 [19], and several efficient IBC-based signature schemes had been found subsequently. However, non-mediated IBC-based encryption (IBE) has proved to be much more challenging, and it is relatively recent that practical IBE schemes were found [8]. The first efficient and secure IBE scheme was given by Boneh and Franklin in 2001, which employs Weil pairing on elliptic curves and is considered more efficient than using regular RSA-based counterparts [20]. Its security is based on the bilinear Diffie-Hellman problem (BDHP), which is considered secure in the random oracle model (ROM) [21]. The Boneh-Franklin (BF-IBE) scheme is semantically secure against chosen ciphertext attacks, even when an adversary has the private-key of any entities other than the one being attacked. Lynn extended the BF-IBE scheme to provide message authenticity without extra computation cost; i.e., receivers can verify the identity of senders and whether the received messages have already been tampered, even without resorting to digital signatures [22].

Based on the latest advances in IBC and related techniques, in the next section, we will design key management schemes to bootstrap secure communications among identifiable peers in wireless ad hoc networks, without PKIs, CAs, key directories, always online authorities, or manually-arranged pairwise preshared secrets among all involved peers.

3. KEY MANAGEMENT

3.1. System Setup

Before an IBC-powered wireless ad hoc network becomes fully functional (i.e., allowing peers to join the system and request keying), an offline PKG first picks a random master-key $x \in \mathcal{Z}_q$ (q is a prime and $\mathcal{Z}_q$ is an algebraic field) and a bilinear mapping $f : \mathcal{G} \times \mathcal{G} \to \mathcal{Z}_q$. f is defined on the points of an elliptic curve (as a group $\mathcal{G}$), and has the following property that for any $P, Q \in \mathcal{G}$ and for any integer a and b,

$$f(aP, bQ) = f(P, bQ)^a = f(aP, Q)^b = f(P, Q)^{ab}. \tag{3}$$

The PKG then picks a random generator P, and publishes P, xP, f and four chosen cryptographic hash functions as the public system parameters. These hash functions, which will be explained shortly, are used to hash an arbitrary identity (e.g., any ASCII strings) to a point on the elliptic curve (H_1), to achieve security against chosen ciphertext attacks (H_2 and H_3), and to encrypt plaintext (H_4), respectively. The PKG should keep x secret, and no one else can derive x even when they have both P and xP.

A lot of offline entities (e.g., the ticketing booth of a recreation park) can assume the role of PKG, as long as they can keep the master-key secret and extract private-keys from the master-key for peers joining the system and requesting to be keyed. Once the private-key is extracted, a peer has no need to communicate with the PKG (nor to keep the PKG online), unless the peer wants to propose a new identity. Also, the offline PKG can key peers in batch (e.g., only during normal business hours), since peers can receive regular, encrypted information even before they request keying. Compared with an online PKI, the offline PKG has many advantages in wireless ad hoc networks. With a PKI, whenever a peer k joins a system, the PKI should verify the binding of the public-key of k and its identity, and broadcast the authenticated public-key to all existing peers, or keep the public-key in a central directory for queries from other peers. No matter when another peer i wants to communicate with k, i has to obtain both the identity and the public-key of k, and i should have a way of verifying the public-key. The complexity of obtaining, verifying and managing public-keys creates considerable overhead in energy-constrained systems that rely on radio technologies to exchange identities, keys and data.

3.2. Peer Keying

When a peer k joins an IBC-powered wireless ad hoc network, k proposes a system-wide unique identity id_k (or the PKG appends a timestamp or sequence number to peer identity). The PKG obtains a corresponding point $Q = H_1(id_k)$ on the elliptic curve by hashing id_k, and extracts k's private-key $pk_k = xQ$ from the master-key x. id_k can be the email address of k, concatenated with temporal or spatial properties (e.g., *a@b.com@date@site*). Identity ownership should be easily verified, e.g., by short-range encounters [23] when peers passing by the PKG or by sending a request-to-confirm email to *a@b.com*. pk_k is conveyed back to k in a secure, out-of-band side channel (e.g., through the ticketing process at a recreation park); the system parameters are periodically broadcasted by the PKG (e.g., through public announcement). To fight against identity theft or spoofing, the PKG should not extract private-keys more than once for the same identity even claimed by the same entity; instead, by using timestamp or sequence number, the entire identity space is always collision-free and forward-secure.

The security of the entire system relies on the master-key x kept by the PKG, since the private-key of all peers in IBC-based wireless ad hoc networks can be derived from x. To reduce the risk of total-exposure even if the PKG is compromised and to address the concern of key escrow for peers with a new PKG, x can be distributed in a t-of-n manner to a group of n PKGs by applying threshold cryptography (TC) techniques [24]. With TC, k thereby derives pk_k alone by combining pk_k^t obtained from any t PKG_t. Unless there are more than t unknowingly-compromised or bogus PKGs, the secrecy of all peers and their private-key are still preserved.

To support a large entity population, Gentry and Silverberg extended the BF-IBE scheme with a hierarchical PKG structure (GS-HIBC), where a lower-level PKG inherits the identities of its ancestors and obtains its master-key from the parent PKG [25]. In HIBC-powered systems, peers are identified by a tuple of identities, corresponding to their location in the PKG hierarchy, which is also their localized public-key. With HIBC, a peer can easily roam from one ad hoc network to another, and communicate with peers in other networks, by just knowing their identities and the system parameters of the root PKG (not the PKG of correspondent peers). For simplicity, here we focus on keying with a single PKG; our schemes can be extended for t-of-n or hierarchical PKGs as well.

3.3. Key Maintenance

In identity-based schemes, the public-key of a peer is exactly its identity or a known transformation of the identity. Hence, a peer can receive regular information encrypted with its identity from other peers even before the peer has obtained its private-key from the PKG. This unique feature allows asynchronous communications in wireless ad hoc networks, where autonomous peers can be in active, idle or sleep state periodically without global synchronization to conserve energy. Also, this feature reduces the cost of operating the offline PKG, since peers can request keying in batch only after they are actively and willingly involved in receiving information from other peers and when the PKG goes online according to its own schedule. In contrast, in SKC or regular PKC systems, peers have to establish pairwise shared-keys or obtain public-key and private-key pairs way before any secure communications can happen; i.e., keying is always mandatory and proactive for all peers, even if they eventually have no secure communications throughout the validity of their keys in these systems.

Once a peer obtains its private-key extracted from its identity and the system parameters, the peer can decrypt received information encrypted with its identity, authenticate itself to other peers, and sign outgoing messages. We will present these procedures in detail in the next section. Also, peers can bootstrap shared-keys or derive session-keys from their identity-based private-keys for symmetric security procedures. Once bootstrapped, symmetric procedures have much less overhead than their asymmetric counterparts. Depending on the definition of peer identity, a peer, as well as the PKG, can determine the lifetime of its private-key. For example, a peer can propose the same identity (e.g., *username*) to systems with different parameters (i.e., the peer will have different private-keys in different systems); even if its private-key is compromised in one system, the information exposure is confined to that system. A peer can propose an ephemeral identity (e.g., *user@time*); even if its private-key is compromised at a certain time, the peer can request a new private-key with a partially-updated identity in *time* portion, without totally losing its identity or forcedly leaving the system. When necessary, a peer can proactively refresh its identity (e.g., *user@date*) with the PKG and remain forward-secure even if its current private-key is captured and compromised by adversaries. To deal with an unknown PKG, a peer can propose a temporary identity (e.g., *user@site*) to a newly-encountered system, while maintaining credentials with other well-known systems. As we mentioned, a peer can request keying with multiple or hierarchical PKGs to reduce its exposure due to compromised PKGs, and to ease its concern of key escrow by untrusted PKGs.

The PKG, on the other hand, can also control the validity of peer identities and extracted private-keys. For example, a peer should have a way of proving its identity ownership (e.g., *a@b.com*) or accept assigned identities (e.g., prepaid personal identification number, PIN). A peer is uniquely identified by its identity, which can be both time and location invariant within the system. No matter how the peer changes its location and status in the system, it solely relies on its identity to receive information and communicate with other peers. In addition, its identity is related to its reputation (e.g., cooperativeness in relaying) and wealth (e.g., collected credits for its cooperation) in the system. If a peer is found greedy and always fails to relay for other peers, this fact can be taken into account when the peer is in need of relaying by other peers. If a peer is found malicious, either persistently or opportunistically, the peer can be excluded from the system by identity blacklisting or key expiring (e.g., the PKG enforces an identity upgrade and refuses to key compromised peers). The PKG can have differentiated policies, e.g., extracting keys of *user@month* for well-established or reputable peers (e.g., a monthly pass to a recreation park) and of *user@day* for new or ill-behaving peers (e.g., a one-time ticket). Certainly, the PKG can enforce a system-wide rekeying after a long time-period by updating the master-key and the system parameters, and peers will need to contact the PKG again to extract their new private-key.

The irreplaceable role of peer identity in wireless ad hoc networks leads to the promotion of identity-based key management schemes in these systems. These key management schemes can effectively and efficiently bootstrap security procedures proposed in Section 4 to ensure the confidentiality, integrity and authenticity of information exchange among peers.

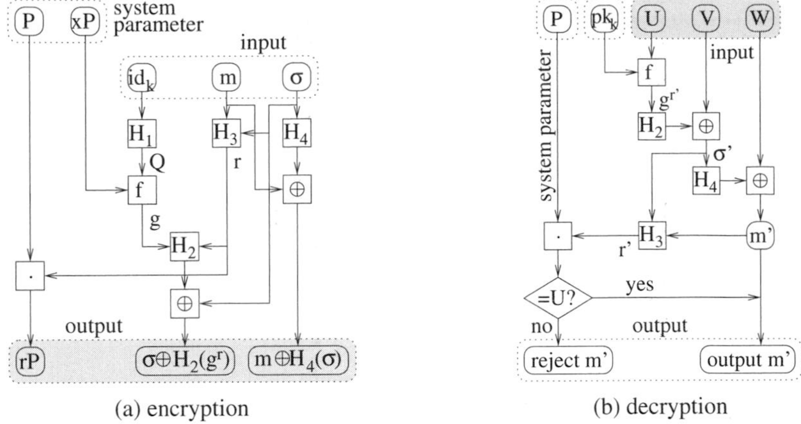

Figure 3. IBC-based encryption and decryption flows.

4. SECURE COMMUNICATIONS

4.1. Information Exchange

Encryption and decryption

Suppose that peer i wants to send a message m to peer k (see Fig. 1). i first picks a random number σ, and obtains $r = H_3(\sigma, m)$. i then employs $g^r = f(xP, Q)^r$ as a session-key for m, where $Q = H_1(id_k)$, and sends rP to k. Consequently, k has

$$g^{r'} = f(rP, pk_k) = f(rP, xQ) = f(P, xQ)^r = g^r, \tag{4}$$

since $f(P, xQ) = f(xP, Q)$ according to the bilinear pairing property of f in (3). With this procedure, both i and k derive the same session key g^r, without knowing the secrecy of their counterpart. Other peers can learn about rP and Q, as well as P and xP, but they cannot obtain r or x; in other words, there is no way for these peers to obtain g^r, nor can they recover the encrypted version of message m.

Fig. 3 gives a detailed illustration of the BF-IBE encryption and decryption flows. Besides f, only hash functions and XOR operations are used, allowing these procedures to be efficiently implemented in resource-constrained peers. Also, there are considerable efforts to implement pairing (e.g., Tate pairing) more efficiently in software and hardware [26]. For a plaintext m, the ciphertext has three parts $\{rP, \sigma \oplus H_2(g^r), m \oplus H_4(\sigma)\}$. When k receives a ciphertext $\{U, V, W\}$, k first recovers $g^{r'}$ from U, with its private-key pk_k, and then recovers σ' from V, with the hashed $H_2(g^{r'})$. m' is recovered from W, with the hashed $H_4(\sigma')$. Finally, $r' = H_3(\sigma', m')$ is recovered. To verify message integrity, k compares $r'P$ with U. If $r'P == U$, m' is accepted as m; otherwise, m' is rejected by k.

Authenticated encryption

If i obtains $g = f(pk_i, H_1(id_k)) = f(xQ_i, Q_k)$, k then has

$$g' = f(H_1(id_i), pk_k) = f(Q_i, xQ_k) = g, \tag{5}$$

since $f(xQ_i, Q_k) = f(Q_i, xQ_k)$ according to (3). With this procedure, both i and k have derived the same shared-key g, even without having any physical communications between them. Also, k knows that only i can create such keys with its own pk_i. Thus, k can be assured that the shared-key g and ciphertext W are indeed created and encrypted by i, respectively. This scheme (IBAE) achieves authenticated encryption for messages between i and k without relying on the digital signature of i or k on each message, which is another advantage for energy-constrained wireless ad hoc networks, since signing digital signatures is an expensive procedure in general.

Signed encryption

Although BF-IBE can verify whether the recovered plaintext should be accepted or rejected, message integrity can be significantly strengthened by applying keyed-hash message authentication code (HMAC) with shared secret (e.g., the authenticated shared-key g), or by applying signed encryption (i.e., signcryption) in asymmetric procedures. Libert and Quisquarter further extended the BF-IBE scheme, and proposed an identity-based signcryption (LQ-IBSC) scheme, by combining the functionality of signature and encryption (but with much less cost than that of a sign-then-encrypt procedure) and offering confidentiality, authenticity, integrity and non-repudiation seamlessly [27]. Message integrity is then verified by applying the same HMAC function with the shared-key derived from the identity of senders and the private-key of receivers, or by applying the unsigncryption procedure in IBSC. When message confidentiality is not a concern, Boneh, Lynn and Shacham proposed an IBC-based short signature scheme (BLS-IBS) that is also based on Weil pairing [28]. With efficient elliptic curve cryptography (ECC) primitives, a BLS-IBS-based signature is only about half the size of a DSA-based signature, but still offers a similar level of security and protection, which is also very attractive for energy-constrained wireless ad hoc networks, where shorter signatures are always preferred.

4.2. Message Routing

With achieved secure information exchange, we can further secure the underlying routing protocol in wireless ad hoc networks. Now, we assume that peers are collaborative once they choose to do so. Designing schemes to stimulate peers to be collaborative and compensate them if they indeed are is one of our future work items.

Route discovery

Here, we want to secure a DSR-like reactive ad hoc routing protocol [29] with identity-based key management. It is feasible to secure other routing protocols with the designed security procedures [30, 31]. In DSR, when a peer i wants to send a message m to another peer k and has no known routes to k in its route cache, i initiates a route discovery procedure by broadcasting a route request message $RREQ\{id_k, rn, id_i\}$ with the identities of i and k, and a sequence number rn to suppress broadcast loops. If a neighboring peer j has a valid route to k in its cache (e.g., $\{j+1, \cdots, k-1, k\}$), j can respond a route reply message $RREP\{id_k, rn, id_i, id_j, id_{j+1}, \cdots, id_k\}$ to i with its identity id_j and the cached route; otherwise, j appends id_j to i's request message and broadcasts the updated $RREQ\{id_k, rn, id_i, id_j\}$. This process is recursive, until the request message reaches k, where a route reply message will be generated and sent back to i by reversing the forward path. With rn, peers never react on duplicated or outdated routing messages, but can learn from bypassed messages.

Obviously, DSR-like routing schemes rely on voluntary peer collaborations, and are highly vulnerable to false routing information corrupted or injected by malicious peers. To fight against these attacks, routing messages should be authenticated by their initiators and verified by their recipients. First, i authenticates its routing request message with its private key pk_i, which is verifiable by all other peers knowing id_i's identity. Similar procedures are required for peers that forward the appended request message. When a routing reply message is generated by j or k, the initiator should also authenticate itself and the route information. Finally, when i receives the reply message, the authenticity of peers (e.g., j and k) among the discovered route is verified by their identities (id_j and id_k). The BLS-IBS-based routing request message arriving at k then has the format

$$RREQ\{\{\{id_k, rn, id_i\}_{pk_i}, id_j\}_{pk_j}, \cdot\cdot, id_{k-1}\}_{pk_{k-1}}, \tag{6}$$

where $\{\cdot\}_{pk_j}$ implies that the message has been authenticated by j's private-key pk_j, and is verifiable by j's identity id_j. En-route peer identity has to appear in routing messages no matter whether systems are powered by IBC. The construction in (6) is similar to that with a regular PKC; however, with IBC, there is no need to obtain the public-key of a peer to verify its messages, which is very attractive for wireless ad hoc networks. The routing reply message generated by k and arriving at i has the format

$$RREP\{\{\{id_k, rn, id_i\}_{pk_i}, id_j\}_{pk_j}, \cdots, id_k\}_{pk_k}. \tag{7}$$

If the reply message is generated by j according to its route cache, it has the following format instead

$$RREP\{\{id_k, rn, id_i\}_{pk_i}, id_j, id_{j+1}, \cdots, id_k\}_{pk_j}. \tag{8}$$

With (7) and (8), i can tell whether a route to k is actually certified by k or just endorsed by j. Hence, no peer can corrupt a relayed routing message without altering message authenticity or revealing its identity.

In the above route discovery procedure, all involved peers have authenticated themselves, so the discovered route is cacheable by all peers, which can reduce the communication cost if these peers also want to obtain the route to k or other downstream peers. However, this procedure requires every en-route peer to sign the hash of the appended request message, which may impose a non-negligible computing overhead in dense ad hoc networks. An alternative is to have all en-route peers authenticate themselves only to the request initiator i, by using a keyed hash of the appended request message with the pairwise shared-key derived from their private-key and i's identity. Accordingly, (6) can be redefined in the following format

$$RREQ\{\{\{id_k, rn, id_i\}_{pk_i}, id_j\}_{sk_{i,j}}, \cdots\}_{sk_{i,k-1}}, \tag{9}$$

where $\{\cdot\}_{sk_{i,j}}$ implies that the message is protected by an HMAC with the shared-key $sk_{i,j}$ defined by (5). Similarly, (7) then is redefined in the following format

$$RREP\{\{\{id_k, rn, id_i\}_{pk_i}, id_j\}_{sk_{i,j}}, \cdot\cdot, id_k\}_{sk_{i,k}}. \tag{10}$$

By doing so, only i can verify the authenticity of all peers that appear in the discovered route, which suggests that the route is not cacheable by peers other than i, unless they have already established trustworthiness with downstream peers. These peers have to initiate their own route discovery if necessary, although some heuristics can help them verify route validity (e.g., i and k exchange data packets successfully after route discovery).

Route Maintenance

Another procedure in DSR is route maintenance. If a peer finds that a route is broken, it notifies the source peer with a route error message, and the source peer initiates another route discovery if there are no alternative routes in its route cache. Apparently, a malicious peer can abuse error report messages and mount DoS attacks. Therefore, report messages should be authenticated by the report initiator with its private-key and be verifiable for everyone knowing its identity (which is included in the message, along with the sequence number and the reversed forward path). The route error message generated by j has the format

$$RERR\{id_k, rn, id_i, \cdots, id_{j-1}, id_j\}_{pk_j}, \tag{11}$$

or alternatively with HMAC,

$$RERR\{id_k, rn, id_i, \cdots, id_{j-1}, id_j\}_{sk_{i,j}}. \tag{12}$$

The source peer and other upstream peers should verify the authenticity and integrity of the message, and update their route cache accordingly. Again, the reporting peer has to trade off computing and communication overhead, by choosing either to sign the hash of the error report message, or to apply a keyed hash on the message to only authenticate with the source peer by using their shared-key.

With identity-based key management schemes, securing information exchange and message routing in wireless ad hoc networks becomes feasible with either asymmetric or symmetric procedures. On the other hand, the irreplaceable role of peer identity in these systems justifies the need and applicability of identity-based key management.

5. FURTHER DISCUSSION

5.1. Practical extensions

Identity-based key management schemes offer many attractive features that are highly desirable in wireless ad hoc networks, in which peer identity usually is the only means to identify autonomous and mobile peers. String-based identity can have very rich semantics (e.g., along with the date and location information). The location-aware identity (e.g., grid-based one) can assist location-aware routing in wireless ad hoc networks: when a peer sends a message to another peer, the routing path for the message is implicitly suggested by their identities. Also, a peer can propose its identity indicating the services (e.g., *email@adhoc.net*) or content (*movie_trailer_title*) provided by itself to assist resource discovery in wireless ad hoc networks. When a peer wants to obtain a specific service or content, it securely solicits the peer identified by the service description or the content hash.

IBC-based schemes with pairing are also very attractive for energy-constrained wireless ad hoc networks. For example, the BF-IBE and follow-on schemes employ bilinear pairings on elliptic curves in ECC, an approach considered much more efficient (in terms of key size and computation complexity) than regular RSA-based PKC procedures. Most operations in these schemes mainly involve hashing and bitwise XOR, and more efficient pairing implementations in software and hardware are appearing as well. In our secure communication schemes, we provide both asymmetric procedures (e.g., BLS-IBS signature) and their symmetric counterparts (HMAC), so that peers can trade off computing and communication overhead properly. Also, IBC-based schemes allow peers to authentically establish shared-key and bootstrap even more efficient symmetric operations without having any physical communications beforehand. In addition, Boyen gave a multipurpose IBC-based signcryption (IBSE) scheme (a.k.a. swiss army knife, since it can be flexibly used for encryption, signing and sign-and-encrypt procedures) with even stronger security properties (i.e., confidentiality, authenticity, integrity, non-repudiation, anonymity and unlinkability) and better runtime efficiency (less ciphertext expansion and fewer high-cost operations) [32]. The Boyen scheme is also based on bilinear pairings, and can be introduced in our identity-based key management schemes. Further, secure IBE schemes without ROM are proposed recently [33], which gives more assurance on adopting them in wireless ad hoc networks.

5.2. Known limitations

In our identity-based key management schemes, peers obtain their private-key from the PKG that oversees the entire system. Therefore, the PKG has total-control over the secrecy and wealth of individual peers. This is not a concern when peers can trust the PKG (e.g., the PKG is the administrator of a managed-open wireless ad hoc network). However, some peers, especially foreign peers, may be concerned about a compromised PKG or an unknown PKG that decrypts messages with their private-key extracted by the PKG, impersonates their identity, and collects their wealth during their tenure in the system. Nevertheless, these concerns also apply to any regular PKC-based systems, in which compromised CAs can always issue false certificates to malicious peers, or bogus PKIs can later reveal public-key and private-key pairs assigned to genuine peers.

There are some identity-based approaches that can alleviate these concerns to some extent. First, the master-key can be distributed to several PKGs that are not under any single administration (e.g., t-of-n PKGs). Therefore, unless the number of compromised or bogus PKGs exceeds a certain threshold, peer secrecy and wealth are still well-preserved. With this approach, peers have to derive their private-key from multiple PKGs, which unavoidably increases their computing cost. Alternatively, peers can resort to hierarchical PKGs when they roam across different systems frequently. Second, the PKG can be required to refresh its master-key and system parameters periodically. Therefore, the vulnerability of a certain master-key and the potential damage of a compromised master-key are limited. With this approach, peers have to inquire the PKG periodically as well to extract their private-key from the latest master-key and system parameters, which increases their communication cost. We argue that the PKG of a wireless ad hoc network usually is the entity, often offline, that enables the system by providing other resources (e.g., the PKG is the ticketing booth of a recreation park), and that peers should have a certain degree of trustworthiness on the PKG while they are willingly in these systems. A visiting peer can propose a PKG-dependent identity to an unknown system, while still maintaining credentials with trusted PKGs in other systems, until the peer has developed trustworthiness with the new PKG.

6. RELATED WORK

Wireless ad hoc networks have attracted intensive research attention in recent years [1, 2, 3, 4, 5, 7]. Their intrinsic vulnerabilities due to the lack of infrastructure, unsecured media, untrusted peers, reliance on relaying, and high system dynamics (e.g., peer membership, working mode and network topology) have geared a considerable amount of research effort toward securing peer communications in these systems [5, 7]. In this section, we briefly review two research topics closely related to our work, and compare reported work with our approach.

Information exchange — Schemes proposed to secure information exchange in wireless ad hoc networks are based on either SKC or PKC. With SKC, pairwise shared-keys, derived from preshared secret or bootstrapped by other means, should be established for all peer pairs beforehand, which is very impractical to achieve for mobile peers. Also, the total number of shared-keys is in the order of N^2, where N is the number of all potential peers and can be very large even in small systems with high membership dynamic. SKC procedures are efficient in general to achieve security properties, but have higher overhead with regard to key management.

Normally, RSA-based PKC procedures are less efficient than those in SKC in achieving the same level of security, but the key management in PKC is has less overhead than SKC, if a PKI or a CA hierarchy has already been built and well-maintained. A distributed certification service is proposed in [24], in which the system private-key used to sign peer public-key certificates is distributed to multiple servers with threshold cryptography. It strengthens the security and reliability of public-key certification, but does not reduce the associated overhead. A self-organizing PGP-like key management is proposed in [34], in which peers probabilistically obtain a certificate chain to other peers by merging their local certificate repositories; however, a roaming peer has difficulty in building its local repository shortly after it joins a foreign system. Random key predistribution has also been attempted in wireless sensor networks [35].

Cryptographically-generated identity [36] is an approach closest to ours. With this approach, peers derive their statistically-unique identity from their public-key (e.g., by hashing), so that the binding between the identity and the public-key of an entity is self-verifiable, which also eliminates the need for public-key certification. However, such identities cannot have any easy-to-understand semantics for its owner and other peers, and additional infrastructures (similar to DNS mapping host name to IP address) may be required to enable distributed applications.

In IBC-based schemes, peers only propose their identity, which is also their public-key, and can potentially have very rich semantics. Therefore, the binding of identity and public-key is intrinsic, and the name-to-identity mapping is unnecessary. This fact reduces the communication and computing overhead for resource-constrained peers in wireless ad hoc networks. For example, sender-only peers have no keying requirement, and peers can request keying even after regular, encrypted information is received. Also, these IBC-based schemes are based on ECC primitives, which are considered more efficient than RSA-based primitives [10, 11, 12, 13]. As we mentioned, BLS-IBS signatures achieve the similar level of security to DSA signatures with a size half of the latter. Further, IBC-based schemes can authentically bootstrap symmetric procedures even without having any physical communications beforehand. All these features are very attractive to resource-constrained peers in wireless ad hoc networks.

Message routing — Many wireless ad hoc routing schemes, no matter reactive or proactive ones, are found vulnerable to corrupted or false routing information. Several security patches have been proposed, which are based on either SKC or regular PKC systems. Broadcast operations often occur in route discovery, while traditional security associations are often based on a point-to-point model. Ariadne is a DSR-like routing scheme, in which message authenticity can be protected by digital signature, preshared secret, or a timed-release hash-chain to allow a group of recipients to verify messages with the same symmetric key (i.e., Tesla keys), without allowing them to forge extra messages [37]. In Ariadne, all peers require loose time synchronization to release key gradually. SRP is another DSR-like routing scheme, where intermediate peers do not perform cryptographic operations and have no *a priori* associations with end-peers [38]; but source and destination peers should have security associations. SAR [39] and SAODV [40] attempt to secure AODV, another on-demand ad hoc routing protocol. SEAD is a DSDV-like routing scheme that employs one-way hash function to protect route update without any asymmetric cryptographic operations [41], but SEAD has to rely on other means to distribute and authenticate the final value (i.e., image) of a hash-chain. ARAN employs PKC to guarantee message authenticity, integrity and non-repudiation, and to prevent modification, impersonation and fabrication attacks [42].

In contrast, IBC-based schemes can be seamlessly integrated with wireless ad hoc routing protocols, and achieve the same level of security more effectively than SKC-based schemes and more efficiently than regular PKC-based schemes. There are other security schemes proposed to defense against more sophisticated attacks such as blackhole, wormhole, rushing and replay attacks in ad hoc networks [14, 17, 18, 16, 15], which are orthogonal to our effort. Further, the identity-based key management schemes proposed in this chapter can help reduce the risk of certain sophisticated attacks associated with forged identities (e.g., Sybil attacks [43]), since malicious peers cannot always request keying from the PKG arbitrarily and then freely spoof their identities to cheat other peers.

7. CONCLUSION

Achieving secure, trustworthy and dependable peer communications imposes a major challenge in the highly-anticipated deployment of large-scale, heterogeneous wireless ad hoc networks. In this chapter, after identifying the irreplaceable role of peer identity in these networks, we promoted identity-based key management schemes, which can effectively and efficiently bootstrap any chosen security procedures in wireless ad hoc networks. In addition, we illustrated secure communication schemes with a security enhancement to a reactive ad hoc routing protocol, and demonstrated that identity-based schemes are intrinsically suitable for and practically capable of ensuring the confidentiality, integrity and authenticity of information exchange among peers.

In this chapter, we assumed that autonomous peers are always collaborative in relaying once they have chosen to do so. Designing accounting and rewarding schemes to stimulate selfish peers to become collaborative and to compensate them if they do so is one of our future work items.

8. REFERENCES

1. C. Perkins (ed). Ad hoc networking. Addison-Wesley, 2001.

2. Z. Haas, J. Deng, B. Liang, P. Papadimitatos, and S. Sajama. Wireless ad hoc networks. in J. Proakis (ed) *Encyclopedia of Telecommunications*, 2002.

3. R. Ramanathan and J. Redi. A brief overview of ad hoc networks: challenges and directions. *IEEE Communications*, 40(5):20–22, 2002.

4. Z. Haas, M. Gerla, D. Johnson, C. Perkins, M. Pursley, M. Steenstrup, and C.-K. Toh (eds). Special issue on wireless ad hoc networks. *IEEE J. on Selected Areas in Communications*, 17(8), 1999.

5. L. Buttyaen and J.-P. Hubaux (eds). Report on a working session on security in wireless ad hoc networks. *Mobile Computing and Communications Review*, 7(1), 2003.

6. S. Capkun and J.-P. Hubaux. BISS: building secure routing out of an incomplete set of secure associations. *Proc. of 2nd ACM Wireless Security (WiSe'03)*, pp. 21–29, 2003.

7. J.-P. Hubaux. What could we submit next year to WiSe? Research challenges in wireless security. *Invited Presentation at 2nd ACM Wireless Security (WiSe'03)*, 2003.

8. M. Gagnee. Identity-based encryption: a survey. *RSA Laboratories Cryptobytes*, 6(1):10–19, 2003.

9. A. Khalili, J. Katz, and W. Arbaugh. Toward secure key distribution in truly ad-hoc networks. *Proc. of IEEE Security and Assurance in Ad-Hoc Networks at Int'l Symp. on Applications and the Internet (SAINT'03)*, pp. 342–346, 2003.

10. G. Appenzeller and B. Lynn. Minimal-overhead IP security using identity based encryption. Available at `http://rooster.stanford.edu/~ben/pubs/ipibe.pdf`, 2002.

11. T. Garefalakis and C. Mitchell. Securing personal area networks. *Proc. of 13th IEEE Personal, Indoor and Mobile Radio Communications (PIMRC'02)*, pp. 1257–1259, 2002.

12. J. Arkko, T. Aura, J. Kempf, V. Mantyla, P. Nikander, and M. Roe. Securing IPv6 neighbor and router discovery. *Proc. 1st ACM Wireless Security (WiSe'01)*, pp. 77–86, 2002.

13. T. Stading. Secure communication in a distributed system using identity based encryption. *Proc. of 3rd IEEE/ACM Cluster Computing and Grid (CCGRID'03)*, pp. 414–420, 2003.

14. H. Deng, W. Li, and D. Agrawal. Routing security in wireless ad hoc networks. *IEEE Communications*, 40(10):70–75, 2002.

15. B. Awerbuch, D. Holmer, C. Nita-Rotaru, and H. Rubens. An on-demand secure routing protocol resilient to byzantine failures. *Proc. of 1st ACM Wireless Security (WiSe'02)*, pp. 21–30, 2002.

16. J. Zhen and S. Srinivas. Preventing replay attacks for secure routing in ad hoc networks. *Proc. of 2nd Ad Hoc Networks & Wireless (ADHOC-NOW'03)*, pp. 140–150, 2003.

17. Y.-C. Hu, A. Perrig, and D. Johnson. Packet leashes: a defense against wormhole attacks in wireless networks. *Proc. of 22nd IEEE Infocom (Infocom'03)*, pp. 1976–1986, 2003.

18. Y. Hu, A. Perrig, and D. Johnson. Rushing attacks and defense in wireless ad hoc network routing protocols. *Proc. of 2nd ACM Wireless Security (WiSe'03)*, pp. 30–40, 2003.

19. A. Shamir. Identity-based cryptosystems and signature schemes. *Proc. of 4th IACR Cryptology (Crypto'84)*, pp. 47–53, 1984.

20. D. Boneh and M. Franklin. Identity-based encryption from the Weil pairing. *Proc. of 21st IACR Cryptology (Crypto'01)*, pp. 213–229, 2001.

21. M. Bellare and P. Rogaway. Random oracle models are practical: a paradigm for designing efficient protocols. *Proc. of 1st ACM Computer and Communications Security (CCS'93)*, pp. 62–73, 1993.

22. B. Lynn. Authenticated identity-based encryption. *Cryptology ePrint Archive*, 2002/072, 2002.

23. S. Capkun, J.-P. Hubaux, and L. Buttyan. Mobility helps security in ad hoc networks. *Proc. of 4th ACM Mobile Ad Hoc Networking and Computing (MobiHoc'03)*, pp. 46–56, 2003.

24. L. Zhou and Z. Haas. Securing ad hoc networks. *IEEE Network*, 13(6):24–30, 1999.

25. C. Gentry and A. Silverberg. Hierarchical ID-based cryptography. *Proc. of 8th IACR AsiaCrypt (AsiaCrypt'02)*, pp. 548–566, 2002.

26. P. Grabher and D. Page. Hardware acceleration of the Tate pairing in characteristic three. *Proc. of 7th IACR Cryptographic Hardware and Embedded Systems (CHES'05)*, pp. 398–411, 2005.

27. B. Libert and J.-J.Quisquarter. New identity based signcryption schemes based on pairings. *Cryptology ePrint Archive*, 2003/023, 2003.

28. D. Boneh, B. Lynn, and H. Shacham. Short signature from the Weil pairing. *Proc. of 7th AsiaCrypt (AsiaCrypt'01)*, pp. 514–532, 2001.

29. D. Johnson. Routing in ad hoc networks of mobile hosts. *Proc. of 1st IEEE Workshop on Mobile Computing Systems and Applications (WMCSA'94)*, pp. 158–163, 1994.

30. E. Royer and C.-K. Toh. A review of current routing protocols for ad hoc mobile wireless networks. *IEEE Personal Communications*, 4(2):46–55, 1999.

31. M. Abolhasan, T. Wysocki, and E. Dutkiewicz. A review of routing protocols for mobile ad hoc networks. *Ad Hoc Networks*, 2:1–22, 2004.

32. X. Boyen. Multipurpose identity-based signcryption: a swiss army knife for identity-based cryptography. *Proc. of 23rd IACR Cryptology (Crypto'03)*, pp. 383–399, 2003.

33. D. Boneh and X. Boyen. Secure identity based encryption without random oracles. *Proc. of 24th IACR Cryptology (Crypto'04)*, 2004.

34. J.-P. Hubaux, L. Buttyaen, and S. Capkun. The quest for security in mobile ad hoc networks. *Proc. of 2nd ACM Mobile Ad Hoc Networking and Computing (MobiHoc'01)*, pp. 146–155, 2001.

35. H. Chan, A. Perrig, and D. Song. Random key predistribution schemes for sensor networks. *Proc. of 24th IEEE Security & Privacy (S&P'03)*, pp. 197–215, 2003.

36. G. Montenegro and C. Castelluccia. Statistically unique and cryptographically verifiable (SUCV) identifiers and addresses. *Proc. of 9th ISOC Network and Distributed Systems Security (NDSS'02)*, 2002.

37. Y.-C. Hu, A. Perrig, and D. Johnson. Ariadne: a secure on-demand routing protocol for ad hoc networks. *Proc. of 8th ACM Mobile Computing and Networking (MobiCom'02)*, pp. 12–23, 2002

38. P. Papadimitratos and Z. Haas. Secure routing for mobile ad hoc networks. *Proc. of 7th SCS Communication Networks and Distributed Systems Modeling and Simulation (CNDS'02)*, 2002.

39. S. Yi, P. Naldurg, and R. Kravets. Security-aware ad hoc routing for wireless networks. *Proc. of 2nd ACM Mobile Ad Hoc Networking and Computing (MobiHoc'01)*, pp. 299–302, 2001.

40. M. Zapata and N. Asokan. Securing ad hoc routing protocols. *Proc. of 1st ACM Wireless Security (WiSe'01)*, pp. 1–10, 2002.

41. Y.-C. Hu, D. Johnson, and A. Perrig. SEAD: secure efficient distance vector routing in mobile wireless ad hoc networks. *Proc. of 4th IEEE Workshop on Mobile Computing Systems and Applications (WMCSA'02)*, pp. 3–13, 2002.

42. K. Sanzgiri, B. Dahill, B. Levine, C. Shields, and E. Belding-Royer. A secure routing protocol for ad hoc networks. *Proc. of 10th IEEE Int'l Conf. on Network Protocols (ICNP'02)*, pp. 78–89, 2002.

43. J. Newsome, E. Shi, D. Song, and A. Perrig. The Sybil attack in sensor networks: analysis & defenses. *Proc. of 3rd IEEE/ACM Information Processing in Sensor Networks (IPSN'04)*, pp. 259–268, 2004.

5

A SURVEY OF ATTACKS AND COUNTERMEASURES IN MOBILE AD HOC NETWORKS

Bing Wu, Jianmin Chen, Jie Wu, Mihaela Cardei
Department of Computer Science and Engineering
Florida Atlantic University
E-mail: {bwu, jchen8}@fau.edu, {jie, mihaela}@cse.fau.edu

Security is an essential service for wired and wireless network communications. The success of mobile ad hoc network (MANET) will depend on people's confidence in its security. However, the characteristics of MANET pose both challenges and opportunities in achieving security goals, such as confidentiality, authentication, integrity, availability, access control, and non-repudiation. We provide a survey of attacks and countermeasures in MANET in this chapter. The countermeasures are features or functions that reduce or eliminate security vulnerabilities and attacks. First, we give an overview of attacks according to the protocol layers, and to security attributes and mechanisms. Then we present preventive approaches following the order of the layered protocol layers. We also put forward an overview of MANET intrusion detection systems (IDS), which are reactive approaches to thwart attacks and used as a second line of defense.

1. INTRODUCTION

A MANET is referred to as a network without infrastructure because the mobile nodes in the network dynamically set up temporary paths among themselves to transmit packets. In a MANET, a collection of mobile hosts with wireless network interfaces form a temporary network without the aid of any fixed infrastructure or centralized administration. Nodes within each other's wireless transmission ranges can communicate directly; however, nodes outside each other's range have to rely on some other nodes to relay messages [22]. Thus, a multi-hop scenario occurs, where several intermediate hosts relay the packets sent by the source host before they reach the destination host. Every node functions as a router. The success of communication highly depends on other nodes' cooperation. At a given time, the system can be viewed as a random graph due to the movement of the nodes, their transmitter/receiver coverage patterns, the transmission power levels, and the co-channel interference levels. The network topology may change with time as the nodes move or adjust their transmission and reception parameters. Thus, a MANET has several salient characteristics [21]:

- Dynamic topology
- Resource constraints
- No infrastructure
- Limited physical security

In 1996, The Internet Engineering Task Force(IETF) created a MANET working group with the goal to standardize IP routing protocol functionality suitable for wireless routing applications within both static and dynamic topologies.

Possible applications of MANET include: soldiers relaying information for situational awareness on the battlefield, business associates sharing information during a meeting, attendees using laptop computers to participate in an interactive conference, and emergency disaster relief personnel coordinating efforts after a fire, hurricane or earthquake. Other possible applications [22] include personal area and home networking, location-based services, and sensor networks.

Security is an essential service for wired and wireless network communications. The success of MANET strongly depends on whether its security can be trusted. However, the characteristics of MANET pose both challenges and opportunities in achieving the security goals, such as confidentiality, authentication, integrity, availability, access control, and non-repudiation.

There are a wide variety of attacks that target the weakness of MANET. For example, routing messages are an essential component of mobile network communications, as each packet needs to be passed quickly through intermediate nodes, which the packet must traverse from a source to the destination. Malicious routing attacks can target the routing discovery or maintenance phase by not following the specifications of the routing protocols. There are also attacks that target some particular routing protocols, such as DSR, or AODV [10] [20]. More sophisticated and subtle routing attacks have been identified in recent published papers, such as the blackhole (or sinkhole) [35], Byzantine [17], and wormhole [15] [32] attacks. Currently routing security is one of the hottest research areas in MANET.

The mobile hosts forming a MANET are normally mobile devices with limited physical protection and resources. Security modules, such as tokens and smart cards, can be used to protect against physical attacks. Cryptographic tools are widely used to provide powerful security services, such as confidentiality, authentication, integrity, and non-repudiation. Unfortunately, cryptography cannot guarantee availability; for example, it cannot prevent radio jamming. Meanwhile, strong cryptography often demands

Table 1. Security Attacks Classification

Passive Attacks	Eavesdropping, traffic analysis, monitoring
Active Attacks	Jamming, spoofing, modification, replaying, DoS

a heavy computation overhead and requires the auxiliary complicated key distribution and trust management services, which mostly are restricted by the capabilities of physical devices (e.g. CPU or battery).

The characteristics and nature of MANET require the strict cooperation of participating mobile hosts. A number of security techniques have been invented and a list of security protocols have been proposed to enforce cooperation and prevent misbehavior, such as 802.11 WEP [47], SEAD [11], ARAN [32], SSL [51], etc. However, none of those preventive approaches is perfect or capable to defend against all attacks. A second line of defense called intrusion detection systems (IDS) is proposed and applied in MANET. IDS are some of the latest security tools in the battle against attacks. Distributed IDS were introduced in MANET to monitor either the misbehavior or selfishness of mobile hosts. Subsequent actions can be taken based on the information collected by IDS.

This chapter is structured as follows. In Section 3, we describe the attacks on each layer of the Internet model: application, transport, network, data link, and physical layer. In Section 4, we overview attack countermeasures, including intrusion detection and co-operation enforcement at different layers of the Internet model. In Section 5, we briefly discuss open challenges and future directions.

2. SECURITY ATTACKS

A variety of attacks are possible in MANET. Some attacks apply to general network, some apply to wireless network and some are specific to MANETs. These security attacks can be classified according to different criteria, such as the domain of the attackers, or the techniques used in attacks. These security attacks in MANET and all other networks can be roughly classified by the following criteria: passive or active, internal or external, different protocol layer, stealthy or non-stealthy, cryptography or non-cryptography related.

- **Passive vs. active attacks:** The attacks in MANET can roughly be classified into two major categories, namely passive attacks and active attacks [9][23]. A passive attack obtains data exchanged in the network without disrupting the operation of the communications, while an active attack involves information interruption, modification, or fabrication, thereby disrupting the normal functionality of a MANET. Table 1 shows the general taxonomy of security attacks against MANET. Examples of passive attacks are eavesdropping, traffic

Table 2. Security Attacks on each layer of the Internet Model

Layer	Attacks
Application layer	Repudiation, data corruption
Transport layer	Session hijacking, SYN flooding
Network layer	Wormhole, blackhole, Byzantine, flooding, resource consumption, location disclosure attacks
Data link layer	Traffic analysis, monitoring, disruption MAC (802.11), WEP weakness
Physical layer	Jamming, interceptions, eavesdropping
Multi-layer attacks	DoS, impersonation, replay, man-in-the-middle

analysis, and traffic monitoring. Examples of active attacks include jamming, impersonating, modification, denial of service (DoS), and message replay.

- **Internal vs. external attacks:** The attacks can also be classified into external attacks and internal attacks, according the domain of the attacks. Some papers refer to outsider and insider attacks [39]. External attacks are carried out by nodes that do not belong to the domain of the network. Internal attacks are from compromised nodes, which are actually part of the network. Internal attacks are more severe when compared with outside attacks since the insider knows valuable and secret information, and possesses privileged access rights.

- **Attacks on different layers of the Internet model:** The attacks can be further classified according to the five layers of the Internet model. Table 2 presents a classification of various security attacks on each layer of the Internet model. Some attacks can be launched at multiple layers.

- **Stealthy vs. non-stealthy attacks:** Some security attacks use stealth [34], whereby the attackers try to hide their actions from either an individual who is monitoring the system or an intrusion detection system (IDS). But other attacks such as DoS cannot be made stealthy.

- **Cryptography vs. non-cryptography related attacks:** Some attacks are non-cryptography related, and others are cryptographic primitive attacks. Table 3 shows cryptographic primitive attacks and the examples.

For the rest of the section, we present a survey of security attacks in MANET on each layer of the Internet model. Physical layer attacks are discussed in Section 3.1, followed

Table 3. Cryptographic Primitive Attacks

Cryptographic Primitive Attacks	Examples
Pseudorandom number attack	Nonce, timestamp, initialization vector (IV)
Digital signature attack	RSA signature, ElGamal signature, digital signature standard (DSS)
Hash collision attack	SHA-0, MD4, MD5, HAVAL-128, RIPEMD

by link layer attacks in Section 3.2; and network layer attacks in Section 3.3. Transport layer attacks are discussed in Section 3.4, application layer attacks are discussed in Section 3.5, and multi-layer attacks are discussed in Section 3.6. Cryptographic primitive attacks are discussed in Section 3.7.

2.1. Physical layer attacks

Wireless communication is broadcast by nature. A common radio signal is easy to jam or intercept. An attacker could overhear or disrupt the service of a wireless network physically.

- **Eavesdropping:** Eavesdropping is the intercepting and reading of messages and conversations by unintended receivers. The mobile hosts in mobile ad hoc networks share a wireless medium. The majorities of wireless communications use the RF spectrum and broadcast by nature. Signals broadcast over airwaves can be easily intercepted with receivers tuned to the proper frequency [47] [48]. Thus, messages transmitted can be overheard, and fake messages can be injected into network.

- **Interference and Jamming:** Radio signals can be jammed or interfered with, which causes the message to be corrupted or lost [47] [48]. If the attacker has a powerful transmitter, a signal can be generated that will be strong enough to overwhelm the targeted signals and disrupt communications. The most common types of this form of signal jamming are random noise and pulse. Jamming equipment is readily available. In addition, jamming attacks can be mounted from a location remote to the target networks.

2.2. Link layer attacks

The MANET is an open multipoint peer-to-peer network architecture. Specifically, one-hop connectivity among neighbors is maintained by the link layer protocols, and the network layer protocols extend the connectivity to other nodes in the network. Attacks may target the link layer by disrupting the cooperation of the layer's protocols.

Wireless medium access control (MAC) protocols have to coordinate the transmissions of the nodes on the common transmission medium. Because a token-passing bus MAC protocol is not suitable for controlling a radio channel, IEEE 802.11 protocol is specifically devoted to wireless LANs. The IEEE 802.11 MAC protocol uses distributed contention resolution mechanisms for sharing the wireless channel. The IEEE 802.11 working group proposed two algorithms for contention resolution. One is a fully distributed access protocol called the distributed coordination function (DCF). The other is a centralized access protocol called the point coordination function (PCF). PCF requires a central decision maker such as a base station. DCF uses a carrier sense multiple access/collision avoidance protocol (CSMA/CA) for resolving channel contention among multiple wireless hosts.

Three values for interframe space (IFS) are defined to provide priority-based access to the radio channel [27]. SIFS is the shortest interframe space and is used for ACK, CTS and poll response frames. DIFS is the longest IFS and is used as the minimum delay for asynchronous frames contending for access. PIFS is the middle IFS and is used for issuing polls by the centralized controller in the PCF scheme. In case there is a collision, the sender waits a random unit of time, based on the binary exponential backoff algorithm, before retransmitting. In Figure 1, node Na and node Nc contend to communicate with node Nb. First node Na gets access and reserves the channel, and then Nc succeeds and reserves the channel while node Na has to back off [30].

Disruption on MAC DCF and backoff mechanism

Current wireless MAC protocols assume cooperative behaviors among all nodes. Obviously the malicious or selfish nodes are not forced to follow the normal operation of the protocols. In the link layer, a selfish or malicious node could interrupt either contention-based or reservation-based MAC protocols.

A malicious neighbor of either the sender or the receiver could intentionally not follow the protocol specifications. For example, the attacker may corrupt the frames easily by introducing some bits or ignoring the ongoing transmission. It could also just wait SIFS or exploit its binary exponential backoff scheme to launch DoS attacks in IEEE 802.11 MAC. The binary exponential scheme favors the last winner amongst the contending nodes. This leads to what is called the capture effect [21]. Nodes that are heavily loaded tend to capture the channel by continually transmitting data, thereby causing lightly loaded neighbors to backoff endlessly. Malicious nodes could take advantage of this capture effect vulnerability. Moreover, a backoff at the link layer can cause a chain reaction in any upper layer protocols that use a backoff scheme, like TCP window management.

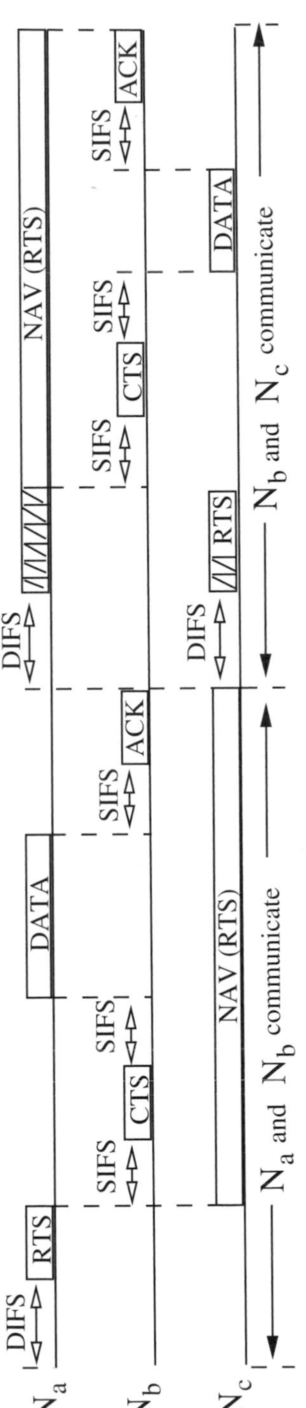

Figure 1. Illustration of Channel Contention in 802.11 MAC

The network allocation vector (NAV) field carried in RTS/CTS frames exposes another vulnerability to DoS attacks in the link layer [21] [29]. Initially the NAV field was proposed to mitigate the hidden terminal problem in the carrier sense mechanism. During the RTS/CTS handshake the sender first sends a small RTS frame containing the time needed to complete the CTS, data, and ACK frames. Each neighbor of the sender and receiver will update the NAV field and defer their transmission for the duration of the future transaction according to the time that they overheard. An attacker may also overhear the NAV information and then intentionally corrupt the link layer frame by interfering with the ongoing transmission.

Weakness of 802.11 WEP

IEEE 802.11 WEP incorporates wired equivalent privacy (WEP) to provide WLAN systems a modest level of privacy by encrypting radio signals. 802.11 WEP standards support WEP cryptographic keys of 40 bits, though some vendors have implemented 104 bits and even 128 bits. It is well known that WEP is broken and WEP is replaced by AES in 802.11i. Some of the weaknesses 802.11 WEP are listed below [27] [28] [47],

- WEP protocol does not specify key management.
- The initialization vector (IV) is a 24-bit field sent in clear and is part of the RC4 encryption key. The reuse of IV and the weakness of RC4 lead to analytic attacks.
- The combined use of a non-cryptographic integrity algorithm, CRC 32, with the stream cipher is a security risk.

2.3. Network layer attacks

Network layer protocols extend connectivity from neighboring 1-hops nodes to all other nodes in MANET. The connectivity between mobile hosts over a potentially multi-hop wireless link relies heavily on cooperative reactions among all network nodes.

A variety of attacks targeting the network layer have been identified and heavily studied in research papers. By attacking the routing protocols, attackers can absorb network traffic, inject themselves into the path between the source and destination, and thus control the network traffic flow, as shown in Figure 2 (a) and (b), where a malicious node M can inject itself into the routing path between sender S and receiver D.

The traffic packets could be forwarded to a non-optimal path, which could introduce significant delay. In addition, the packets could be forwarded to a nonexistent path and get lost. The attackers can create routing loops, introduce severe network congestion, and channel contention into certain areas. Multiple colluding attackers may even prevent a source node from finding any route to the destination, causing the network to partition, which triggers excessive network control traffic, and further intensifies network congestion and performance degradation.

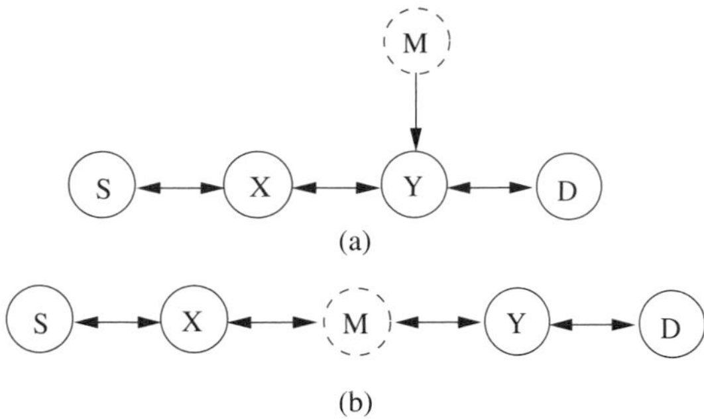

Figure 2. Illustration of Routing Attack

Attacks at the routing discovery phase

There are malicious routing attacks that target the routing discovery or maintenance phase by not following the specifications of the routing protocols. Routing message flooding attacks, such as hello flooding, RREQ flooding, acknowledgement flooding, routing table overflow, routing cache poisoning, and routing loop are simple examples of routing attacks targeting the route discovery phase [6] [35]. Proactive routing algorithms, such as DSDV [22] and OLSR [45], attempt to discover routing information before it is needed, while reactive algorithms, such as DSR [22] and AODV [22], create routes only when they are needed. Thus, proactive algorithms performs worse than on-demand schemes because they do not accommodate the dynamic of MANETs, clearly proactive algorithms require many costly broadcasts. Proactive algorithms are more vulnerable to routing table overflow attacks. Some of these attacks are listed below.

- **Routing table overflow attack:** A malicious node advertises routes that go to non-existent nodes to the authorized nodes present in the network. It usually happens in proactive routing algorithms, which update routing information periodically. The attacker tries to create enough routes to prevent new routes from being created. The proactive routing algorithms are more vulnerable to table overflow attacks because proactive routing algorithms attempt to discover routing information before it is actually needed. An attacker can simply send excessive route advertisements to overflow the victim's routing table.

- **Routing cache poisoning attack:** In route cache poisoning attacks, attackers take advantage of the promiscuous mode of routing table updating, where a node overhearing any packet may add the routing information contained in that packet header to its own route cache, even if that node is not on the path. Suppose a malicious node M wants to poison routes to node X. M could broadcast spoofed packets with source route to X via M itself; thus, neighboring nodes that overhear the packet may add the route to their route caches.

Attacks at the routing maintenance phase

There are attacks that target the route maintenance phase by broadcasting false control messages, such as link-broken error messages, which cause the invocation of the costly route maintenance or repairing operation. For example, AODV and DSR implement path maintenance procedures to recover broken paths when nodes move. If the destination node or an intermediate node along an active path moves, the upstream node of the broken link broadcasts a route error message to all active upstream neighbors. The node also invalidates the route for this destination in its routing table. Attackers could take advantage of this mechanism to launch attacks by sending false route error messages.

Attacks at data forwarding phase

Some attacks also target data packet forwarding functionality in the network layer. In this scenario the malicious nodes participate cooperatively in the routing protocol routing discovery and maintenance phases, but in the data forwarding phase [18] [33] they do not forward data packets consistently according to the routing table. Malicious nodes simply drop data packets quietly, modify data content, replay, or flood data packets; they can also delay forwarding time-sensitive data packets selectively or inject junk packets.

Attacks on particular routing protocols

There are attacks that target some particular routing protocols. In DSR, the attacker may modify the source route listed in the RREQ or RREP packets. It can delete a node from the list, switch the order, or append a new node into the list. In AODV, the attacker may advertise a route with a smaller distance metric than the actual distance, or advertise a routing update with a large sequence number and invalidate all routing updates from other nodes.

Other advanced attacks

More sophisticated and subtle routing attacks have been identified in recent research papers. The blackhole (or sinkhole), Byzantine, and wormhole attacks are the typical examples, which are described in detail below.

- **Wormhole attack:** An attacker records packets at one location in the network and tunnels them to another location. Routing can be disrupted when routing control messages are tunneled. This tunnel between two colluding attackers is referred as a wormhole [8] [32]. Wormhole attacks are severe threats to MANET routing protocols. For example, when a wormhole attack is used against an on-demand routing protocol such as DSR or AODV, the attack could prevent the discovery of any routes other than through the wormhole.

- **Blackhole attack:** The blackhole attack has two properties. First, the node exploits the mobile ad hoc routing protocol, such as AODV, to advertise itself as having a valid route to a destination node, even though the route is spurious, with the intention of intercepting packets. Second, the attacker consumes the intercepted packets without any forwarding. However, the attacker runs the risk that neighboring nodes will monitor and expose the ongoing attacks. There is a more subtle form of these attacks when an attacker selectively forwards packets. An attacker suppresses or modifies packets originating from some nodes, while leaving the data from the other nodes unaffected, which limits the suspicion of its wrongdoing.

- **Byzantine attack:** A compromised intermediate node works alone, or a set of compromised intermediate nodes works in collusion and carry out attacks such as creating routing loops, forwarding packets through non-optimal paths, or selectively dropping packets, which results in disruption or degradation of the routing services [17].

- **Rushing attack:** Two colluded attackers use the tunnel procedure to form a wormhole. If a fast transmission path (e.g. a dedicated channel shared by attackers) exists between the two ends of the wormhole, the tunneled packets can propagate faster than those through a normal multi-hop route. This forms the rushing attack [19]. The rushing attack can act as an effective denial-of-service attack against all currently proposed on-demand MANET routing protocols, including protocols that were designed to be secure, such as ARAN and Ariadne [20].

- **Resource consumption attack:** This is also known as the sleep deprivation attack. An attacker or a compromised node can attempt to consume battery life by requesting excessive route discovery, or by forwarding unnecessary packets to the victim node.

- **Location disclosure attack:** An attacker reveals information regarding the location of nodes or the structure of the network. It gathers the node location information, such as a route map, and then plans further attack scenarios. Traffic analysis, one of the subtlest security attacks against MANET, is unsolved. Adversaries try to figure out the identities of communication parties and analyze traffic to learn the network traffic pattern and track changes in the traffic pattern. The leakage of such information is devastating in security-sensitive scenarios.

2.4. Transport layer attacks

The objectives of TCP-like Transport layer protocols in MANET include setting up of end-to-end connection, end-to-end reliable delivery of packets, flow control, congestion control, and clearing of end-to-end connection. Similar to TCP protocols in the Internet, the mobile node is vulnerable to the classic SYN flooding attack or

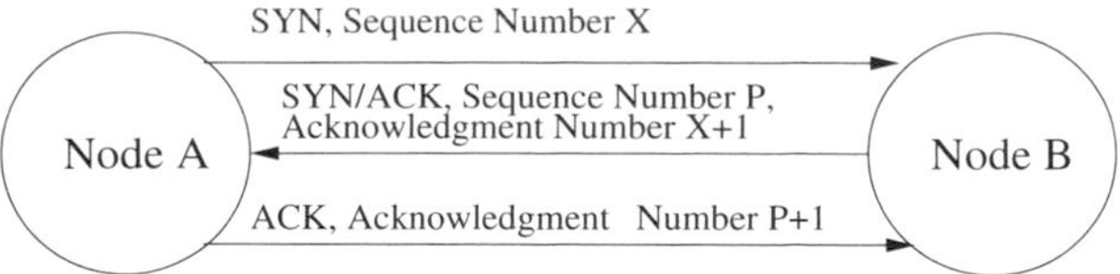

Figure 3. TCP Three-way Handshake

session hijacking attacks. However, a MANET has a higher channel error rate when compared with wired networks. Because TCP does not have any mechanism to distinguish whether a loss was caused by congestion, random error, or malicious attacks, TCP multiplicatively decreases its congestion window upon experiencing losses, which degrades network performance significantly [49].

- **SYN flooding attack:** The SYN flooding attack is a denial-of-service attack. The attacker creates a large number of half-opened TCP connections with a victim node, but never completes the handshake to fully open the connection.

  For two nodes to communicate using TCP, they must first establish a TCP connection using a three-way handshake. The three messages exchanged during the handshake, illustrated in Figure 3, allow both nodes to learn that the other is ready to communicate and to agree on initial sequence numbers for the conversation.

  During the attack, a malicious node sends a large amount of SYN packets to a victim node, spoofing the return addresses of the SYN packets. The SYN-ACK packets are sent out from the victim right after it receives the SYN packets from the attacker and then the victim waits for the response of ACK packet. Without receiving the ACK packets, the half-open data structure remains in the victim node. If the victim node stores these half-opened connections in a fixed-size table while it awaits the acknowledgement of the three-way handshake, all of these pending connections could overflow the buffer, and the victim node would not be able to accept any other legitimate attempts to open a connection. Normally there is a time-out associated with a pending connection, so the half-open connections will eventually expire and the victim node will recover. However, malicious nodes can simply continue sending packets that request new connections faster than the expiration of pending connections.

- **Session hijacking:** Session hijacking takes advantage of the fact that most communications are protected (by providing credentials) at session setup, but not thereafter. In the TCP session hijacking attack, the attacker spoofs the victim's IP address, determines the correct sequence number that is expected by the target, and then performs a DoS attack on the victim. Thus the attacker impersonates the victim node and continues the session with the target.

  The TCP ACK storm problem, illustrated in Figure 4, could be created when an attacker launches a TCP session hijacking attack. The attacker sends injected

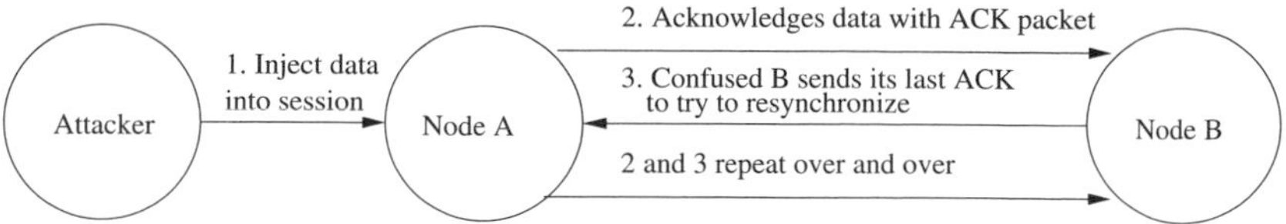

Figure 4. TCP ACK Storm

session data, and node A will acknowledge the receipt of the data by sending an ACK packet to node B. This packet will not contain a sequence number that node B is expecting, so when node B receives this packet, it will try to resynchronize the TCP session with node A by sending it an ACK packet with the sequence number that it is expecting. The cycle goes on and on, and the ACK packets passing back and forth create an ACK storm. Hijacking a session over UDP is the same as over TCP, except that UDP attackers do not have to worry about the overhead of managing sequence numbers and other TCP mechanisms. Since UDP is connectionless, edging into a session without being detected is much easier than the TCP session attacks.

2.5. Application layer attacks

The application layer communication is also vulnerable in terms of security compared with other layers. The application layer contains user data, and it normally supports many protocols such as HTTP, SMTP, TELNET, and FTP, which provide many vulnerabilities and access points for attackers. The application layer attacks are attractive to attackers because the information they seek ultimately resides within the application and it is direct for them to make an impact and reach their goals.

- **Malicious code attacks:** Malicious code, such as viruses, worms, spywares, and Trojan Horses, can attack both operating systems and user applications. These malicious programs usually can spread themselves through the network and cause the computer system and networks to slow down or even damaged. In MANET, an attacker can produce similar attacks to the mobile system of the ad hoc network.

- **Repudiation attacks:** In the network layer, firewalls can be installed to keep packets in or keep packets out. In the transport layer, entire connections can be encrypted, end-to-end. But these solutions do not solve the authentication or non-repudiation problems in general. Repudiation refers to a denial of participation in all or part of the communication. For example, a selfish person could deny conducting an operation on a credit card purchase, or deny any on-line bank transaction, which is the prototypical repudiation attack on a commercial system.

2.6. Multi-layer attacks

Some security attacks can be launched from multiple layers instead of a particular layer. Examples of multi-layer attacks are denial of service (DoS), man-in-the-middle, and impersonation attacks.

- **Denial of service:** Denial of service (DoS) attacks could be launched from several layers. An attacker can employ signal jamming at the physical layer, which disrupts normal communications. At the link layer, malicious nodes can occupy channels through the capture effect, which takes advantage of the binary exponential scheme in MAC protocols and prevents other nodes from channel access. At the network layer, the routing process can be interrupted through routing control packet modification, selective dropping, table overflow, or poisoning. At the transport and application layers, SYN flooding, session hijacking, and malicious programs can cause DoS attacks.
- **Impersonation attacks:** Impersonation attacks are launched by using other node's identity, such as MAC or IP address. Impersonation attacks sometimes are the first step for most attacks, and are used to launch further, more sophisticated attacks.
- **Man-in-the-middle attacks:** An attacker sits between the sender and the receiver and sniffs any information being sent between two ends. In some cases the attacker may impersonate the sender to communicate with the receiver, or impersonate the receiver to reply to the sender.

2.7. Cryptographic primitive attacks

Cryptography is an important and powerful security tool. It provides security services, such as authentication, confidentiality, integrity, and non-repudiation. In all likelihood, there exist attacks on many cryptographic primitives that have not yet been discovered. There could be new attacks designed and developed for hash functions, digital signatures, both block and stream ciphers. Most security holes are due to poor implementation, i.e. weakness in security protocols. For example, authentication protocols and key exchange protocols are often the target of malicious attacks. Cryptographic primitives are considered to be secure, however, recently some problems were discovered, such as collision attacks on hash function, e.g. SHA-1 [46]. Pseudorandom number attacks [51], digital signature attacks [14], and hash collision attacks [46] are discussed as following.

- **Pseudorandom number attacks:** To make packets fresh, a timestamp or random number (nonce) is used to prevent a replay attack [51]. The session key is often generated from a random number. In the public key infrastructure the shared secret key can be generated from a random number too. The conventional random number generators in most programming languages are designed for statistical randomness, not to resist prediction by cryptanalysts. In the optimal case, random numbers are generated based on physical sources of randomness that cannot be predicted. The noise from an electronic device or the position of a pointer device is a source of such randomness. However, true random numbers are difficult to generate. When true physical randomness is not available, pseudorandom numbers must be used. Cryptographic pseudorandom generators typically have a large pool (seed value) containing randomness. New environmental noise should be mixed into the pool to prevent others from determining previous or future values. The design and implementation of cryptographic pseudorandom generators could easily become the weakest point of the system.

- **Digital signature attacks:** The RSA public key algorithm can be used to generate a digital signature. The signature scheme has one problem: it could suffer the blind signature attack. The user can get the signature of a message and use the signature and the message to fake another message's signature. The ElGamal signature is based on the difficulty in breaking the discrete log problem. Digital Signature Algorithm (DSA) is an updated version of the ElGamal digital signature scheme published in 1994 by FIPS, and was chosen as the digital signature standard (DSS) [14]. The attack models for digital signature can be classified into known-message, chosen-message, and key-only attacks. In the known-message attack, the attacker knows a list of messages previously signed by the victim. In the chosen-message attack, the attacker can choose a specific message that it wants the victim to sign. But in the key-only attack, the adversary only knows the verification algorithm, which is public. Very often the digital signature algorithm is used in combination with a hash function. The hash function needs to be collision resistant.

- **Hash collision attacks:** The goal of a collision attack is to find two messages with the same hash, but the attacker cannot pick what the hash will be. Collision attacks were announced in SHA-0, MD4, MD5, HAVAL-128, and RIPEMD. The collisions against MD4, MD5, HAVAL-128, and RIPEMD were found recently. A successful attack against SHA-1 [46] was found, and the collisions in SHA-1 can be found with an estimated effort of 269 hash computations .

  Normally all major digital signature techniques (including DSA and RSA) involve first hashing the data and then signing the hash value. The original message data is not signed directly by the digital signature algorithm for both performance and security reasons. Collision attacks could be used to tamper with existing certificates. An adversary might be able to construct a valid certificate corresponding to the hash collision.

- **Key management vulnerability:** Key management protocols deal with the key generation, storage, distribution, updating, revocation, and certificate service. Attackers can launch attacks to disclose the cryptographic key at the local host or during the key distribution procedure. The lack of a central trusted entity in MANET makes it more vulnerable to key management attacks [5] [7] [9]

[24]. For example, the man-in-the-middle attack is a design pitfall of the Diffie-Hellman (DH) key exchange protocol. For key management protocols that rely on a trusted key distribution center or certificate authority, the trusted central entity becomes the focus of attacks.

3. SECURITY ATTACK COUNTERMEASURES

Security is essential for the widespread of MANET. However, the characteristics of MANET pose both challenges and opportunities in achieving the security goals, such as confidentiality, authentication, integrity, availability, access control, and non-repudiation.

The attacks countermeasures presentation is as follows. An overview of security attributes and security mechanisms is presented in Sections 3.1 and 3.2, respectively. We describe the attack countermeasures by different network layers. Physical layer defense is discussed in Section 3.3, link layer defense is discussed in Section 3.4, and network layer defense is discussed in Section 3.5. Transport layer defense and application layer defense are discussed in Section 3.6 and Section 3.7 respectively. Multi-layer defense is in Section 3.8. Defense against key management attacks is in Section 3.9, and MANET intrusion detection systems are discussed in 3.10.

3.1. Security attributes

Security is the combination of processes, procedures, and systems used to ensure confidentiality, authentication, integrity, availability, access control, and non-repudiation.

- **Confidentiality:** The goal of confidentiality is to keep the information sent unreadable to unauthorized users or nodes. MANET uses an open medium, so usually all nodes within the direct transmission range can obtain the data. One way to keep information confidential is to encrypt the data, and another technique is to use directional antennas.

- **Authentication:** The goal of authentication is to be able to identify a node or a user, and to be able to prevent impersonation. In wired networks and infrastructure-based wireless networks, it is possible to implement a central authority at a point such as a router, base station, or access point. But there is no central authority in MANET, and it is much more difficult to authenticate an entity. Authentication can be achieved by using message authentication code (MAC) [62].

- **Integrity:** The goal of integrity is to be able to keep the message sent from being illegally altered or destroyed in the transmission. When the data is sent through the wireless medium, the data can be modified or deleted by malicious attackers. The malicious attackers can also resend it, which is called a replay attack. The integrity can be achieved by hash functions.

- **Non-repudiation:** The goal of non-repudiation is related to a fact that if an entity sends a message, the entity cannot deny that the message was sent by it. By producing a signature for the message, the entity cannot later deny the message. In public key cryptography, a node A signs the message using its private key. All other nodes can verify the signed message by using A's public key, and A cannot deny that its signature is attached to the message.

- **Availability:** The goal of availability is to keep the network service or resources available to legitimate users. It ensures the survivability of the network despite malicious incidents.

- **Access control:** The goal of access control is to prevent unauthorized use of network services and system resources. Obviously, access control is tied to authentication attributes. In general, access control is the most commonly thought of service in both network communications and individual computer systems.

3.2. Security mechanisms

A variety of security mechanisms have been invented to counter malicious attacks. The conventional approaches such as authentication, access control, encryption, and digital signature provide a first line of defense. As a second line of defense, intrusion detection systems and cooperation enforcement mechanisms implemented in MANET can also help to defend against attacks or enforce cooperation, reducing selfish node behavior.

- **Preventive mechanism:** The conventional authentication and encryption schemes are based on cryptography, which includes asymmetric and symmetric cryptography. Cryptographic primitives such as hash values (message digests) are sufficient in providing data integrity in transmission as well. Threshold cryptography can be used to hide data by dividing it into a number of shares. Digital signatures can also be used to achieve data integrity and authentication services.

  It is also necessary to consider the physical safety of mobile devices, since the hosts are normally small devices, which are physically vulnerable. For example, a device could easily be stolen, lost, or damaged. In the battlefield they are at risk of being hijacked. The protection of the sensitive data on a physical device can be enforced by some security modules, such as tokens or a smart card that is accessible through PIN, passphrases, or biometrics.

- **Reactive mechanism:** A number of malicious attacks could bypass the preventive mechanisms due to its design, implementation, or restrictions. An intrusion detection system provides a second line of defense. There are widely used to detect misuse and anomalies. A misuse detection system attempts to define improper behavior based on the patterns of well-known attacks, but it lacks the ability to detect any attacks that were not considered during the creation of

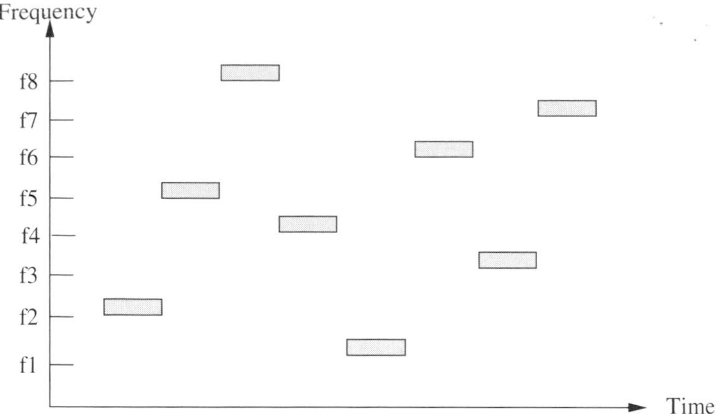

Figure 5. Illustration of Frequency Hopping Spread Spectrum

the patterns; Anomaly detection attempts to define normal or expected behavior statistically. It collects data from legitimate user behavior over a period of time, and then statistical tests are applied to determine anomalous behavior with a high level of confidence. In practice, both approaches can be combined to be more effective against attacks. Some intrusion detection systems for MANET have been proposed in recent research papers.

3.3. Physical layer defense

Spread spectrum technology, such as frequency hopping (FHSS) [27] or direct sequence (DSSS) [27], can make it difficult to detect or jam signals. It changes frequency in a random fashion to make signal capture difficult or spreads the energy to a wider spectrum so the transmission power is hidden behind the noise level. Directional antennas can also be deployed due to the fact that the communication techniques can be designed to spread the signal energy in space.

- **FHSS:** The signal is modulated with a seemingly random series of radio frequencies, which hops from frequency to frequency at fixed intervals. The receiver uses the same spreading code, which is synchronized with the transmitter, to recombine the spread signals into their original form. Figure 5 shows an example of a frequency-hopping signal.

  With the transmitter and the receiver synchronized properly, data is transmitted over a single channel. However, the signal appears to be unintelligible duration impulse noise for the eavesdroppers. Meanwhile, interference is minimized as the signal is spread across multiple frequencies.

- **DSSS:** Each data bit in the original signal is represented by multiple bits in the transmitted signal, using a spreading code. The spreading code spreads the

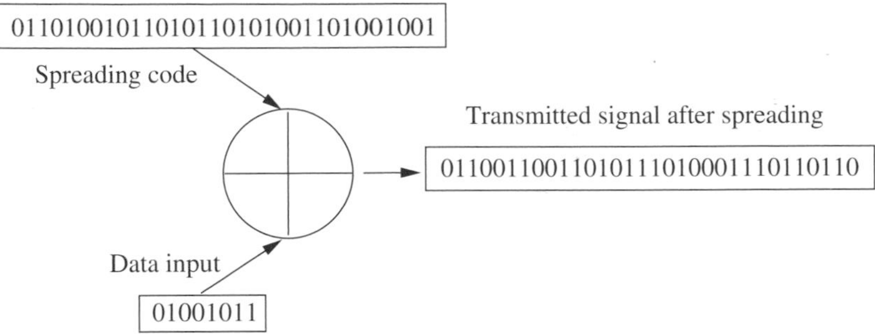

Figure 6. Illustration of Direct Sequence Spread Spectrum

signal across a wider frequency band in direct proportion to the number of bits used. The receiver can use the spreading code with the signal to recover the original data. Figure 6 illustrates that each original bit of data is represented by 4 bits in the transmitted signal. The first bit of data, a 0 is transmitted as 0110 which is first 4 bits of spreading code. The second bit, 1, is transmitted as 0110 which is bit-wise complement of the second 4 bits of spreading code. In turn, each input bit is combined, using exclusive-or, with four bits of the spreading code.

Both FHSS and DSSS pose difficulties for outsiders attempting to intercept the radio signals. The eavesdropper must know the frequency band, spreading code, and modulation techniques in order to accurately read the transmitted signals. The property that spread spectrum technologies do not interoperate with each other further adds difficulties to the eavesdropper. Spread spectrum technology also minimizes the potential for interference from other radios and electromagnetic devices. Despite the capability of spread spectrum technology, it is secure only when the hopping pattern or spreading code is unknown to the eavesdroppers.

3.4. Link layer defense

There are malicious attacks that target the link layer by disrupting the cooperative nature of link layer protocols. Link layer protocols help to discover 1-hop neighbors, handle fair channel access, frame error control, and maintain neighbor connections. Selfish nodes could disobey the channel access rule, manipulate the NAV field, cheat backoff values in order to maximize their own throughput. Neighbors should monitor these misbehaviors. Although it is still an open challenge to prevent selfishness, some schemes have been proposed, such as ERA-802.11 [12], where detection algorithms are proposed. Traffic analysis is prevented by encryption at data link layer.

WEP encryption scheme defined in the IEEE 802.11 wireless LAN standard uses link encryption to hide the end-to-end traffic flow information. However, WEP has been widely criticized for its weaknesses [28] [47]. Some secure link layer protocols have been proposed in recent research, such as LLSP.

In MANET, some papers propose to create a security cloud, construct a traffic cover mode or dynamic mix method, or use traditional traffic padding and traffic rerouting techniques to prevent traffic analysis. A security cloud means that each node under the security cloud is identical in terms of traffic generation. A traffic cover mode hides the changes of an end-to-end flow traffic pattern, because certain tactical information might be inferred from the unusual changes in the traffic pattern. A dynamic mix method is used to hide the source and destination information during message delivery via a cryptographic method and to "mix" nodes in the network.

3.5. Network layer defense

The passive attack on routing information can be countered with the same methods that protect data traffic. Some active attacks, such as illegal modification of routing messages, can be prevented by mechanisms such as source authentication and message integrity. DoS attacks on a routing protocol could take many forms. DoS attacks can be limited by preventing the attacker from inserting routing loops, enforcing the maximum route length that a packet should travel, or using some other active approaches. The wormhole attack can be detected by an unalterable and independent physical metric, such as time delay or geographical location. For example, packet leashes are used to combat wormhole attacks [15].

In general, some kind of authentication and integrity mechanism, either the hop-by-hop or the end-to-end approach, is used to ensure the correctness of routing information. For instance, digital signature, one-way hash function, hash chain, message authentication code (MAC), and hashed message authentication code
(HMAC) are widely used for this purpose. IPsec and ESP are standards of security protocols on the network layer used in the Internet that could also be used in MANET, in certain circumstances, to provide network layer data packet authentication, and a certain level of confidentiality; in addition, some protocols are designed to defend against selfish nodes, which intend to save resources and avoid network cooperation. Some secure routing protocols have been proposed in MANET in recent papers. We outline those defense techniques at below sections.

Section 4.5.1 describes the proposed defense against wormhole attacks. Section 4.5.2 outlines the defense against blackhole attacks. Section 4.5.3 presents the defense against impersonation and repudiation attacks. Section 4.5.4 talks about the defense against modification attacks.

Defense against wormhole attacks

A packet leash protocol [15] is designed as a countermeasure to the wormhole attack. The SECTOR mechanism [52] is proposed to detect wormholes without the need of clock synchronization. Directional antennas [42] are also proposed to prevent wormhole attacks.

In the wormhole attack, an attacker receives packets at one point in the network, tunnels them to another point in the network, and then replays them into the network from that point. To defend against wormhole attacks, some efforts have been put into hardware design and signal processing techniques. If data bits are transferred in some special modulating method known only to the neighbor nodes, they are resistant to closed wormholes. Another potential solution is to integrate the prevention methods into intrusion detection systems. However, it is difficult to isolate the attacker with a software-only approach, since the packets sent by the wormhole are identical to the packets sent by legitimate nodes.

- **Packet leashes [15]:** The Packet leashes are proposed to detect wormhole attacks. A leash is the information added into a packet to restrict its transmission distance. A temporal packet leash sets a bound on the lifetime of a packet, which adds a constraint to its travel distance. A sender includes the transmission time and location in the message. The receiver checks whether the packet has traveled the distance between the sender and itself within the time frame between its reception and transmission. Temporal packet leashes require tightly synchronized clocks and precise location knowledge. In geographical leashes, location information and loosely synchronized clocks together verify the neighbor relation.

- **SECTOR [52]:** The SECTOR mechanism is based primarily on distance-bounding techniques, one-way hash chains, and the Merkle hash tree. SECTOR can be used to prevent wormhole attacks in MANET without requiring any clock synchronization or location information. SECTOR can also be used to help secure routing protocols in MANET using last encounters, and to help detect cheating by means of topology tracking.

- **Directional antennas [42]:** Directional antennas are also proposed as a countermeasure against wormhole attacks. This approach does not require either location information or clock synchronization, and is more efficient with energy.

Defense against blackhole attacks

Some secure routing protocols, such as the security-aware ad hoc routing protocol (SAR) [54], can be used to defend against blackhole attacks. The security-aware ad hoc routing protocol is based on on-demand protocols, such as AODV or DSR. In SAR, a security metric is added into the RREQ packet, and a different route discovery procedure is used. Intermediate nodes receive an RREQ packet with a particular security metric or trust level. At intermediate nodes, if the security metric or trust level is satisfied, the node will process the RREQ packet, and it will propagate to its neighbors using controlled flooding. Otherwise, the RREQ is dropped. If an end-to-end path with the required security attributes can be found, the destination will generate a RREP packet with the specific security metric. If the destination node fails to find a route with the required security metric or trust level, it sends a notification to the sender and allows the sender to adjust the security level in order to find a route.

To implement SAR, it is necessary to bind the identity of a user with an associated trust level. To prevent identity theft, stronger access control mechanisms such as authentication and authorization are required. In SAR, a simple shared secret is used to generate a symmetric encryption/decryption key per trust level. Packets are encrypted using the key associated with the trust level; nodes belonging to different levels cannot read the RREQ or RREP packets. It is assumed that an outsider cannot obtain the key.

In SAR, a malicious node that interrupts the flow of packets by altering the security metric to a higher or lower level cannot cause serious damage because the legitimate intermediate or destination node is supposed to drop the packet, and the attacker is not able to decrypt the packet. SAR provides a suite of cryptographic techniques, such as digital signature and encryption, which can be incorporated on a need-to-use basis to prevent modification.

Defense against impersonation and repudiation attacks

ARAN [32] is one example to provide authentication and non-repudiation, however this does not need to be part of a routing protocol. There are several other solutions, each with its own weaknesses. Here routing protocol ARAN is used as a case study to defend against impersonation and repudiation attacks at network layer. ARAN provides authentication and non-repudiation services using predetermined cryptographic certificates for end-to-end authentication. In ARAN, each node requests a certificate from a trusted certificate server. Route discovery is accomplished by broadcasting a route discovery message RDP from the source node. The reply message REP is unicast from the destination to the source. The routing messages are authenticated at each intermediate hop in both directions.

Routing discovery authentication at each hop is illustrated in Figure 7. The RDP packet includes $[RDP, IP_X, Cert_A, N_A, t]K_{A-}$, where RDP is a packet identifier, A is the source node, IP_X is the destination node X's IP address, N_A is a nonce, $Cert_A$ is A's certificate, t is the current time, and K_{A-} after the packet RDP, IP_X, $Cert_A$, N_A, t means the packet was signed with A's private key. If the intermediate node B is the first hop from node A, after validating A's signature and checking its certificate for expiration, it will decide to sign the packet by adding its own signature and certificate, and then it will forward $[[RDP, IP_X, Cert_A, N_A, t]K_{A-}]K_{B-}$, Cert$_B$ to all its neighbors. Each hop verifies the signature of the previous hop and replaces it with its own. The destination node X unicasts a REP packet $[REP, IP_A, Cert_X, N_A, t]K_{X-}$ back to source A.

Because RDPs do not contain a hop count or specific recorded source route, and because messages are signed at each hop, malicious nodes have no chance to form a routing loop by redirecting traffic or using impersonation to instantiate routes. The disadvantage of ARAN is that it uses hop-by-hop authentication, which incurs a large computation overhead. Meanwhile, each node needs to maintain one table entry per source-destination pair that is currently active.

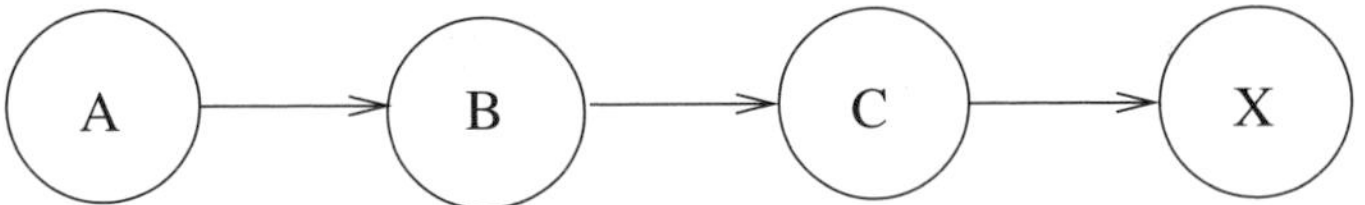

$A \longrightarrow$ broadcast: $[RDP, IP_X, cert_A, N_A, t]K_{A-}$

$B \longrightarrow$ broadcast: $[[RDP, IP_X, cert_A, N_A, t]K_{A-}]\ K_{B-}, cert_B$

$C \longrightarrow$ broadcast: $[[RDP, IP_X, cert_A, N_A, t]K_{A-}]\ K_{C-}, cert_C$

Figure 7. Illustration of ARAN Routing Discovery Authentication at Each Hop

Defense against modification attacks

The security protocol SEAD [11] is used here as an example of a defense against modification attacks at network layer. Similar to a packet leash [15], the SEAD protocol utilizes a one-way hash chain to prevent malicious nodes from increasing the sequence number or decreasing the hop count in routing advertisement packets. In SEAD, nodes need to authenticate neighbors by using TESLA [12] broadcast authentication or a symmetric cryptographic mechanism. Specifically, in SEAD, a node generates a hash chain and organizes the chain into segments of m elements as $(h_0, h_1, ..., h_{m-1})$, ..., $(h_{km}, h_{km+1}, ..., h_{km+m-1})$, ..., h_n, where $k = \frac{n}{m} - i$, m is the maximum network diameter, and i is the sequence number.

Illustrated in table 4, the network diameter is 5, the length of hash chain n's value is 20, i is the sequence number, and j is the metric, which is number of hops to destination. Because h_i=$H(h_{i-1})$, given h_i it is easy to verify the authenticity of h_j, as long as $j<i$. Given h_i, h_j cannot be derived for $j<i$, but h_j can be derived for $j>i$. Because different hash function is used for different i and j and used by the order showed in table 4, the attacker can never forge lower metric value, or greater sequence value. Because, after receiving a routing update in routing protocol DSDV, a node updates its advertised routing table when the sequence number is greater or when the sequence number is the same but the metric is lower, SEAD prevents malicious nodes from decreasing the hop count value or increasing the sequence number based on the design of DSDV.

3.6. Transport layer defense

In MANET, like TCP protocols in the Internet, nodes are vulnerable to the classic SYN flooding attack, or session hijacking attack.

Point-to-point or end-to-end encryption provides message confidentiality at or above the transport layer in two end systems. TCP is a connection-oriented reliable transport layer protocol. Because TCP does not perform well in MANET, TCP feedback (TCP-F) [49], TCP explicit failure notification (TCP-ELFN) [49], ad hoc transmission control protocol (ATCP) [49], and ad hoc transport protocol (ATP) [49] have been invented, but none of these protocols are designed with security in mind.

Table 4. SEAD Example: Hash Function used for Message Authentication, i is sequence number, j is metric, the network diameter (m) is 5, the length of hash chain (n) is 20

	j=0	1	2	3	4
i=1	h_{15}	h_{16}	h_{17}	h_{18}	h_{19}
2	h_{10}	h_{11}	h_{12}	h_{13}	h_{14}
3	h_5	h_6	h_7	h_8	h_9
4	h_0	h_1	h_2	h_3	h_4

Secure Socket Layer (SSL) [51], Transport Layer Security (TLS) [51], and Private Communications Transport (PCT) [51] protocols were designed for secure communications and are based on public key cryptography. TLS/SSL can help secure data transmission. It can also help to protect against masquerade attacks, man-in-the-middle (or bucket brigade) attacks, rollback attacks, and replay attacks. TLS/SSL is based on public key cryptography, which is CPU-intensive and requires comprehensive administrative configuration. Therefore, the application of these schemes in MANET is restricted. TLS/SSL has to be modified in order to address the special needs of MANET. Some firewall at a higher level can be configured to defend against SYN flooding attacks.

3.7. Application layer defense

Like the other protocol layers, the application layer also needs to be secured. In a network with a firewall installed, the firewall can provide access control, user authentication, packet filtering, and a logging and accounting service. Application layer firewalls can effectively prevent many attacks, and application-specific modules, for example, spyware detection software, have also been developed to guard mission-critical services. However, a firewall is mostly restricted to basic access control and is not able to solve all security problems. For example, it is not effective against attacks from insiders. Because of MANET's lack of infrastructure, a firewall is not particularly useful.

In MANET, an Intrusion Detection System (IDS) can be used as a second line of defense. Intrusion detection can be installed at the network layer, but in the application layer it is not only feasible, but also necessary. Certain attacks, such as an attack that tries to gain unauthorized access to a service, may seem legitimate to the lower layers, such as the MAC protocols. Also some attacks may be more obvious in the application layer. For instance, the application layer can detect a DoS attack more quickly than the lower layers when a large number of incoming service connections have no actual operations, since low layers need more time to recognize it.

3.8. Defense against multi-layer attacks

The DoS attacks, impersonation attacks, man-in-the-middle attacks, and many other attacks can target multiple layers. The countermeasures for these attacks need to be implemented at different layers. For example, directional antennas [52] are used at the media access layer to defend against wormhole attacks, and packet leashes [15] are used as a network layer defense against wormhole attacks. The countermeasures for multi-layer attacks can also be implemented in an integrated scheme. For example, if a node detects a local intrusion at a higher layer, lower layers are notified to do further investigation.

As an example, we give a detailed description about the defense against DoS attacks.

- **Defense against DoS attacks:** In MANET, two types of DoS attacks [55] are quite common. One is at the routing layer, and another is at the MAC layer. Attacks at the routing layer could consist of but is not limited to the following:

    1. The malicious node participates in a route but simply drops some of the data packets.
    2. The malicious node transmits falsified route updates.
    3. The malicious node could potentially replay stale updates.
    4. The malicious node reduces the TTL (time-to-live) field in the IP header so that the packet never reaches the destination.

    If end-to-end authentication is enforced, attacks by independent malicious node of types (2) and (3) may be thwarted. An attack of type (1) may be handled by assigning confidence levels to nodes and using routes that provide the highest level of confidence. An attack of type (4) may be countered by making it mandatory that a relay node ensures that the TTL field is set to a value greater than the hop count to the intended destinations.

    If nodes collude, the authentication mechanisms fail and it is an open problem to provide protection against such routing attacks.

    At the MAC layer DoS attacks could include, among others, the following misbehaviors:

    1. Keeping the channel busy in the vicinity of a node leads to a denial of service attack at that node.
    2. By using a particular node to continually relay spurious data, the battery life of that node may be drained.

    End-to-end authentication may prevent the above two cases from succeeding. If the node does not have a certificate of authentication, it may be prevented from accessing the channel. Usually the nodes are outsiders. However, if nodes collude, and the colluding nodes include the sending node and the destination, MAC layer attacks are very feasible.

3.9. Defense against key management attacks

Cryptographic algorithms are security primitives, which are widely used for the purposes of authentication, confidentiality, integrity, and non-repudiation. Most cryptographic systems include the underlining secure, robust, and efficient key management system. Key management is in the central part of any secure communication, and is the weak point of system security and protocol design. A key is a piece of input information for cryptographic algorithms. If the key were released, the encrypted information would be disclosed. The secrecy of the symmetric key and private key must be assured locally. The Key Encryption Key (KEK) approach [62] could be used at local hosts to build a line of defense.

Key distribution and key agreement over an insecure channel are at high risk and suffer from potential attacks. In the traditional digital envelop approach, a session key is generated at one side and is encrypted by the public-key algorithm. Then it is delivered and recovered at the other end. In the Diffie-Hellman (DH) scheme [62], the communication parties at both sides exchange some public information and generate a session key on both ends. Several enhanced DH schemes have been invented to counter man-in-the-middle attacks. In addition, a multi-way challenge response protocol, such as Needham-Schroeder [62], can also be used. Kerberos [62], which is based on a variant of Needham-Schroeder, is an authentication protocol used in many real systems including Microsoft Windows.

Key integrity and ownership should be protected from advanced key attacks. Digital signature, hash function, and hash function based on message authentication code (HMAC) [62] are techniques used for data authentication or integrity purposes. Similarly, public key is protected by the public-key certificate, in which a trusted entity called the certification authority (CA) in PKI [62] vouches for the binding of the public key with the owner's identity. In systems lacking a trusted third party (TTP) [62], the public-key certificate is vouched for by peer nodes in a distributed manner, such as pretty good privacy (PGP) [62]. In some distributed approaches, the system secret is distributed to a subset or all of the network hosts based on threshold cryptography. Obviously, a certificate cannot prove whether an entity is "good" or "bad", but can prove ownership of a key. Mainly it is for key authentication.

A cryptographic key could be compromised or disclosed after a certain period of usage. Since the key should no longer be useable after its disclosure, some mechanism is required to enforce this rule. In PKI, this can be done implicitly or explicitly. The certificate contains the lifetime of validity-it is not useful after expiration. But in some cases, the private key could be disclosed during the valid period, in which case certification authority (CA) needs to revoke a certificate explicitly and notify the network by posting it onto the certificate revocation list (CRL) to prevent its usage.

Currently there are three types of key management on MANET: the first one is virtual CA approach [3], the second one is certificate chaining [57], and the third one is composite key management, which combines the first two [9].

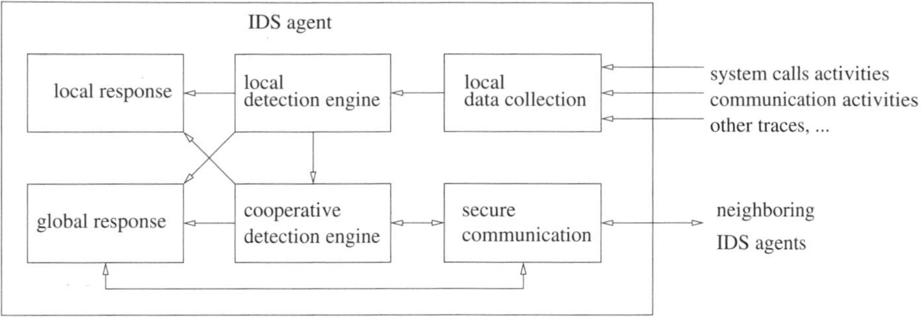

Figure 8. A Conceptual Model for an IDS Agent in MANET

3.10. MANET intrusion detection systems (IDS)

Because MANET has features such as an open medium, dynamic changing topology, and the lack of a centralized monitoring and management point, many of the intrusion detection techniques developed for a fixed wired network are not applicable in MANET. Zhang [37] gives a specific design of intrusion detection and response mechanisms for MANET. Marti [36] proposes two mechanisms: watchdog and pathrater, which improve throughput in MANET in the presence of nodes that agree to forward packets but fail to do so. In MANET, cooperation is very important to support the basic functions of the network so the token-based mechanism, the credit-based mechanism, and the reputation-based mechanism were developed to enforce cooperation. Each mechanism is discussed in this chapter.

MANET IDS agent conceptual architecture

The basic approach in MANET [36] is that each mobile node runs an IDS agent independently. It has to observe the behavior of neighboring nodes, detect local intrusion, cooperate with neighboring nodes, and, if needed, make decisions and take actions. An IDS agent has data collection, a local detection engine, local response, a cooperative detection engine, global response, and secure communication with neighboring IDS agents. Figure 8 is a conceptual model of an IDS agent.

Approaches to detect routing misbehavior

Watchdog and pathrater [36] are proposed for the DSR routing protocol. It is assumed that wireless links are bi-directional; wireless interfaces support promiscuous mode operation, which means that if a node A is within the transmission range of B, it can overhear communications to and from B even if those communications do not directly involve A.

The watchdog methods detect misbehaving nodes. A node may measure a neighboring node's frequency of dropping or misrouting packets, or its frequency of invalid routing information advertisements. The implementation of a watchdog maintains a buffer of recently sent packets and compares each overheard packet with the packets in the buffer to see if there is a match. If there is a match, the node removes the packet from the buffer; otherwise if a packet has remained in the buffer for longer than a certain timeout, the watchdog increments a failure tally for the neighboring node. If the tally exceeds a certain threshold bandwidth, it sends a message to the source notifying it of the misbehaving node. The weaknesses of watchdog are that it might not detect a misbehaving node because of ambiguous collisions, receiver collisions, limited transmission power, false behavior, collusion, and partial dropping.

In another scheme, pathrater is run by each node. Each node keeps track of the trustworthiness rating of every known node, including calculating path metrics by averaging the node ratings in the path to each known node. If there are multiple paths to the same destination, then according to standard DSR routing protocol the shortest path in the route cache is chosen, but when using pathrater the path with the highest metric is chosen.

Cooperation enforcement

Generally, there are two kinds of misbehaving nodes: one is the selfish node, and the other is the malicious node. Selfish nodes don't cooperate for selfish reasons, such as saving power. Even though the selfish nodes do not intend to damage other nodes, the main threat from selfish nodes is the dropping of packets, which may affect the performance of the network severely. Malicious nodes have the intention to damage other nodes, and battery saving is not a priority. Without any incentive for cooperating, network performance can be severely degraded. The mechanisms to enforce cooperating are currently split into three research areas: token-based, micro-payment, and reputation-based. Yang [58] proposed a token-based scheme. Buttyan [59] proposed the nuglets scheme. The nuglets scheme is micro-payment scheme. Buchegger's CONFIDANT [41], Michiard's CORE [60], and Bansel's OCEAN [61] are reputation-based schemes.

- **Token-based mechanism:** The token-based scheme [58] is a unified network-layer security solution in MANET based on the AODV protocol. In this scheme, each node carries a token in order to participate network operations, and its local neighbors collaboratively monitor any misbehavior in routing or packet forwarding services. The approach is different from a watchdog, which monitors neighbors alone, not collaboratively.

  Nodes without a valid token are isolated in the network, and all of their legitimate neighbors will not interact with them in routing and forwarding services. Upon expiration of the token, each node renews its token via its neighbors. The lifetime of a token is related to the node's behavior. A well-behaving node with a good record needs to renew its token less often.

  This approach uses asymmetric cryptographic primitives such as RSA. There is a global secret key and public key pair. Each legitimate node carries a token stamped with an expiration time and marked with a signature. The design is based on several assumptions to simplify the mechanism:

1. Any two nodes within wireless transmission range may monitor each other.
2. The approach is only based on network-layer security, not physical-layer or link-layer issues.
3. Only the secure route for data forwarding between the source and destination is discussed, not data packet confidentiality and integrity.
4. Each node has a unique ID.
5. Multiple attackers are possible, but there is a limit to attackers in any neighborhood.
6. Every legitimate node has a token signed with the private key, which can be verified by its neighbors.

- **Credit-based mechanism:** The nuglets scheme [59] is an approach analogous to virtual currency. A node that consumes a service must pay the nodes that provide the service in nuglets. The combination of watchdog and pathrater cannot hold any misbehaving nodes accountable, and misbehaving nodes are still able to send and receive packets. However, in the nuglets scheme, a misbehaving node will be locked out by its neighbors. That is much better in fairness.

  Nuglets are designed to simulate packet forwarding. The nuglets are related to the counters in the nodes. The counter is maintained by a trusted and tamper-resistant hardware module at each node. A packet purse holds nuglets, which are contained in the packet. The packet purse is protected from unauthorized modification and detachment from the original packet by cryptographic mechanisms. The packet forward protocol is designed on fixed per hop charges.

- **Reputation-based mechanism:** CONFIDANT [41] presents an extension to the routing protocol in order to detect and isolate misbehaving nodes. The protocol is designed to be able to make cooperation fair. With CONFIDANT, each node has four components: a monitor, a reputation system, a trust manager, and a path manager.

  The CONFIDANT approach copes with MANET security, robustness, and fairness by retaliating for malicious behavior and warning affiliated nodes to avoid bad experiences. Nodes learn not only from their own experience, but also from observing the neighborhood and from the experience of their friends.

4. OPEN CHALLENGES AND FUTURE DIRECTIONS

Security is such an important feature that it could determine the success and wide deployment of MANET. A variety of attacks have been identified. Security countermeasures either currently used in wired or wireless networking or newly designed specifically for MANET are presented in the above sections. Security must be ensured for the entire system at all levels since overall security level is determined by the system's weakest point.

The research on MANET is still in an early stage. Existing papers are typically based on one specific attack. They could work well in the presence of designated attacks, but there are many unanticipated or combined attacks that remain undiscovered. Research is still being performed and will result in the discovery of new threats as well as the creation of new countermeasures. More research is needed on robust key management system, trust-based protocols, integrated approaches to routing security, and data security at different layers. Here are some research topics and future work in the area:

- **Key management:** Cryptography is the fundamental security technique used in almost all aspects of security. The strength of any cryptographic system depends on proper key management. The public-key cryptography approach relies on the centralized CA entity, which is a security weak point in MANET. Some papers propose to distribute CA functionality to multiple or all network entities based on a secret sharing scheme, while some suggest a fully distributed trust model, in the style of PGP. Symmetric cryptography has computation efficiency, yet it suffers from potential attacks on key agreement or key distribution. Many complicated key exchange or distribution protocols have been designed, but for MANET, they are restricted by a node's available resources, dynamic network topology, and limited bandwidth. Efficient key agreement and distribution in MANET is an ongoing research area.

- **Trust-based system:** Most of the current work is on preventive methods with intrusion detection as the second line of defense. One interesting research issue is to build a trust-based system so that the level of security enforcement is dependant on the trust level. Building a sound trust-based system and integrating it into the current preventive methods can be done in future research.

- **Multi-fence solution:** Since most attacks are unpredictable, a resiliency-oriented security solution will be more useful, which depends on a multi-fence security solution. Cryptography-based methods offer a subset of solutions. Other solutions will be in future research.

5. ACKNOWLEDGEMENT

This work was supported in part by NSF grants CCR 0329741, CNS 0422762, CNS 0434533, ANI 0073736, EIA 0130806, and by a federal earmark project on Secure Telecommunication Networks.

6. REFERENCES

1. A. Salomaa, *Public-Key Cryptography*, Springer-Verlag, 1996.
2. A. Tanenbaum, *Computer Networks*, PH PTR, 2003.
3. L. Zhou and Z. Haas, Securing Ad Hoc Networks, *IEEE Network Magazine* Vol.13 No.6 (1999) pp. 24-30.
4. S. Yi, P. Naldurg, and R. Kravets, Security Aware Ad hoc Routing for Wireless Networks. Report No.UIUCDCS-R-2002-2290, UIUC, 2002.
5. H. Luo and S. Lu, URSA: Ubiquitous and Robust Access Control for Mobile Ad-Hoc Networks, *IEEE/ACM Transactions on Networking* Vol.12 No.6 (2004) pp. 1049-1063.
6. W. Lou and Y. Fang, A Survey of Wireless Security in Mobile Ad Hoc Networks: Challenges and Available Solutions. *Ad Hoc Wireless Networks*, edited by X. Chen, X. Huang and D. Du. Kluwer Academic Publishers, pp. 319-364, 2003.
7. S. Burnett and S. Paine, *RSA Security's Official Guide to Cryptography*, RSA Press, 2001.
8. M. Ilyas, *The Handbook of Ad Hoc Wireless Networks*, CRC Press, 2003.
9. S. Yi and R. Kravets, Composite Key Management for Ad Hoc Networks. *Proc. of the 1st Annual International Conference on Mobile and Ubiquitous Systems: Networking and Services* (MobiQuitous'04), pp. 52-61, 2004.
10. M. Zapata, Secure Ad Hoc On-Demand Distance Vector (SAODV). Internet draft, draft-guerrero-manet-saodv-01.txt, 2002.
11. Y. Hu, D. Johnson, and A. Perrig, SEAD: Secure Efficient Distance Vector Routing in Mobile Wireless Ad-Hoc Networks. *Proc. of the 4th IEEE Workshop on Mobile Computing Systems and Applications* (WMCSA'02), pp. 3-13, 2002.
12. A. Perrig, R. Canetti, J. Tygar, and D. Song, The TESLA Broadcast Authentication Protocol. Internet Draft, 2000.
13. P. Papadimitratos and Z. Haas, Secure Routing for Mobile Ad Hoc Networks. *Proc. of the SCS Communication Networks and Distributed Systems Modeling and Simulation Conference* (CNDS 2002), 2002.
14. W. Mehuron, Digital Signature Standard (DSS). U.S. Department of Commerce, National Institute of Standards and Technology (NIST), Information Technology Laboratory (ITL). FIPS PEB 186, 1994.
15. Y. Hu, A. Perrig, and D. Johnson, Packet Leashes: A Defense Against Wormhole Attacks in Wireless Ad Hoc Networks. *Proc. of IEEE INFORCOM*, 2002.
16. H. Deng, W. Li, and D. P. Agrawal, Routing Security in Wireless Ad Hoc Networks. *IEEE Communications Magazine*, vol. 40, no. 10, 2002.
17. B. Awerbuch, D. Holmer, C. Nita-Rotaru, and H. Rubens, An On-demand Secure Routing Protocol Resilient to Byzantine Failures. *Proceedings of the ACM Workshop on Wireless Security*, pp. 21-30, 2002.
18. P. Papadimitratos and Z. Haas, Secure Data Transmission in Mobile Ad Hoc Networks. *Proc. of the 2003 ACM Workshop on Wireless Security*, pp. 41-50, 2003.
19. Y. Hu, A. Perrig, and D. Johnson, Rushing Attacks and Defense in Wireless Ad Hoc Network Routing Protocols. *Proc. of the ACM Workshop on Wireless Security* (WiSe), pp. 30-40, 2003.
20. Y. Hu, A. Perrig, and D. Johnson, Ariadne: A Secure On-Demand Routing for Ad Hoc Networks. *Proc. of MobiCom 2002*, Atlanta, 2002.

21. H. Yang, H. Luo, F. Ye, S. Lu, and L. Zhang, Security in Mobile Ad Hoc Networks: Challenges and Solutions. *IEEE Wireless Communications*, pp. 38-47, 2004.

22. C. Perkins, *Ad Hoc Networks*, Addison-Wesley, 2001.

23. R. Oppliger, *Internet and Intranet Security*, Artech House, 1998.

24. B. Wu, J. Wu, E. Fernandez, S. Magliveras, and M. Ilyas, Secure and Efficient Key Management in Mobile Ad Hoc Networks. *Proc. of 19th IEEE International Parallel & Distributed Processing Symposium*, Denver, 2005.

25. L. Buttyan and J. Hubaux, Report on Working Session on Security in Wireless Ad Hoc Networks. *Mobile Computing and Communications Review*, vol. 6, 2002.

26. S. Ravi, A. Raghunathan, and N. Potlapally, Secure Wireless Data: System Architecture Challenges. *Proc. of International Conference on System Synthesis*, 2002.

27. W. Stallings, *Wireless Communication and Networks*, Pearson Education, 2002.

28. N. Borisov, I. Goldberg and D.Wagner, Interception Mobile Communications: The Insecurity of 802.11. *Conference of Mobile Computing and Networking*, 2001.

29. P. Kyasanur and N. Vaidya, Detection and Handling of MAC Layer Misbehavior in Wireless Networks. *Proc. of the International Conference on Dependable Systems and Networks*, pp. 173-182, 2003.

30. A. Crdenas, S. Radosavac, and J. Baras, Detection and Prevention of MAC layer Misbehavior in Ad Hoc Networks. *Proc. of the 2nd ACM Workshop on Security of Ad Hoc and Sensor Networks*, pp. 17-22, 2004.

31. C. Murthy and B. Manoj, *Ad Hoc Wireless Networks: Architectures and Protocols*, Prentice Hall PTR, 2005.

32. K. Sanzgiri, B. Dahill, B. Levine, C. Shields, and E. Belding-Royer, A Secure Routing Protocol for Ad Hoc Networks. *Proc. of IEEE International Conference on Network Protocols* (ICNP), pp. 78-87, 2002

33. K. Ng and W. Seah, Routing Security and Data Confidentiality for Mobile Ad Hoc Networks. *Proc. of Vehicular Technology Conference*(VTC), Jeju, Korea, 2003.

34. M. Jakobsson, S. Wetzel, and B. Yener, Stealth Attacks on Ad Hoc Wireless Networks. *Proc. of IEEE Vehicular Technology Conference* (VTC), 2003.

35. Y. Hu and A. Perrig, A Survey of Secure Wireless Ad Hoc Routing. *IEEE Security & Privacy*, pp. 28-39, 2004.

36. S. Marti, T. Giuli, K. Lai, and M. Baker, Mitigating Routing Misbehavior in Mobile Ad Hoc Networks, *Proc. of the Sixth Annual International Conference on Mobile Computing and Networking* (MOBICOM), Boston, 2000.

37. Y. Zhang and W. Lee, Intrusion Detection in Wireless Ad-hoc Networks, *Proc. of the Sixth Annual International Conference on Mobile Computing and Networking* (MOBICOM), Boston, 2000.

38. P. Kyasanur and N. Vaidya, Detection and Handling of MAC Layer Misbehavior in Wireless Networks, *Proc. of Dependable Computing and Communications Symposium (DCC) at the International Conference on Dependable Systems and Networks (DSN)*, 2003.

39. A. Cardenas, N. Benammar, G. Papageorgiou, and J. Baras, Cross-Layered Security Analysis of Wireless Ad Hoc Networks, *Proc. of 24th Army Science Conference*, 2004.

40. H. Yang, X. Meng, and S. Lu, Self-Organized Network-Layer Security in Mobile Ad Hoc Networks. *Proc. of ACM MOBICOM Wireless Security Workshop* (WiSe'02), Atlanta, 2002.

41. S. Buchegger and J. Boudec, Nodes Bearing Grudges: Towards Routing Security, Fairness, and Robustness in Mobile Ad Hoc Networks, *Proc. of the 10th Euromicro Workshop on Parallel, Distributed and Network-based Processing*, Canary Islands, Spain, 2002.

42. L. Hu and D. Evans, Using Directional Antennas to Prevent Wormhole Attacks. *Proc. of Networks and Distributed System Security Symposium* (NDSS), 2004.

43. P. Ning and K. Sun, How to Misuse AODV: A Case Study of Inside Attacks against Mobile Ad-Hoc Routing Protocols, *Proceedings of the 2003 IEEE Workshop on Information Assurance, United States Military Academy*, West Point, NY, 2003.

44. V. Park and S. Corson, Temporally-Ordered Routing Algorithm (TORA) Ver. 1 Functional Specification, IETF draft, 2001.

45. T. Clausen and P. Jacquet, Optimized Link State Routing Protocol (OLSR) Project, Hipercom, INRIA, www.ietf.org/rfc/rfc3626.txt, RFC-3626, 2003.

46. X. Wang, D. Feng, X. Lai, and H. Yu, Collisions for Hash Functions MD4, MD5, HAVAL-128 and RIPEMD, *Cryptology ePrint Archive*, Report 2004/199, http://eprint.iacr.org/, 2004.

47. T. Karygiannis and L. Owens, Wireless Network Security-802.11, Bluetooth and Handheld Devices. National Institute of Standards and Technology. Technology Administration, U.S Department of Commerce, *Special Publication* 800-848, 2002.

48. R. Nichols and P. Lekkas, *Wireless Security-Models, Threats, and Solutions*, McGraw-Hill, Chapter 7, 2002.

49. H. Hsieh and R. Sivakumar, Transport Over Wireless Networks. *Handbook of Wireless Networks and Mobile Computing*, Edited by Ivan Stojmenovic. John Wiley and Sons, Inc., 2002.

50. N. Weaver, V. Paxson, S. Staniford, and R. Cunningham, "A Taxonomy of Computer Worms", *First Workshop on Rapid Malcode* (WORM), 2003.

51. C. Kaufman, R. Perlman, and M. Speciner, *Network Security Private Communication in a Public World*, Prentice Hall PTR, A division of Pearson Education, Inc., 2002

52. S. Capkun, L. Buttyan, and J. Hubaux, Sector: Secure Tracking of Node Encounters in Multi-hop Wireless Networks. *Proc. of the ACM Workshop on Security of Ad Hoc and Sensor Networks*, 2003.

53. W. Wang, B. Bhargava, Y. Lu, and X. Wu, Defending Against Wormhole Attacks in Mobile Ad Hoc Networks, under review at Wiley Journal Wireless Communication and Mobile Computing (WCMC).

54. S. Yi, P. Naldurg, and R. Kravets, Security-Aware Ad-hoc Routing for Wireless Networks. Report No.UIUCDCS-R-2002-2290, UIUC, 2002.

55. V. Gupta, S. V. Krishnamurthy, and M. Faloutsos, Denial of Service Attacks at the MAC Layer in Wireless Ad Hoc Networks. *In Proc. of MILCOM*, 2002.

56. I. Aad, J. Hubaux, and E. W. Knightly, Denial of Service Resilience in Ad Hoc Networks, *In Proc. of 10th Ann. Int'l Conf. Mobile Computing and Networking* (MobiCom 2004), pp. 202 - 215, ACM Press, 2004.

57. J. Hubaux, L. Buttyan, and S. Capkun, The Quest for Security in Mobile Ad Hoc Networks, *In Proc. of the ACM Symposium on Mobile Ad Hoc Networking & Computing* (MobiHoc 2001), Long Beach, CA, Oct. 2001.

58. H. Yang, X. Meng, and S. Lu, Self-organized Network Layer Security in Mobile Ad Hoc Networks, *ACM MOBICOM Wireless Security Workshop* (WiSe'02).

59. L. Buttyan and J. Hubaux, Nuglets: A Virtual Currency to Simulate Cooperation in Self-organized Ad Hoc Networks. Technial Report DSC/2001/001, Swiss Federal Institute of Technology - Lausanne, 2001.

60. P. Michiardi and R. Molva, Core: A Collaborative Reputation Mechanism to Enforce Node Cooperation in Mobile Ad Hoc Networks, *IFIP-Communication and Multimedia Security Conference 2002*.

61. S. Bansal and M. Baker, Observation-based Cooperation Enforcement in Ad Hoc Networks, http://arxiv.rog/pdf/cs.NI/0307012, July 2003.

62. A. Menezes, P. Oorschot, and S. Vanstone, Handbook of Applied Cryptography, CRC Press, 1996.

6

SECURE ROUTING IN WIRELESS AD-HOC NETWORKS

Venkata C. Giruka
Department of Computer Science
University of Kentucky, Lexington KY, 40506 USA.
E-mail: venkata@cs.uky.edu

Mukesh Singhal
Department of Computer Science
University of Kentucky, Lexington KY 40506 USA.
E-mail: singhal@cs.uky.edu

Routing in wireless ad-hoc networks is one of the fundamental tasks which helps nodes send and receive packets. Traditionally, routing protocols for wireless ad-hoc networks assume a non-adversarial and a cooperative network setting. In practice, there may be malicious nodes that may attempt to disrupt the network communication by launching attacks on the network or the routing protocol itself. In this chapter, we present several routing protocols for ad-hoc networks, the security issues related to routing, and securing routing protocols for mobile wireless ad-hoc networks.

1. INTRODUCTION

Wireless ad-hoc networks are rapidly deployable networks in which nodes with wireless radios form a network on-the-fly without the need of any fixed infrastructure. Two main features of ad-hoc networks, namely, connecting without cables and user mobility, provide a powerful combination that enables networking in situations where it is not feasible to establish and maintain a network. Ad-hoc networks were of primary interest in military communications and disaster relief because of their "infrastructure-less" nature. However, over the past decade these networks gained popularity in the form of personal-area networks [6] and civilian networks [23].

One of the basic functions of an ad-hoc network is routing, which enables nodes to send and receive packets. Due to the limited transmission range (typically 250m or less) of nodes, routing in ad-hoc networks generally involves multiple hops. Thus, each node acts as a router as well as an end-node to relay or receive packets in the network. Routing in ad-hoc networks is challenging because of node mobility, lack of predefined infrastructure, peer-to-peer mode of communication and limited radio range. Several protocols [3, 9, 10, 11, 12, 13, 17, 20] have been proposed in the literature for routing in ad-hoc networks, each with its own niche of applicability. At the core, all these routing protocols try to find a 'good path' from the source to the destination, assuming that nodes in the network are 'friendly' and cooperative. If we relax the assumption of node cooperation for routing and take into account the presence of malicious nodes, then it adds a new dimension to the problem, viz., security.

Security in ad-hoc networks is an essential component that safeguards the proper functioning of the network and underlying protocols. In general, securing ad-hoc networks is a nontrivial task due to lack of a pre-existing infrastructure, wireless nature of communication links, and frequently changing network topology. Unlike wired networks where the attacker needs to gain access to the physical medium to launch any kind of attack, in ad-hoc networks, an intruder can easily eavesdrop on the on-going traffic. Further, lack of infrastructural support impedes the use of well known (those used in wired-networks) security architectures/protocols to detect and thwart intruders in the context of multi-hop wireless networks. However, there are several protocols for wireless networks in the literature that enforce/implement security at different layers and at different levels in ad-hoc networks. In this chapter, we focus on secure-routing protocols for ad-hoc networks.

An enthusiastic reader may ask: "Is a secure-routing protocol a new routing protocol that is designed from the scratch with security as one of its goal or is it a secure extension of an existing routing protocol?" Our answer is: It can be either. The former approach still makes sense, since there is no standard routing protocol for ad-hoc networks as of the time of writing this book. However, authors believe that one may end-up re-inventing something similar to one of the existing routing techniques plus security extensions, in designing a new secure-routing protocol. The later approach is motivated by the fact that routing in ad-hoc networks, which is challenging in itself, has some of approaches like AODV [20], DSR [11], and OLSR [10], which IETF MANET group is considering for standardization. Thus, securing such a routing protocol requires assessing attacks specific to that protocol and securing them accordingly.

In the rest of the chapter, we concentrate on secure versions existing single path routing protocols. To this end, we present a classification and a brief review of few well known routing protocols for ad-hoc networks in Section 2. We discuss possible attacks on routing protocols in Section 3. Section 4 presents secure topology-based routing protocols for ad-hoc networks. Section 5 presents security issues and counter measures in position-based routing, and we summarize the chapter in Section 6.

2. ROUTING IN AD-HOC NETWORKS

Routing in ad-hoc networks involves finding a path from the source to the destination, and delivering packets to the destination nodes while nodes in the network are moving freely. Due to node mobility, a path established by a source may not exist after a short interval of time. To cope with node mobility nodes need to maintain routes in the network. Depending on how nodes establish and maintain paths, routing protocols for ad-hoc networks broadly fall into pro-active [17], reactive [11, 20], hybrid [9], and location-based [3, 12, 13] categories.

2.1. Proactive Routing Protocols

Pro-active routing protocols are table-driven protocols that maintain up-to-date routing table using the routing information learnt from the neighbors on a continuous basis. Routing in such protocols involves selecting a path form the source to the destination, where the source node and each intermediate node selects a next hop, by routing table look up, and forwarding the packet to next hop until destination receives the packet. A drawback of such protocols is the proactive overhead due to route maintenance and frequent route updates to cope with node mobility. An example of this class is the DSDV [17].

DSDV: The Destination-Sequenced Distance-Vector Routing protocol (DSDV) is an enhanced version of distributed Bellman-Ford algorithm, for mobile ad-hoc networks. In this protocol, each node maintains a routing table that contains an entry for every node in the network. Each entry in the routing table consists of the destination ID, the next hop ID, a hop count, and a sequence number for that destination. The sequence number helps nodes maintain a fresh route to the destination(s) and avoid routing loops. To cope with frequently changing network topology, nodes periodically broadcast routing table updates thought-out the network.

When a node receives a route-update packet, it changes its routing table entries if the sequence number of the destination in the update packet is higher (fresh) than the one in its routing table. If the sequences numbers are the same, then the node selects a route with smaller metric (hop count). To reduce the network traffic due to large update packets, DSDV employs two types of updates –full dump and incremental. A full dump packet generated by a node contains all entries in its routing table. Whereas an incremental packet contains only the routing table entries that are changed by the node since the last full dump. A node triggers an update when either the metric for a destination changes or when the sequence number changes. In the later case, it is called DSDV-SQ.

2.2. Reactive Routing Protocols

Reactive routing protocols are demand-driven protocols that find path on-the-fly as and when necessary. In such protocols, establishing a new route involves a route discovery phase consisting of route request (flooding) and a route reply (by the destination node). Nodes maintain only the active routes until a desired period or until destination becomes inaccessible along every path from the source node. A drawback of such protocols is the delay due to route discovery on-the-fly. We briefly discuss the AODV and DSR protocols next.

AODV: In Ad-hoc On-demand Distance Vector Routing (AODV), a node discovers and maintains a route to the destination as and when necessary. Nodes maintain a routing table containing routes towards source(s)-destination(s) that are actively communicating with each other. Each entry in the routing table consists of the destination ID, the next hop ID, a hop count, and a sequence number for that destination (the same as one in DSDV). The sequence number helps nodes maintain a fresh route to the destination(s) and avoid routing loops. Thus, each node maintains a sequence number for itself and the respective source(s) and destination(s). A node increments its sequence number if it initiates a new route request or if it detects a link-break with one of its neighbors.

To establish a path to the destination, a source node broadcasts a route request (RREQ) packet. The RREQ packet contains the source ID, the destination ID, sequence number of the source, and the latest sequence number of the destination node that is known to the source node. When a node receives a RREQ packet, it makes an entry for the route request in the route-request cache, and stores the address of the node from which it received the request as the next hop towards the source in its routing table. If receiving node is the destination or it has a fresh route to that destination[1], then it responds with a route reply (RREP). Otherwise, it rebroadcasts the RREQ to its neighbors. When a node receives a RREP, it stores the address of the node from which it received RREP as the next hop towards the destination in its routing table and unicast the RREP to the next hop towards the source node.

Once the source receives the RREP packet, it starts transmitting data packets along the path traced by the RREP packet. Due to the node mobility, path(s) established by a source node may break. A node detects a path break if it attempts to forward a data packet and receives a packet-drop notification from the media access control (MAC) layer. When a node detects a path-break, it drops the packet for the destination and generates a route error (RERR) packet for the destination and sends the RERR to the source. Upon receiving a RERR, the source node buffers data packets for the destination and tries to re-establish a path to the destination.

DSR: Dynamic Source Routing (DSR) [11] was one of the first reactive routing protocols for ad-hoc networks. In DSR, nodes use RREQ, RREP, and RERR packets to establish and maintain paths to the destination. However, unlike AODV, RREQ packet accumulates a list of node IDs along the path from the source to the destination and the corresponding RREP packet carries this list of IDs back to the source. Once the source node receives RREP packet, it starts transmitting data packets to the destination

[1] A node determines the freshness of its route table entry (provided such an entry exists) for that destination by comparing the destination sequence number in the RREQ with that of its route table entry.

by embedding the route from the source to the destination in the packet header. The path in the data packet header is referred to as the "source route".

Every node in the network stores route to other nodes in the network by maintaining a dynamic route cache. A node learns routes to other nodes when it initiates a RREQ to a particular destination or when the node lies on an active path to that destination. In addition to these, a node may also learn a route by overhearing transmissions (in the promiscuous mode) along the routes of which it is not a part.

2.3. Hybrid Routing Protocols

Hybrid protocols combine the advantages of various approaches of routing protocols into a single protocol. The Zone Routing Protocol (ZRP) [9], is one such hybrid protocol that combines both the proactive and reactive routing approaches. ZRP takes advantage of pro-active discovery within a node's local neighborhood, and uses a reactive protocol for communication between these neighborhoods. The local neighborhoods are called Zones, and each node may be within multiple overlapping zones. ZRP is motivated by the fact that " the most communication takes place between nodes close to each other. Changes in the topology are most important in the vicinity of a node - the addition or the removal of a node on the other side of the network has only limited impact on the local neighborhoods". The performance of ZRP depends on choosing a radius, which decides the transition from pro-active to reactive behavior. With a carefully chosen radius, ZRP can achieve better efficiency and scalability over both pro-active and reactive routing protocols.

2.4. Position-based Routing Protocols

Position-based routing protocols utilize position of nodes in the network and make the least use of the topology information. Routing protocols using such a scheme eliminate drawbacks due to frequently changing network topology. DREAM [3], GPSR [12], and LAR [13] are some of the examples of position-based routing protocols.

In Position-based routing protocols nodes maintain local (one or two hop) topology information with the help of a hello protocol. To route a packet to the destination, the source node uses a greedy-forwarding to select a next hop towards the destination. In greedy-forwarding, a node selects a next-hop towards the destination that is geographically closest to the destination among its neighboring nodes. Since there is no pre-established route from a source to the destination, each packet may follow a different path depending on the network topology.

There are two parts to position-based routing: (a) given the position of the source, the position of the destination, and a local neighbor table of each node, delivering packets from the source to the destination, and (b) given that each node can determine its own position, using some positioning system like GPS, obtaining the position of any other node in the system. The former part is the position-based routing, examples include GFG [5], GPSR [12]. Position-based routing is typically greedy-forwarding along with a recovery mechanism to circumvent local optima due to greedy-forwarding, a condition where there is no node close to an intermediate node in its neighborhood than the node itself. The later part is called the location service. Some of the examples of location-service protocols are GLS [15], DLM [27], and RLS [22]. Interestingly, most location-service protocols including GLS and DLM, rely on the underlying greedy forwarding algorithm (although there are few other variants of greedy forwarding [26] exists) to send and receive control packets like location updates and location queries.

The advantage of these protocols is that nodes need not establish, maintain routes, and these protocols are more scalable compared to reactive and pro-active routing protocols.

3. POSSIBLE ATTACKS ON ROUTING PROTOCOLS

Having explained functioning of some routing protocols in the previous sections, we now present possible attacks on routing protocols. Attacks on routing protocols can be both active and passive. In passive attacks an attacker does not actively participate in bringing the network down. Attackers are typically involved in unauthorized listening to routing packets. An attacker just eavesdrops on the network traffic as to determine which nodes are trying to establish routes to which other nodes, which nodes are the center of the network and so on. A major advantage for the attacker is that passive attacks are usually impossible to detect and hence makes defending against such attacks extremely difficult. Further, routing information can reveal relationships between nodes or disclose their addresses. If a route to a particular node is requested more often than to other nodes, the attacker might expect that the node is important for the functioning of the network, and disabling it could bring the entire network down. Such attacks can be prevented mostly by applying cryptographic techniques on messages, to protect the message contents from being exposed to the attacker.

Active attacks involves modification, fabrication of messages, or preventing the network from functioning properly. Further, active attacks can be due to an external attacker(s) and an internal attacker(s). External attackers are unauthorized nodes without a shared cryptography key in the network. Internal attackers are authorized but compromised nodes and are more dangerous and hard to detect as they are in the network and own the necessary cryptography keys. Active attacks can be classified into packet-dropping, modification, fabrication, and other miscellaneous attacks.

3.1. Packet Dropping

Malicious nodes may ensure that certain messages are not transmitted by simply forwarding few packets and dropping the remaining one. By dropping packets, an attacker succeeds in disrupting the network operation. Such misbehavior can be hard to detect as valid nodes may, from time to time, drop packets due to congestion/collision. Depending on the strategy of dropping packets, there are two types of attacks:
Black holes: The attacker injects falsified routing packets to attract traffic. The attacker intercepts or drops control as well as data packets to deny services to authentic nodes.

This attack can be prevented by establishing routes free of such nodes or by removing them from existing routes.

Gray holes: The attacker drops data packets but not control packets. This attack is difficult to detect. A promiscuous mode operation within the routing protocol is required to detect such an attack.

3.2. Modification

Most routing protocols assume that nodes do not alter fields of the protocol messages. The protocol messages, or control packets, carry important routing information that governs the behavior of their transmission. Since the level of trust in a traditional ad-hoc network cannot be measured or enforced, malicious nodes may participate directly in route discovery and may intercept and disrupt communication. They can easily cause redirection of network traffic and denial of service attacks by simply altering fields in protocol messages. These attacks can be classified as follows [24]:

Remote redirection with modified route sequence number: A malicious node uses the routing protocol to advertise itself as having the shortest path to destination whose packets it wants to intercept. Typically, routing protocols maintain routes using monotonically increasing sequence numbers for each destination. A malicious node may divert traffic through itself by advertising a route to a node with a destination sequence number greater than the authentic value.

Redirection with modified hop count: In some protocols such as AODV, the route length is represented in the message by a hop count field. A malicious node can succeed in diverting all the traffic to a particular destination through itself by advertising a shortest route (with a very low hop count) to that destination.

Denial of service with modified source routes: DSR routing protocol explicitly states routes in data packets called the source route. In the absence of any integrity checks on the source route, a malicious node can modify this source route and hence succeed in creating loops in the network or launching a simple denial of service attack.

3.3. Fabrication

Fabrication of messages means generating false routing messages. Such attacks are difficult to detect. There are three types of such attacks.

Falsifying route error messages: AODV and DSR have measures to handle broken routes when constituent nodes move or fail. If the destination node or an intermediate node along an active path moves or fails, the node, which precedes the broken link, broadcasts a route error message to all active neighbors which precede the broken link. The nodes then invalidate the route for this destination in their routing tables. A malicious node can succeed in launching a denial of service attack against a benign node by sending false route error messages against this benign node.

Route cache poisoning: In DSR, a node can learn routing information by overhearing transmissions on routes of which it is not a part. The node then adds this information to its own cache. An attacker can easily exploit this method of learning and poison route caches. If a malicious node, M, wants to launch a denial of service attack on node X, it can simply broadcast spoofed packets with source routes to X via itself. Any neighboring nodes that overhear the packet transmission may add the route to their route cache.

Routing table overflow attack: A malicious node may attempt to overwhelm the protocol by initiating route discovery to non-existent nodes. The logic behind this is to create so many routes that no further routes could be created as the routing tables of nodes are already overflowing.

3.4. Other Attacks

Impersonation: A malicious node masquerades as another node. It does this by misrepresenting its identity by changing its own IP or MAC address to that of some other node, thereby masquerading as that node. Using stronger authentication procedures can prevent this type of attack.

Sybil attack: In the Sybil attack, an adversary presents multiple identities to other nodes in the network. This attack disrupts routing protocols by causing nodes to appear to be "in more than one place at once" [14]. This reduces the diversity of routes available in the network. It also diminishes the effectiveness of fault-tolerant schemes such as distributed storage, disparity, multi-path routing, and topology maintenance.

Wormhole attacks: The attacker receives packets at one point in the network and tunnels them to another part of the network. It then replays them into the network from that point onwards. This kind of attack does not require the attacker to have any knowledge of cryptographic keys. Using packet leashes can prevent these attacks [25].

Location spoofing attacks: Apart from the usual attacks on routing protocols, position-based protocols face a new attack, viz., the position spoofing attack. In the position spoofing attack, a malicious node aims to disrupt the normal functioning of greedy forwarding by fabricating its position information in favor of itself. A selfish node may declare a selected position (e.g., away from the destinations' position) to stay away from forwarding data packets. On the other hand, a malicious node can declare a false position (e.g., closest to the destination) to attract the traffic so that it can launch attacks. An attack combining Sybil attack and position spoofing can construct a wall around a node and control all traffic from that node.

4. SECURE TOPOLOGY-BASED ROUTING PROTOCOLS

In this section we present secure proactive and reactive routing protocols. The SEAD protocol that is described in the next subsection is a secure extension of DSDV and it is a proactive secure routing protocol. The rest of the protocols in this section are reactive protocols, and are secure extensions of DSR or AODV routing protocols.

4.1. SEAD

The Secure Efficient Ad hoc Distance vector routing protocol (SEAD) [7] is a secure routing protocol based on the DSDV-SQ protocol described in Section 2.1. Recall that in DSDV-SQ, nodes send both periodic routing updates as well as triggered routing updates. These updates can be either the whole routing table (full dump) or only those table entries that correspond to the destinations for which route has changed since the last full dump. A node sends a triggered update, if the node receives a new metric value for the destination or if it receives a new sequence number for the destination, to communicate such changes to the other nodes.

A malicious node may send updates advertising lower hop count for certain destinations to its neighboring nodes. The neighbors would be fooled into believing that this malicious node has the shortest path to those destinations, and so they would make this malicious node as the next hop for routes to those destinations. Thus, this malicious node would be able to launch denial of service attacks against those destinations by having all routes to go through itself. It can then selectively drop packets and wreck havoc in the network. In SEAD protocol, nodes prevent such modification attacks by authenticating the routing update packets.

The SEAD protocol assumes that the network diameter has a value of at most $m-1$, where m is a positive integer. Thus all metrics in any routing update are less than m. SEAD uses one-way hash chains, which are computationally efficient as compared to public key cryptography or secret key cryptography paradigms, for authenticating update packets from a given node. To generate a hash chain, a node picks a ρ bit long random number x, and generates the values $h_0, h_1, h_2 \cdots, h_n$, where $h_0 = x$, $h_i = H(h_{i-1})$ $1 \leq i \leq n$, n is divisible by m, H is a one-way hash function like SHA-1 [16] or MD5 [21]. Given the authentic element h_n, a node can authenticate h_{n-3} by computing $H(H(H(h_{n-3})))$ and comparing with h_n. If they are equal then message carrying h_{n-3} is authentic, else it is not authentic. SEAD assumes some mechanism for a node to distribute its authentic h_n element of the hash chain that can be used by other nodes to authenticate all other elements of the hash chain of that node.

There are two parts for authenticating an entry in a routing update, the sequence number and the metric value for that entry. SEAD uses a single element from the hash chain of the node, corresponding an entry in the route update to authenticate the route update entry. Note that for a given sequence number i, the corresponding metric value can be a number j, $0 \leq j < m$. For this reason, a node generates a hash chain $h_0, h_1, h_2 \cdots, h_n$, such that n is divisible by m. The number n/m represents the maximum value of sequence number for a node. For each sequence number there is a group of m elements in the hash chain, one for each metric value. A node X releases hash values in the reverse order of their generation for the purpose of authentication. For the routing update entry with a sequence number i and a metric value of j, X uses h_{km+j} to authenticate itself, where $k = n/m - i$.

When a node sends a routing update, the node includes one hash value with every entry in that update. If a node lists an entry for some other destination in its update, it sets the destination address to that nodes' address, the metric and the sequence number to the corresponding values in its routing table for that destination node and the hash value is set to the hash of the hash value of the routing update entry from which it learned the route to that destination. If the node lists an entry for itself in that update, it sets the address in that entry to its own node address, the metric to 0, the sequence number to its own next sequence number, and a hash value its own hash chain elements corresponding to that sequence number and metric as explained before. The role played by sequence number and metric in selecting the hash value for routing update entry prevents any node from advertising a route to a destination claiming a greater sequence number than that destination's current sequence number due to one-way hash functions.

Suppose an attacker receives a routing update having metric j for a particular entry. The attacker decides to decrease the metric for that entry to say $j - 1$, then the attacker will have to authenticate the entry with hash chain element h_{km+j-1}. However, this chain element cannot be calculated from h_{km+j} as the hash function cannot be inverted. Hence, any attempt to decrease the metric of a particular routing table entry would be thwarted as the attacker cannot generate the necessary hash chain element to authenticate the resulting metric.

When a node receives a routing update, depending upon the sequence number and metric in the received entry and the sequence number and metric of the prior authentic hash value for that destination, it decides how many times the hash value in the newly received update entry needs to be hashed so that it should be the same as the prior authentic hash value. If the two hash values are found to be equal, the entry is authentic and the node processes the update, else it drops the update packet.

SEAD, however, cannot prevent the same distance attack where a node receives an advertisement for a particular sequence number and metric and then it re-advertises the same sequence number and metric. This is because SEAD only secures the lower bound on the metric ensuring the node does not reduce the metric.

4.2. SRP

The Secure Routing Protocol (SRP) [19] is an extension to reactive routing protocols like Dynamic Source Routing (DSR), which helps nodes defend against attacks that disrupt the route establishment phase. SRP attempts to guarantee that the node initiating the route discovery will be able to differentiate between the legitimate replies and the replies meant to provide false topological information and can discard such malevolent replies. The protocol assumes that there is a security association (SA) between the source node S and the destination T. By using the SA, a source/destination pair that participated in the route establishment verify each other. The source and destination share a secret key K_{ST}, which is negotiated by the SA.

In SRP, a source node adds an additional header, called SRP header, to the underlying routing protocol packet. The SRP header contains a query sequence number, a random query identifier, and a Message Authentication Code (MAC), called SRP

MAC, generated by the source using the shared key K_{ST}. The Query Sequence Number, Q_{SEQ}, is a monotonically increasing 32 bit sequence number maintained by the source node S for each destination T it has a security association with. Q_{SEQ} increases monotonically for every route request generated by S for T, thus allowing T to detect outdated/replayed requests. Q_{SEQ} is initialized at the establishment of the SA and is generally not allowed to wrap around. The Query Identifier Q_{ID} is a random 32 bit identifier generated by S and is used by the intermediate nodes as a means to identify the request. Since Q_{ID} is an output of a secure pseudo-random number generator and is unpredictable by an adversary, it provides protection against attackers who fabricate requests only to cause subsequent requests to be dropped. SRP MAC is a 96 bit value calculated using the shared key K_{ST} over IP addresses of the source S and target T and the two identifiers Q_{SEQ} and Q_{ID}. It not only validates the integrity of the request but also authenticates the origin of the packet to the target, as the MAC could have been calculated only by the source or the destination node which have the knowledge of K_{ST}.

When an intermediate node receives a route request, and if an SRP header is not present in the route request packet, it drops the packet. Otherwise, the node extracts the IP address of the source and destination as well as the Q_{ID} from the request and creates an entry for the request in the query table. If an entry already exists for that source destination pair with the same Q_{ID}, the request is dropped by the node. Otherwise, the node appends its IP address to the request and rebroadcasts the request. Thus IP addresses of the intermediate nodes keep on accumulating on the route request.

The above situation warrants that the Q_{ID} should be sufficiently random and an adversary with finite computation capacity should not be able to predict it. Otherwise, the attacker can prevent route from being established between the given source and the destination pair, as it would fabricate request packets with this Q_{ID} and the intermediate nodes will not forward the legitimate requests, as an entry already exists in the query table for that particular Q_{ID}.

When the destination T receives this request packet, it verifies that the packet originated at the node with which it has SA. The destination compares the Q_{SEQ} with S_{MAX}, the maximum query sequence number received from S. If $Q_{SEQ} \leq S_{MAX}$, the request is outdated/replayed and the destination discards the packet. Else, it calculates the keyed hash of the request field and matches against the SRP MAC. The equality validates the integrity of the request as well as the authenticity of the sender.

The destination broadcasts a route reply to its one-hop neighbors in order to thwart a potentially malicious neighbor from controlling multiple replies. For each valid request, the destination puts the accumulated route in the form of IP addresses of intermediate nodes into the route reply packet. The Q_{SEQ} and Q_{ID} fields from the route request are copied into the corresponding fields of the reply packet. MAC is calculated to preserve the integrity of the packet in transit. The Q_{SEQ} and Q_{ID} fields verify the freshness of the packet to the source.

When the source S receives the route reply packet, it checks source and destination addresses, Q_{ID} and Q_{SEQ} and discards the reply if it does not correspond to the currently pending query. Otherwise, it compares the reply IP source-route with the reverse of the route carried in the reply payload. If the two routes match, MAC is calculated using the replied route, the SRP header fields, and K_{ST}. The successful verification confirms that the request did indeed reach the intended destination T and the reply was not corrupted on the way back from T to S. Furthermore, since the reply packet has been routed and successfully received over the reverse of the route it carries, the routing information has not been compromised during the request propagation.

Intermediate nodes also measure the frequency of queries received from their neighbors. Intermediate nodes maintain a priority ranking of their neighbors - highest priority to nodes generating requests at the lowest rate and the lowest rating for nodes generating requests with highest rate. In case two packets arrive at the same time, the neighbor whose ranking is higher, is given priority in routing over the one with the lower ranking.

The secure routing protocol guarantees the discovery of a correct route, even in the presence of malicious nodes. The protocol obviates the need of a certification authority, thereby suiting itself to the ad-hoc paradigm. The protocol does not necessitate the knowledge of keys of all member nodes. The only requirement of this protocol is that there should be a prior security association between the source and the destination nodes. This kind of a security association is realized through shared secret keys between any two pair of nodes. However, when malicious nodes succeed in subverting benign nodes, the malicious nodes could easily gain access to the shared secret keys. The malicious node can then masquerade as the subverted node and initiate communication with other good nodes with whom the subverted node has a security association.

4.3. ARIADNE

Ariadne [8] is an on-demand secure routing protocol based on the DSR protocol. Ariadne prevents attackers or compromised nodes from tampering with uncompromised routes consisting of benign nodes. It is based on efficient symmetric cryptographic primitives and prevents several types of denial of service attacks. Unlike SRP, Ariadne uses a broadcast authentication protocols TESLA [18], which enables a node to verify that a broadcast packet (like RREQ) received by the node is indeed generated by the initiator of the message. Such a broadcast authentication is essential in defending against impersonation and denial of service attacks. The basic idea of the Ariadne protocol is to insure that the destination node can authenticate the source node, the source node can authenticate every intermediate node on the path from the source to the destination (received by the source in RREP), and malicious nodes cannot tamper with routes in RREQ or RREP by inserting dummy IDs or removing benign node IDs.

The idea behind the TESLA protocol is to have a random initial key (k_n) for each node from which each node generates a one-way key chain by repeated computation of a one-way hash function (H) such that $k_{n-1} = H(k_n)$ and in general for any $j < i$, $k_j = H_{i-j}(k_i)$. A node discloses each key of its one-way key chain in an order that is exactly the reverse of the order in which the node generates the keys. Further, a node publishes its key k_i at a time $T_0 + i * t$, where T_0 is the time at which k_0 is published, and t is the key publication interval. The rationale behind having a reverse key disclosure schedule is that using a previously known hash chain element, like k_j, any other node can authenticate subsequent elements, k_i, $i > j$, from a nodes hash chain by using the equation $k_j = H_{i-j}(k_i)$. However, other nodes cannot generate k_i due to the one-way property of the hash function.

For broadcast authentication using TESLA, a node generates a broadcast packet, adds a Message Authentication Code (MAC) of the packet generated by the node using its future (next in its schedule) TESLA key and then releases the key used in MAC at a later time. A node receiving the packet verifies the TESLA *security condition* that the key k_i used to authenticate the packet has not yet been released by the nodes[2]. If the condition holds, then the receiving node waits for the TESLA key to be released by the sender and verifies the key (using the one-way hash function) and the MAC of the packet. If they are authentic, then the receiver accepts the packet, else it drops the packet.

The Protocol: Ariadne protocol assumes that the source and the destination share a secret key K_{ST} that allows them to authenticate each other. To establish a secure route to the destination, the source node floods a RREQ packet that has eight fields <ROUTE REQUEST, initiator, target, id, time interval, hash chain, node list, MAC list>. The initiator and the target are set to the source ID and the destination ID, respectively. The 'id' is an identifier that has not been recently used in route discovery. The 'time interval' is set to TESLA time interval at the pessimistic arrival time of the request at the target, with maximum possible clock offset/skew and maximum transmission delay. The hash chain field is initialized by the initiator to the MAC calculated over initiator, target, id, time interval, using the key K_{ST} ($MAC_{K_{ST}}$(initiator, target, id, time interval)). The node list and MAC list are empty initially and will be filled by the intermediate and target nodes.

When an intermediate node, A, receives a RREQ, the node checks its local table for the (initiator, id) entry. If it finds an entry for the same route discovery, it discards the RREQ, else the node verifies the time interval of the RREQ. If the time interval is too much in the future or the key corresponding to it has been disclosed, the RREQ is discarded. Otherwise, the node appends its address to the node list in the RREQ packet, and replaces the hash chain field with $H(A, oldhashchain)$. The node then appends a MAC of the entire request to the MAC list, where the MAC is calculated using key k_i corresponding to the time interval in the RREQ. The node then rebroadcasts the modified RREQ.

When the destination node receives the RREQ, it determines whether the keys corresponding to the time interval mentioned in the RREQ have not been disclosed yet, and the hash chain field is equal to

$$H(I_n, H(I_{n-1}, H(\cdots, H(I_1, MAC_{K_{ST}}(\text{initiator, target, id, time interval}))\cdots))),$$

where I_i is the intermediate node at position i and n is the number of nodes in the node list. If both the conditions hold, then the destination is assured that the RREQ is valid, and it constructs a RREP packet.

[2] This is because if the key is released it is also known to malicious nodes.

The RREP packet consists of target, initiator, time interval, node list, MAC list (which correspond to fields from the corresponding RREQ), target MAC and key list. Target MAC is a MAC calculated by the destination over first five fields with the key K_{ST}. Key list is left empty to be initialized by the intermediate nodes, along the reverse route in the RREQ. The destination sends the RREP to the initiator along the source route which is the reverse of the sequence of hops in the node list in the RREQ. The node forwarding the route RREP waits until it is able to disclose the key for the specified time interval. The node then appends the key to the key list field in the RREP and forwards the RREP to the next hop towards the source. The waiting delays do not add significant computation overhead but adds to storage overheads. When the initiator receives the RREP, it checks if the keys in the key list are valid, target MAC is valid and each MAC in the MAC list is valid. If all these are valid only then will it accept the RREP.

One-way hash chain in RREQ/RREP ensures that no hop is omitted by some malicious node. To change or remove a previous hop form the RREQ/RREP, the attacker must be able to invert the one-way hash function, which is computationally infeasible. However, a malicious node might succeed in removing the address of any previous node from the node list, but won't be able to remove that node's address from the hash chain field. Such a fabrication would be easily detected by the destination/source, since the computed hash chain field won't be the same as the hash chain in the received packet and hence the RREQ/RREP would be discarded.

When an intermediate node detects a route break, i.e., it is unable to deliver the packet to the next hop after a fixed finite number of retransmissions, it generates a route error RERR and sends it to the source node. To deal with false RERR messages, the protocol requires the source to authenticate the RERR messages using TESLA. If the authentication succeeds, then the source tries to reestablish a route to the destination, else it drops the RERR. However, the protocol does not guard against attackers intentionally dropping genuine RERR messages.

4.4. ARAN

Authenticated Routing for Ad-hoc Networks (ARAN) [24] detects and protects against malicious actions by third parties and peers in an ad-hoc environment. ARAN assumes a managed-open environment, meaning that there is an opportunity for pre-deployment of certain security infrastructure. Using such an infrastructure helps nodes exchange initialization parameters before hand through a trusted third party like a certification authority. With the help of initialization parameters, like a certificate from a trusted server, ARAN provides authentication, message integrity, and non-repudiation in an ad-hoc environment. Table 1 presents the notations used in the rest of this section.

ARAN protocol assumes a trusted certification server T, whose public key is known to all the valid nodes in the networks. The protocol consists of three stages, the preliminary certification stage, the authenticated route discovery phase, and the authenticated route setup phase. In the preliminary certification stage, each node obtains a certificate, $cert_A$, from the server T. The certificate of a node, $cert_A = [IP_A, K_{A+}, t, e]K_{T-}$, contains the IP address of A, the public key of A, a timestamp t of the time the certifi-

Table 1. Notations used in ARAN

K_{A+}	Public-key of node A.
K_{A-}	Private-key of node A.
$\{d\}K_{A+}$	Encryption of data d with key K_{A+}.
$\{d\}K_{A-}$	Data d digitally signed by node A
$cert_A$	Certificate belonging to node A.
N_a	Nonce issued by node A.
IP_A	IP address of node A.
RDP	Route Discovery Packet identifier.
REP	REPly packet identifier.
ERR	ERRor packet identifier.

cate was generated by the server, and a time e at which the certificate expires. These variables are concatenated and signed by the server. Nodes maintain a fresh certificate issued to them by the trusted server, which helps them authenticate themselves to the other nodes during the exchange of control messages.

The authenticated route discovery phase provides end-to-end authentication, in which the source node verifies that the intended destination is reached. The source node, A, initiates a route discovery for the destination X, by broadcasting a route discovery packet (RDP). The broadcast message, $[RDP, IP_X, N_A]K_{A-}, cert_A$, includes a packet type identifier (RDP), the IP address of the destination (IP_X), A's certificate ($cert_A$), and a monotonically increasing nonce N_A, the all signed with A's private key.

Upon receiving an RDP packet, an intermediate node stores (IP_A, N_A) of the RDP packet. If an intermediate node has already seen the (IP_A, N_A) tuple, it drops the RDP packet. Otherwise, it keeps track of the predecessor node from which it received the RDP packet, validates the signature with the given certificate, removes A's certificate from the RDP, and rebroadcasts the RDP packet by signing it. For instance, if B is a neighbor of A, then B broadcasts

$$[[RDP, IP_X, N_A]K_{A-}]K_{B-}, cert_A, cert_B.$$

Such message signing prevents spoofing attacks that may alter the route or form loops. When node C receives the broadcast packet, C validates signatures of A and B using their respective certificates in the RDP packet. C removes B's signature and certificate, records B as its predecessor node, signs the contents $[[RDP, IP_X, N_A]K_{A-}]$ with its private key and appends its own certificate, and broadcasts the RDP.

$$C \rightarrow broadcast : [[RDP, IP_X, N_A]K_{A-}]K_{C-}, cert_A, cert_C.$$

Each intermediate node repeats the same process as node C, and eventually the destination receives the RDP packet.

The destination replies to the first RDP packet it receives. Note that such an RDP packet may not have traversed the shortest path due to network congestion, or due to the presence of malicious nodes. The rationale behind choosing such a path is to prefer them over a congested least-hop path that reduces the end-to-end delay. As a response to the RDP, the destination generate a Reply packet (REP), and unicasts it along the reverse path to the source node. If D is the next node towards the source from the destination X, then X unicasts a REP ($[REP, IP_A, N_A]K_{X-}, cert_X$) packet to D, where REP in the packet is a packet-type identifier. Since nodes keep track of predecessor nodes during the RDP phase, an intermediate node forwards the REP to the predecessor. Each intermediate node along the reverse path signs the REP and appends its own certificate before forwarding the REP. For instance, if C is be the predecessor of D, then D unicasts

$$[[REP, IP_A, N_A]K_{X-}]K_{D-}, cert_X, cert_D.$$

to C. Node C validates D's signature on the received message, removes the signature and certificate, then signs the contents of the message and appends its own certificate before unicasting the REP to its predecessor. This avoids impersonation and replay of the message sent by X. When the source receives the REP, it verifies the destination's signature and the nonce returned by the destination. If they are valid, then it starts transmitting the date along the established path.

Route maintenance: In ad-hoc networks, routes may not be used actively by the source node for a long time or they may break due to the node mobility. In ARAN, nodes purge route table entries that are not used by the source-destination for a predetermined time period (route's lifetime). An intermediate node generates an error message (ERR) if there is no active route towards the destination in its route table, or if the node finds a link break due to node mobility. If a node B finds a route break, then it generates and sends a ERR ($[ERR, IP_A, IP_X, N_B]K_{B-}, cert_B,$) message to its predecessor node C. The ERR message is forwarded along the path towards the source without modification. The nonce N_B ensures the ERR message is fresh. Because messages are signed, malicious nodes cannot generate ERR messages for other nodes. Non-repudiation provided by the signed ERR message allows a node to be verified as the source of each ERR message that it sends. A node which transmits a large number of ERR messages, whether the ERR messages are valid or fabricated, should be avoided.

Key revocation: ARAN attempts a best effort key revocation that is backed with limited time certificates. In the event of a certificate revocation, the trusted certificate server, T, sends a broadcast message ($[revoke, Cert_R]K_{T-}$) to the ad-hoc group that announces the revocation. Any node receiving this message re-broadcasts it and stores the message until the revoked certificate expires normally. Neighbors of the node with the revoked certificate need to reform routes as necessary to avoid transmission through such nodes.

If an untrusted node whose certificate is being revoked, is the only link between two partitions of an ad-hoc network, it may not propagate the revocation message to the other part - leading to a partitioned network. Such a partition may last until the untrusted node is no longer the sole connection between the two partitions. Thus, this method is not fail-safe. To detect such situations and to hasten the propagation of revocation notices, nodes exchange a summary of its revocation notices with new neighbors (as and when discovered). If these summaries do not match, then nodes that detect inconsistency rebroadcasts signed notices to restart propagation of the notice.

4.5. Coping with Byzantine Failures

The secure routing protocols described so far assume that nodes in the network do not collude to attack the network. However, in realistic networks attacks can be due to an individual malicious node or due to colluding malicious nodes. Baruch et al.[1] proposed an on-demand protocol to provide resilience to Byzantine failures caused by individual or colluding nodes. In this protocol [1], the emphasis is on survivability of routes under situations where an intermediate node or group of nodes are known to be malicious and may attempt 'Byzantine' attacks such as creation of routing loops, misrouting of packets along non-optimal (unnecessarily long) paths or selective dropping of packets (black or gray holes).

Instead of laying the blame of a route failure on a single misbehaving node, the protocol [1] takes into account a pair of nodes that share a link in the network. Such an approach can ameliorate routing misbehaviors wherein two adjacent nodes are colluding with each other and dropping packets. Each link between two adjacent nodes has certain weight associated with it. When a node detects a link to be faulty, it increases the weight associated with multiplicatively. When multiple routes are discovered for a particular destination, the initiating node selects the route that has the least sum of link weights. The least sum of link weights of a route implies that the route has the least likelihood of having a faulty link on it. The protocol consists of the following three phases. (a) Route Discovery with fault avoidance, (b) Byzantine fault detection, and (c) link weight management.

Route Discovery with Fault Avoidance: A source node initiates a Route Discovery by generating a route REQUEST packet, digitally signing it using its private key, and flooding the REQUEST packet in the network. The request consists of the source ID, the destination ID, a sequence number and link weight list. Digital signature helps intermediate nodes to authenticate the source, and to safeguard against malicious nodes trying to initiate route discovery and consume valuable network resources.

When an intermediate node receives a route request, it checks its valid request list to see if there is a matching request in the list for the same source. If there is no matching request and the source's signature is valid, it rebroadcasts the request, else the request is dropped. When the destination receives a request from the source for the first time, it checks the source signature on the request. If the signature is valid, it generates and signs the response consisting of source, destination, a response sequence number and the weight list from the request packet. Unlike DSR, intermediate nodes do not cache routes and respond to the source node.

When an intermediate node receives a response, it computes the total weight of the path by summing weights of all the links, which constitute the path. If the total weight is less than any of the previous responses for that particular request, it checks the signature on the response header and every hop listed on the packet. If each element of the packet is verified, the node appends its identifier to the end of the packet, signs the new packet, and broadcasts it.

When the source receives the response, it verifies the digital signature of intermediate nodes. If the path is better than the best path received so far, the source updates the route used to send packets to the particular destination. This type of route discovery attempts to find the route having lowest sum of link weights, thereby selecting a route which is least likely to have a faulty link on it. Faulty links have more link weight and get automatically precluded from route discovery.

In spite of this fault avoiding route discovery, there may still be a faulty link along a route because no alternate routes with lower link weights were discovered. To detect a faulty link, nodes invoke a Byzantine fault detection mechanism that uses an adaptive probing technique.

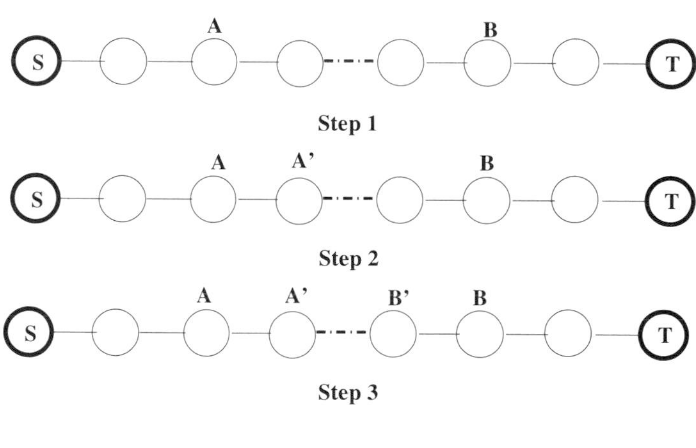

Figure 1. Fault Detection.

Byzantine Fault Detection: For the purpose of Byzantine fault detection, the protocol requires the destination to return an ACK-message to the source for every successfully received data packet. If the source node does not receives valid ACKs during the timeout, it assumes that the packets were lost in transit due to the presence of malicious nodes or because the destination is unreachable due to a network partition. For each destination, the source node selects an ACK loss rate less than a fixed threshold as tolerable, and this may vary with every route. The source keeps track of number of losses on a path. If this number exceeds the threshold, the source node initiates a binary search on the path, assuming a faulty link exists on the source-destination route, in an attempt to locate the faulty link.

The fault detection mechanism is best explained by an example shown in Figure 1. The source specifies two random intermediate nodes, A and B, on the route called probes, each of which must send an ack for the successfully received packet. The probes divide the route into non-overlapping continuous segments. In the example, probes A and B divide the path into SA, AB, and BD. Due to the presence of the faulty link, S does not receives an ack from node B. Thus S determines a fault on the segment AB. S inserts a new probe A' in between that segment. The probe insertion and interval subdivision continues until the faulty interval narrows down to a single faulty link. in the example it is the link A'B'. Due to binary search, the source detects a faulty link after log(L) steps, L being the total number of nodes on the route.

Link Weight Management: When a node detects a faulty link, it uses a multiplicative increase scheme to double its weight. The higher the weight, the lower the probability of that link being on any further routes.

Thus using these techniques, route discovery with fault avoidance, Byzantine fault detection, and link weight management, nodes establish routes that are free of nodes known to be malicious and may attempt 'Byzantine' attacks.

5. SECURE POSITION-BASED ROUTING

Security in position-based routing is a relatively new area, and to the best of authors knowledge, there is no secure position-based routing protocol in the literature so far. Thus, to keep the presentation simple, we discuss security issues related to position-based greedy forwarding and some possible counter measures to fight attacks in greedy-forwarding. Fundamental to greedy-forwarding is a neighbor discovery or a hello protocol using which nodes exchange their ID and position information periodically. However, malicious node may not follow the protocol properly, and may try to spoof their ID or location as explained in Section 3.

To prevent external attacks, nodes may employ an authentication mechanism like TESLA broadcast authentication as explained in Section 4.3, along with digital signatures to avoid attacks due to unauthorized external nodes. On the other hand, compromised internal nodes can pose severe threats to the greedy-forwarding. Zhou et al. [29] identified location spoofing, traffic abusing and forwarding misbehavior as three main internal attacks, and proposed the following counter measures.

Defense against location spoofing: A possible way to defend against location spoofing is to use the Time of Flight (ToF) of the message and the speed of signal to estimate the distance between the two nodes. Precisely, if t is the round-trip time and s is the speed of the signal, then the distance d between two communicating nodes should be less than $(t \times s)/2$, i.e., $d \leq (t \times s)/2$. However, this method does not provide an upper bound on the distance, as a malicious node can hold the probe message for a arbitrary time to increase the ToF value. By doing this, a malicious node succeeds in claiming a farther position than its true position.

To mitigate this problem, the basic distance estimation method described above can be augmented by using a neighbor monitoring scheme along with voting. The idea depends on the fact that a false-position reported by a node tends to be inconsistent among neighbors. However, the success of this method depends on the ability of the voting system to cope with false accusations.

Defense against traffic abusing: Traffic abusing may range from dropping packets to flooding the network with junk or meaningful data at high-rate. By doing this, an attacker may attempt to exhaust network resources or overwhelm a node to do lot of packet-processing. To mitigate this problem, one can use the following observation: when an attacker abuses a node X with traffic, neighboring nodes of X experience anomalous traffic even before X. Thus, neighboring nodes may choose to drop such packets to save the attacked node.

Further, nodes can choose an upper bound and lower bound on the traffic intensity to detect anomalous traffic behavior. If a node experiences a traffic intensity above a preset lower bound, then the node may simply stop processing packets. This method works even if a node is surrounded by a group of colluding malicious nodes.

Defense against forwarding misbehaviors: Another common problem in secure-routing is to deal with forwarding misbehavior. Forwarding misbehaviors are more serious due to compromised internal nodes or due to 'selfish' or malicious nodes. Such nodes may want to gain services from network, but may not want to 'give' services to save their limiting resources like battery. Note that a 'selfish node' may be not malicious because a selfish node may not harm the network. To keep up with our discussion, we consider malicious nodes for forwarding misbehaviors. However, readers interested in dealing with selfish nodes are referred to [4, 28] for more details.

A simple way to work around forwarding misbehaviors is to use multiple paths. Multi-path approach mitigates packet delivery failure, but incurs control overhead to have multiple paths. Another approach is to maintain two-hop neighbor table, in contrast to one-hop neighbor table that is maintained in most position-base protocols, at each node, and employ a neighbor monitoring mechanism to verify the next hop transmission. For this approach nodes need to work in the promiscuous mode. In the promiscuous mode a node can overhear transmission for other nodes within its radio range. When a node A selects a next hop B using greedy forwarding, it starts a timer to check if B forwarded the packet correctly to one of its neighbors C selected using the greedy forwarding. If the timer expires before A hears a transmission from B, then A suspects B and takes necessary actions (like flooding an accusation message). Else, if A hears a transmission from B, it checks if B selected a proper next hop. Since A, as well as other nodes, maintains a two-hop neighbor table, it can verify the next hop selection of B. However, the neighbor monitoring in promiscuous mode is prone to error, and sometime malicious node may attempt to falsely accuse benign nodes. Thus, protocols that deal with such errors and false accusations [2] may help mitigate the problem.

6. SUMMARY

Ad-hoc networks are potential enablers of networking any-where and any-time concept, which is the current trend in this information-sharing age. While these networks are rapidly deployable and do not need an infrastructure to operate, they are very vulnerable to attacks from both inside and outside of the network. As explained in this chapter, even the fundamental task of routing becomes non-trivial in presence of malicious node. Especially, when the number of malicious nodes cross beyond certain threshold, routing becomes impossible. Another extreme is a case where there is a single malicious node that connects two part of the network. In such cases, excluding malicious node renders the network partitioned in to two or parts. In this chapter, we presented a brief description of routing protocols for ad-hoc networks, possible attacks of routing protocols, and various secure routing protocols that establish secure paths from a source to the destination. Further, we discussed some security counter measures for position-based routing. Secure routing in ad-hoc network, as of now, is an active area of research. Coming up with an efficient and secure routing protocol under a robust security model with provable security is still an open problem.

7. REFERENCES

1. Baruch Awerbuch, David Holmer, Cristina Nita Rotaru and Herbert Rubens. An On Demand Secure Routing Protocol Resilient to Byzantine Failures. *In ACM Workshop on Wireless Security (WiSe)*, Atlanta, Georgia, September 28 2002.

2. Sonja Buchegger and Jean-Yves Le Boudec. Performance Analysis of the CONFIDANT Protocol (Cooperation Of Nodes: Fairness In Dynamic Ad-hoc NeTworks). *In proceedings of the 3rd ACM international symposium on Mobile ad hoc networking & computing (MOBIHOC) 2002. Lausanne, Switzerland.*

3. S. Basagni, I. Chlamtac, V. Syroutik, and B. Woodward. A distance effect routing algorithm for mobility (DREAM). *In proceedings of the 4th annual ACM/IEEE Int. Conf. on Mobile Computing and networking (MOBICOM)*, pages 76-84, Dallas, TX, USA, 1998.

4. Levente Buttyán and Jean-Pierre Hubaux. Stimulating cooperation in self-organizing mobile ad hoc networks.*Mob. Netw. Appl.*, 8(5), 579–592, Kluwer Academic Publishers 2003.

5. Prosenjit Bose, Pat Morin, Ivan Stojmenovic, and Jorge Urrutia. Routing with Guaranteed Delivery in Ad Hoc Wireless Networks. *Wireless Networksi* 7, 609-616, Kluwer Academic Publishers 2001.

6. http://www.bluetooth.com

7. Yih-Chun Hu, David B. Johnson and Adrian Perrig, SEAD: Secure Efficient Distance Vector Routing in Mobile Wireless Ad-hoc Networks. *Fourth IEEE Workshop on Mobile Computing Systems and Applications WMCSA '02.*

8. Yih-Chun Hu, Adrian Perrig and David B. Johnson. Ariadne: A secure on-demand routing protocol for ad-hoc networks. *The 8th ACM International Conference on Mobile Computing and Networking, September 2002.*

9. Z. Haas and M. Pearlman. The performance of query control scheme for the zone routing protocol. *ACM/IEEE Transactions on Networking, 9(4) pages 427-438*, August 2001.

10. P. Jacquet, P. Muhlethaler, T. Clausen, A. Laouiti, A. Qayyum, L. Viennot. Optimized Link State Routing Protocol. *IEEE INMIC Pakistan 2001.*

11. D. Johnson and D. Maltz. Mobile Computing. *chapter 5. Dynamic Source Routing*, pages 153-181. Kulwer Academic Publishers, 1996.

12. Karp, B., and Kung. H. T. GPSR: Greedy Perimeter Stateless Routing for Wireless Networks. *Proc. 6th Annual International Conference on Mobile Computing and Networking* (MOBICOM 2000), 243-254.

13. Young-Bae Ko , Nitin H. Vaidya. Location-aided routing (LAR) in mobile ad-hoc networks. *ACM/Blatzer Wireless Networks journal*, 6(4) pages 307-321, 2000.

14. Chris Karlof and David Wagner. Secure routing in sensor networks: Attacks and countermeasures. *In Proceedings of the IEEE International Workshop on Sensor Network Protocols and Applications (SNPA-03)*, May 2003.

15. Jinyang Li, John Jannotti, Douglas S. J. De Couto, David R. Karger, Robert Morris. A Scalable Location Service for Geographic Ad Hoc Routing. *Proceedings of 6th ACM International Conference on Mobile Computing and Networking (MOBICOM)* 2000.

16. National Institute of Standards and Technology (NIST). Secure Hash Standard, May 1993. *Federal Information Processing Standards (FIPS) Publication 180-1.*

17. C.Perkins and P. Bhagwat. Highly dynamic destination sequenced distance-vector routing for mobile computers. *Computer Communication Review*, pages 234-244, October 94.

18. Adrian Perrig, Ran Canetti, Dawn Song, and J. D. Tygar. Efficient and Secure Source Authentication for Multicast. *In Network and Distributed System Security Symposium, NDSS 01*, pages 35-46, February 2001.

19. P.Papadimitratos and Z.J. Haas. Secure Routing for Mobile Ad hoc Networks. *In the proceedings of SCS Communication Networks and Distributed Systems Modeling And Simulation Conference (CNDS 2002)*, San Antonio, TX, January 27-31, 2002.

20. C. Perkins and E. Royer. Ad-hoc on-demand distance vector routing. *In Proc. Of the 2nd IEEE Workshop on Mobile Computing Systems and Applications*, pages 90-100, Feb 1999.

21. Ronald L. Rivest. The MD5 Message-Digest Algorithm. *RFC 1321, April 1992.*

22. Michael Käsemann, Holger Füßler, Hannes Hartenstein, Martin Mauve. A Reactive Location Service for Mobile Ad Hoc Networks. *TR-14-2002, Department of Computer Science, University of Mannheim, November 2002.*

23. http://www.terminodes.org

24. K. Sanzgiri, B. Dahill, B.N. Levine, C. Shields, E.M. Belding-Royer. An Authenticated Routing Protocol for Secure Ad Hoc Networks. *IEEE Journal on Selected Areas in Communication, special issue on Wireless Ad hoc Networks, 23(3)* pages 598-610, March 2005.

25. Ning Song, Lijun Qian, Xiangfang Li. Wormhole Attacks Detection in Wireless Ad Hoc Networks: A Statistical Analysis Approach. *19th IEEE International Parallel and Distributed Processing Symposium (IPDPS'05)* - Workshop 17, 2005.

26. Ivan Stojmenovic. Position based routing in ad-hoc networks. *IEEE Communications Magazine*,40(7), pages 128-134, 2002.

27. Y Xue, B Li and K Nahrstedt. A scalable location management scheme in mobile ad-hoc networks. *26th Annual IEEE Conference on Local Computer Networks* (LCN 2001).

28. Yongwei Wang, Venkata C. Giruka and Mukesh Singhal. A Fair Distributed Solution for Selfish Nodes Problem in Wireless Ad Hoc Networks. *In Proceedings of Third International Conference, ADHOC-NOW 2004,* Vancouver, Canada, July 22-24, 2004. Proceedings.

29. Zhi Zhou and Kin Choong Yow. Geographic Ad Hoc Routing Security: Attacks and Countermeasures. *Ad Hoc & Sensor Wireless Networks, vol.1, number 1* pp 235-253, 2005.

7

A SURVEY ON INTRUSION DETECTION IN MOBILE AD HOC NETWORKS

Tiranuch Anantvalee
Department of Computer Science and Engineering
Florida Atlantic University, Boca Raton, FL 33428
E-mail: tanantva@fau.edu

Jie Wu
Department of Computer Science and Engineering
Florida Atlantic University, Boca Raton, FL 33428
E-mail: jie@cse.fau.edu

In recent years, the use of mobile ad hoc networks (MANETs) has been widespread in many applications, including some mission critical applications, and as such security has become one of the major concerns in MANETs. Due to some unique characteristics of MANETs, prevention methods alone are not sufficient to make them secure; therefore, detection should be added as another defense before an attacker can breach the system. In general, the intrusion detection techniques for traditional wireless networks are not well suited for MANETs. In this paper, we classify the architectures for intrusion detection systems (IDS) that have been introduced for MANETs. Current IDS's corresponding to those architectures are also reviewed and compared. We then provide some directions for future research.

1. INTRODUCTION

A mobile ad hoc network (MANET) is a self-configuring network that is formed automatically by a collection of mobile nodes without the help of a fixed infrastructure or centralized management. Each node is equipped with a wireless transmitter and receiver, which allow it to communicate with other nodes in its radio communication range. In order for a node to forward a packet to a node that is out of its radio range, the cooperation of other nodes in the network is needed; this is known as multi-hop communication. Therefore, each node must act as both a host and a router at the same time. The network topology frequently changes due to the mobility of mobile nodes as they move within, move into, or move out of the network.

A MANET with the characteristics described above was originally developed for military purposes, as nodes are scattered across a battlefield and there is no infrastructure to help them form a network. In recent years, MANETs have been developing rapidly and are increasingly being used in many applications, ranging from military to civilian and commercial uses, since setting up such networks can be done without the help of any infrastructure or interaction with a human. Some examples are: search-and-rescue missions, data collection, and virtual classrooms and conferences where laptops, PDA or other mobile devices share wireless medium and communicate to each other. As MANETs become widely used, the security issue has become one of the primary concerns. For example, most of the routing protocols proposed for MANETs assume that every node in the network is cooperative and not malicious [1]. Therefore, only one compromised node can cause the failure of the entire network.

There are both passive and active attacks in MANETs. For passive attacks, packets containing secret information might be eavesdropped, which violates confidentiality. Active attacks, including injecting packets to invalid destinations into the network, deleting packets, modifying the contents of packets, and impersonating other nodes violate availability, integrity, authentication, and non-repudiation. Proactive approaches such as cryptography and authentication [10, 11, 12, 13] were first brought into consideration, and many techniques have been proposed and implemented. However, these applications are not sufficient. If we have the ability to detect the attack once it comes into the network, we can stop it from doing any damage to the system or any data. Here is where the intrusion detection system comes in.

Intrusion detection can be defined as a process of monitoring activities in a system, which can be a computer or network system. The mechanism by which this is achieved is called an intrusion detection system (IDS). An IDS collects activity information and then analyzes it to determine whether there are any activities that violate the security rules. Once an IDS determines that an unusual activity or an activity that is known to be an attack occurs, it then generates an alarm to alert the security administrator. In addition, IDS can also initiate a proper response to the malicious activity.

Although there are several intrusion detection techniques developed for wired networks today, they are not suitable for wireless networks due to the differences in their characteristics. Therefore, those techniques must be modified or new techniques must be developed to make intrusion detection work effectively in MANETs.

In this paper, we classify the architectures for IDS in MANETs, each of which is suitable for different network infrastructures. Current intrusion detection systems corresponding to those architectures are reviewed and compared.

The rest of the paper is structured as follows. Section 2 describes the background on intrusion detection systems. Intrusion detection in MANETs - how it differs from intrusion detection in wired networks - is also presented in this section. In Section 3, architectures that have been introduced for IDS in MANETs are presented. Some of current intrusion detection systems for MANETs are given in Section 4. Then, some of the intrusion detection techniques for node cooperation are reviewed and compared in Section 5. Finally, the conclusion and future directions are given in Section 6.

2. BACKGROUND

2.1. Intrusion Detection System (IDS)

Many historical events have shown that intrusion prevention techniques alone, such as encryption and authentication, which are usually a first line of defense, are not sufficient. As the system become more complex, there are also more weaknesses, which lead to more security problems. Intrusion detection can be used as a second wall of defense to protect the network from such problems. If the intrusion is detected, a response can be initiated to prevent or minimize damage to the system.

Some assumptions are made in order for intrusion detection systems to work [1]. The first assumption is that user and program activities are observable. The second assumption, which is more important, is that normal and intrusive activities must have distinct behaviors, as intrusion detection must capture and analyze system activity to determine if the system is under attack.

Intrusion detection can be classified based on audit data as either host-based or network-based. A network-based IDS captures and analyzes packets from network traffic while a host-based IDS uses operating system or application logs in its analysis. Based on detection techniques, IDS can also be classified into three categories as follows [2].

- *Anomaly detection systems*: The normal profiles (or normal behaviors) of users are kept in the system. The system compares the captured data with these profiles, and then treats any activity that deviates from the baseline as a possible intrusion by informing system administrators or initializing a proper response.
- *Misuse detection systems*: The system keeps patterns (or signatures) of known attacks and uses them to compare with the captured data. Any matched pattern is treated as an intrusion. Like a virus detection system, it cannot detect new kinds of attacks.
- *Specification-based detection*: The system defines a set of constraints that describe the correct operation of a program or protocol. Then, it monitors the execution of the program with respect to the defined constraints.

2.2. Intrusion Detection in MANETs

Many intrusion detection systems have been proposed in traditional wired networks, where all traffic must go through switches, routers, or gateways. Hence, IDS can be added to and implemented in these devices easily [17, 18]. On the other hand, MANETs do not have such devices. Moreover, the medium is wide open, so both legitimate and malicious users can access it. Furthermore, there is no clear separation between normal and unusual activities in a mobile environment. Since nodes can move arbitrarily, false routing information could be from a compromised node or a node that has outdated information. Thus, the current IDS techniques on wired networks cannot be applied directly to MANETs. Many intrusion detection systems have been proposed to suit the characteristics of MANETs, some of which will be discussed in the next sections.

3. ARCHITECTURES FOR IDS IN MANETS

The network infrastructures that MANETs can be configured to are either flat or multi-layer, depending on the applications. Therefore, the optimal IDS architecture for a MANET may depend on the network infrastructure itself [9]. In a flat network infrastructure, all nodes are considered equal, thus it may be suitable for applications such as virtual classrooms or conferences. On the contrary, some nodes are considered different in the multi-layered network infrastructure. Nodes may be partitioned into clusters with one clusterhead for each cluster. To communicate within the cluster, nodes can communicate directly. However, communication across the clusters must be done through the clusterhead. This infrastructure might be well suited for military applications.

3.1. Stand-alone Intrusion Detection Systems

In this architecture, an intrusion detection system is run on each node independently to determine intrusions. Every decision made is based only on information collected at its own node, since there is no cooperation among nodes in the network. Therefore, no data is exchanged. Besides, nodes in the same network do not know anything about the situation on other nodes in the network as no alert information is passed. Although this architecture is not effective due to its limitations, it may be suitable in a network where not all nodes are capable of running an IDS or have an IDS installed. This architecture is also more suitable for flat network infrastructure than for multi-layered network infrastructure. Since information on each individual node might not be enough to detect intrusions, this architecture has not been chosen in most of the IDS for MANETs.

3.2. Distributed and Cooperative Intrusion Detection Systems

Since the nature of MANETs is distributed and requires cooperation of other nodes, Zhang and Lee [1] have proposed that the intrusion detection and response system in MANETs should also be both distributed and cooperative as shown in Figure 1. Every node participates in intrusion detection and response by having an IDS agent running on them. An IDS agent is responsible for detecting and collecting local events and data to identify possible intrusions, as well as initiating a response independently. However, neighboring IDS agents cooperatively participate in global intrusion detection actions when the evidence is inconclusive. Similarly to stand-alone IDS architecture, this architecture is more suitable for flat network infrastructure, not multi-layered one.

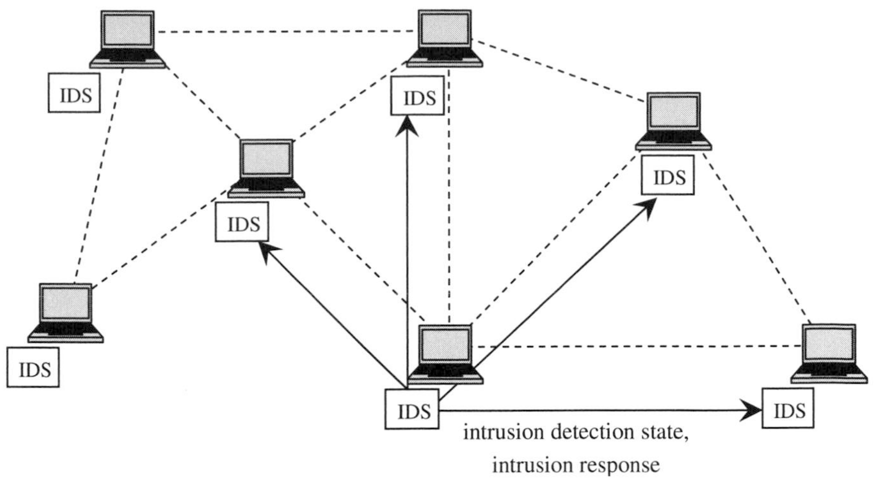

Figure 1. Distributed and Cooperative IDS in MANETs proposed by Zhang and Lee [1]

3.3. Hierarchical Intrusion Detection Systems

Hierarchical IDS architectures extend the distributed and cooperative IDS architectures and have been proposed for multi-layered network infrastructures where the network is divided into clusters. Clusterheads of each cluster usually have more functionality than other members in the clusters, for example routing packets across clusters. Thus, these clusterheads, in some sense, act as control points which are similar to switches, routers, or gateways in wired networks. The same concept of multi-layering is applied to intrusion detection systems where hierarchical IDS architecture is proposed. Each IDS agent is run on every member node and is responsible locally for its node, i.e., monitoring and deciding on locally detected intrusions. A clusterhead is responsible locally for its node as well as globally for its cluster, e.g. monitoring network packets and initiating a global response when network intrusion is detected.

3.4. Mobile Agent for Intrusion Detection Systems

A concept of mobile agents has been used in several techniques for intrusion detection systems in MANETs. Due to its ability to move through the large network, each mobile agent is assigned to perform only one specific task, and then one or more mobile agents are distributed into each node in the network. This allows the distribution of the intrusion detection tasks.

There are several advantages for using mobile agents [2]. Some functions are not assigned to every node; thus, it helps to reduce the consumption of power, which is scarce in mobile ad hoc networks. It also provides fault tolerance such that if the network is partitioned or some agents are destroyed, they are still able to work. Moreover, they are scalable in large and varied system environments, as mobile agents tend to be

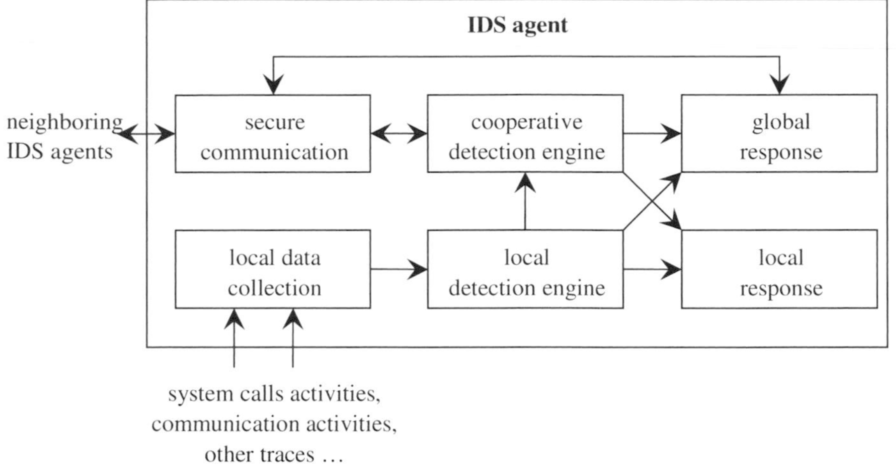

Figure 2. A Model for an IDS Agent [1]

independent of platform architectures. However, these systems would require a secure module where mobile agents can be stationed to. Additionally, mobile agents must be able to protect themselves from the secure modules on remote hosts as well.

Mobile-agent-based IDS can be considered as a distributed and cooperative intrusion detection technique as described in Section 3.2. Moreover, some techniques also use mobile agents combined with hierarchical IDS, for example, what will be described in Section 4.3.

4. SAMPLE INTRUSION DETECTION SYSTEMS FOR MANETS

Since the IDS for traditional wired systems are not well-suited to MANETs, many researchers have proposed several IDS especially for MANETs, which some of them will be reviewed in this section.

4.1. Distributed and Cooperative IDS

As described in Section 3.2, Zhang and Lee also proposed the model for a distributed and cooperative IDS as shown in Figure 2 [1].

The model for an IDS agent is structured into six modules. The *local data collection* module collects real-time audit data, which includes system and user activities within its radio range. This collected data will be analyzed by the *local detection engine* module for evidence of anomalies. If an anomaly is detected with strong evidence, the IDS agent can determine independently that the system is under attack and initiate a response through the *local response* module (i.e., alerting the local user) or the *global response* module (i.e., deciding on an action), depending on the type of intrusion, the type of network protocols and applications, and the certainty of the evidence. If an anomaly is detected with weak or inconclusive evidence, the IDS agent can request the cooperation of neighboring IDS agents through a *cooperative detection engine* module, which communicates to other agents through a *secure communication* module.

4.2. Local Intrusion Detection System (LIDS)

Albers *et al.* [3] proposed a distributed and collaborative architecture of IDS by using mobile agents. A Local Intrusion Detection System (LIDS) is implemented on every node for local concern, which can be extended for global concern by cooperating with other LIDS. Two types of data are exchanged among LIDS: security data (to obtain complementary information from collaborating nodes) and intrusion alerts (to inform others of locally detected intrusion). In order to analyze the possible intrusion, data must be obtained from what the LIDS detects, along with additional information from other nodes. Other LIDS might be run on different operating systems or use data from different activities such as system, application, or network activities; therefore, the format of this raw data might be different, which makes it hard for LIDS to analyze. However, such difficulties can be solved by using SNMP (Simple Network Management Protocol) data located in MIBs (Management Information Base) as an audit data source. Such a data source not only eliminates those difficulties, but also reduces the increase in using additional resources to collect audit data if an SNMP agent is already run on each node.

To obtain additional information from other nodes, the authors proposed mobile agents to be used to transport SNMP requests to other nodes. In another words, to distribute the intrusion detection tasks. The idea differs from traditional SNMP in that the traditional approach transfers data to the requesting node for computation while this approach brings the code to the data on the requested node. This is motivated by the unreliability of UDP messages used in SNMP and the dynamic topology of MANETs. As a result, the amount of exchanged data is tremendously reduced. Each mobile agent can be assigned a specific task which will be achieved in an autonomous and asynchronous fashion without any help from its LIDS.

The LIDS architecture is shown in Figure 3, which consists of

- **Communication Framework**: To facilitate for both internal and external communication with a LIDS.
- **Local LIDS Agent**: To be responsible for local intrusion detection and local response. Also, it reacts to intrusion alerts sent from other nodes to protect itself against this intrusion.
- **Local MIB Agent**: To provide a means of collecting MIB variables for either mobile agents or the Local LIDS Agent. Local MIB Agent acts as an interface with SNMP agent, if SNMP exists and runs on the node, or with a tailor-made agent developed specifically to allow updates and retrievals of the MIB variables used by intrusion detection, if none exists.

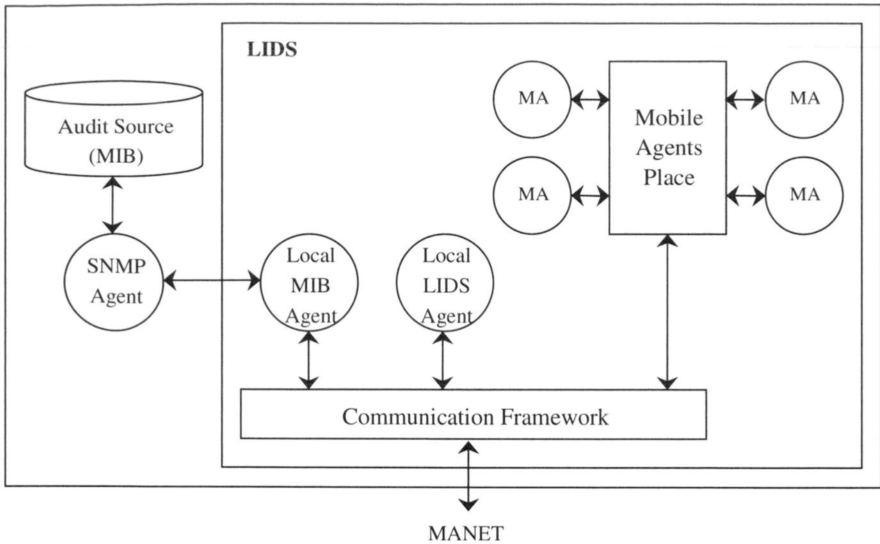

Figure 3. LIDS Architecture in A Mobile Node [3]

- **Mobile Agents (MA)**: They are distributed from its LID to collect and process data on other nodes. The results from their evaluation are then either sent back to their LIDS or sent to another node for further investigation.
- **Mobile Agents Place**: To provide a security control to mobile agents.

For the methodology of detection, Local IDS Agent can use either anomaly or misuse detection. However, the combination of two mechanisms will offer the better model. Once the local intrusion is detected, the LIDS initiates a response and informs the other nodes in the network. Upon receiving an alert, the LIDS can protect itself against the intrusion.

4.3. Distributed Intrusion Detection System Using Multiple Sensors

Kachirski and Guha [4] proposed a multi-sensor intrusion detection system based on mobile agent technology. The system can be divided into three main modules, each of which represents a mobile agent with certain functionality: monitoring, decision-making or initiating a response. By separating functional tasks into categories and assigning each task to a different agent, the workload is distributed which is suitable for the characteristics of MANETs. In addition, the hierarchical structure of agents is also developed in this intrusion detection system as shown in Figure 4.

- **Monitoring agent**: Two functions are carried out at this class of agent: network monitoring and host monitoring. A host-based monitor agent hosting system-level sensors and user-activity sensors is run on every node to monitor within

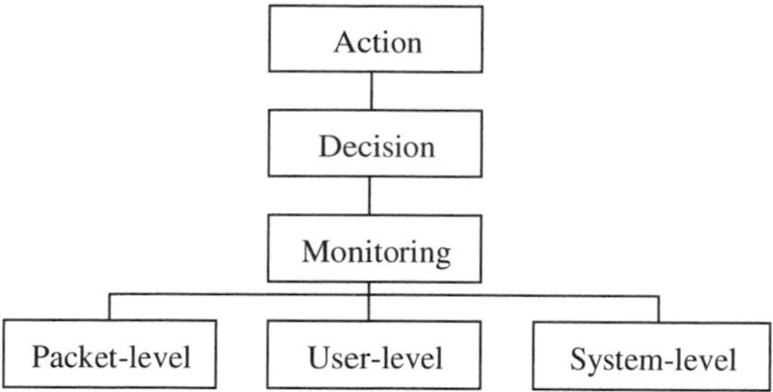

Figure 4. Layered Mobile Agent Architecture proposed by Kachirski and Guha [4]

the node, while a monitor agent with a network monitoring sensor is run only on some selected nodes to monitor at packet-level to capture packets going through the network within its radio ranges.

- **Action agent**: Every node also hosts this action agent. Since every node hosts a host-based monitoring agent, it can determine if there is any suspicious or unusual activities on the host node based on anomaly detection. When there is strong evidence supporting the anomaly detected, this action agent can initiate a response, such as terminating the process or blocking a user from the network.

- **Decision agent**: The decision agent is run only on certain nodes, mostly those nodes that run network monitoring agents. These nodes collect all packets within its radio range and analyze them to determine whether the network is under attack. Moreover, from the previous paragraph, if the local detection agent cannot make a decision on its own due to insufficient evidence, its local detection agent reports to this decision agent in order to investigate further. This is done by using packet-monitoring results that comes from the network-monitoring sensor that is running locally. If the decision agent concludes that the node is malicious, the action module of the agent running on that node as described above will carry out the response.

The network is logically divided into clusters with a single clusterhead for each cluster. This clusterhead will monitor the packets within the cluster and only packets whose originators are in the same cluster are captured and investigated. This means that the network monitoring agent (with network monitoring sensor) and the decision agent are run on the clusterhead.

In this mechanism, the decision agent performs the decision-making based on its own collected information from its network-monitoring sensor; thus, other nodes have no influence on its decision. This way, spoofing attacks and false accusations can be prevented.

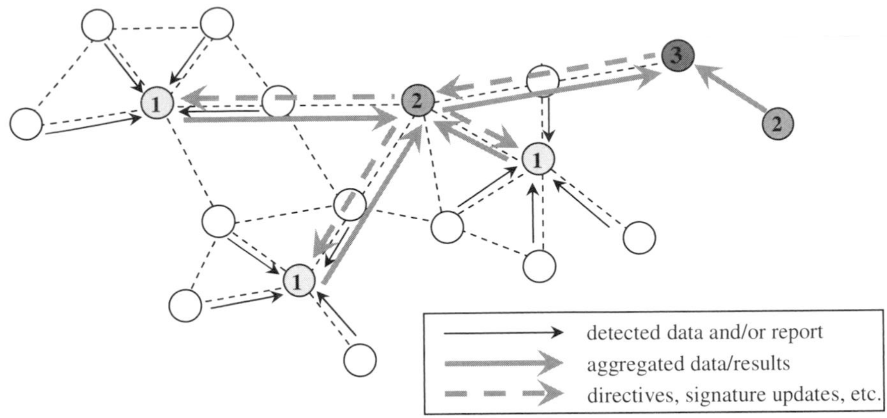

Figure 5. Dynamic Intrusion Detection Hierarchy [16]

4.4. Dynamic Hierarchical Intrusion Detection Architecture

Since nodes move arbitrarily across the network, a static hierarchy is not suitable for such dynamic network topology. Sterne *et al.* [16] proposed a dynamic intrusion detection hierarchy that is potentially scalable to large networks by using clustering like those in Section 4.3 and 5.5. However, it can be structured in more than two levels as shown in Figure 5. Nodes labeled "1" are the first level clusterheads while nodes labeled "2" are the second level clusterheads and so on. Members of the first level of the cluster are called leaf nodes.

Every node has the responsibilities of monitoring (by accumulating counts and statistics), logging, analyzing (i.e., attack signature matching or checking on packet headers and payloads), responding to intrusions detected if there is enough evidence, and alerting or reporting to clusterheads. Clusterheads, in addition, must also perform:

- **Data fusion/integration and data reduction**: Clusterheads aggregate and correlate reports from members of the cluster and data of their own. Data reduction may be involved to avoid conflicting data, bogus data and overlapping reports. Besides, clusterheads may send the requests to their children for additional information in order to correlate reports correctly.

- **Intrusion detection computations**: Since different attacks require different sets of detected data, data on a single node might not be able to detect the attack, e.g., DDoS attack, and thus clusterheads also analyze the consolidated data before passing to upper levels.

- **Security Management**: The uppermost levels of the hierarchy have the authority and responsibility for managing the detection and response capabilities of the clusters and clusterheads below them. They may send the signatures update, or directives and policies to alter the configurations for intrusion detection and response. These update and directives will flow from the top of the hierarchy to the bottom.

To form the hierarchical structure, every node uses clustering, which is typically used in MANETs to construct routes, to self-organize into local neighborhoods (first level clusters) and then select neighborhood representatives (clusterheads). These representatives then use clustering to organize themselves into the second level and select the representatives. This process continues until all nodes in the network are part of the hierarchy. The authors also suggested criteria on selecting clusterheads. Some of these criteria are:

- *Connectivity*: the number of nodes within one hop
- *Proximity*: members should be within one hop of its clusterhead
- *Resistance to compromise (hardening)*: the probability that the node will not be compromised. This is very important for the upper level clusterheads.
- *Processing power, storage capacity, energy remaining, bandwidth capabilities*

Additionally, this proposed architecture does not rely solely on promiscuous node monitoring like many proposed architectures, due to its unreliability as described in [5]. Therefore, this architecture also supports direct periodic reporting where packet counts and statistics are sent to monitoring nodes periodically.

4.5. Zone-Based Intrusion Detection System (ZBIDS)

Sun et al. [24] has proposed an anomaly-based two-level nonoverlapping Zone-Based Intrusion Detection System (ZBIDS). By dividing the network in Figure 6 into nonoverlapping zones (zone A to zone I), nodes can be categorized into two types: the intrazone node and the interzone node (or a gateway node). Considering only zone E, node 5, 9, 10 and 11 are intrazone nodes, while node 2, 3, 6, and 8 are interzone nodes which have physical connections to nodes in other zones. The formation and maintenance of zones requires each node to know its own physical location and to map its location to a zone map, which requires prior design setup.

Each node has an IDS agent run on it which the model of the agent is shown in Figure 7. Similar to an IDS agent proposed by Zhang and Lee (Figure 2), the *data collection module* and the *detection engine* are responsible for collecting local audit data (for instance, system call activities, and system log files) and analyzing collected data for any sign of intrusion respectively. In addition, there may be more than one for each of these modules which allows collecting data from various sources and using different detection techniques to improve the detection performance. The *local aggregation and correlation (LACE)* module is responsible for combining the results of these local detection engines and generating alerts if any abnormal behavior is detected. These alerts are broadcasted to other nodes within the same zone. However, for the *global*

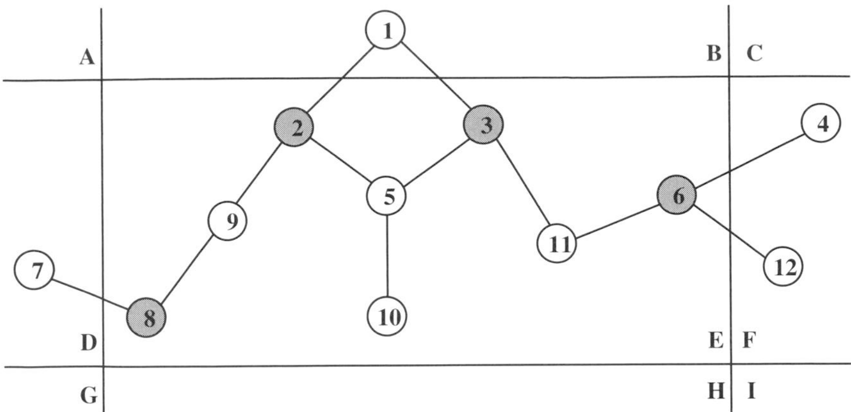

Figure 6. ZBIDS for MANETs [24]

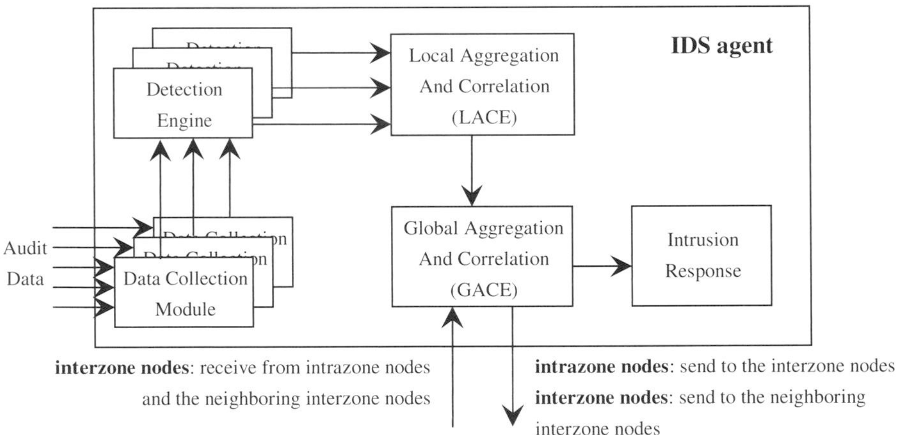

Figure 7. An IDS agent in ZBIDS [24]

aggregation and correlation (GACE), its functionality depends on the type of the node. As described in Figure 7, if the node is an intrazone node, it only sends the generated alerts to the interzone nodes. Whereas, if the node is an interzone node, it receives alerts from other intrazone nodes, aggregates and correlates those alerts with its own alerts, and then generates alarms. Moreover, the GACE also cooperates with the GACEs of the neighboring interzone nodes to have more accurate information to detect the intrusion. Lastly, the *intrusion response* module is responsible for handling the alarms generated from the GACE.

The local aggregation and correlation algorithm used in ZBIDS is based on a local Markov chain anomaly detection. An IDS agent first creates a normal profile by

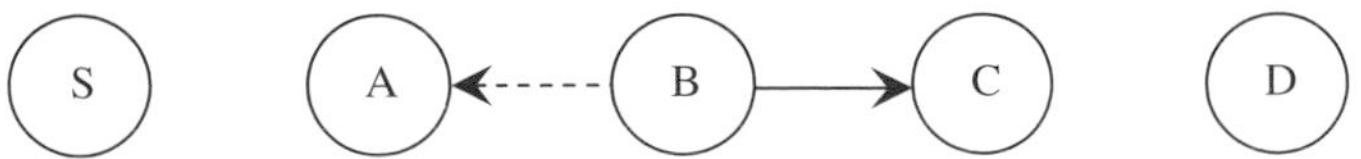

Figure 8. How watchdog works: Although node B intends to transmit a packet to node C, node A could overhear this transmission

constructing a Markov chain from the routing cache. A valid change in the routing cache can be characterized by the Markov chain detection model with probabilities, otherwise, it's considered abnormal, and the alert will be generated. For the global aggregation and correlation algorithm, it's based on information provided in the received alerts containing the type, the time, and the source of the attacks.

5. INTRUSION DETECTION TECHNIQUES FOR NODE COOPERATION IN MANETS

Since there is no infrastructure in mobile ad hoc networks, each node must rely on other nodes for cooperation in routing and forwarding packets to the destination. Intermediate nodes might agree to forward the packets but actually drop or modify them because they are misbehaving. The simulations in [5] show that only a few misbehaving nodes can degrade the performance of the entire system. There are several proposed techniques and protocols to detect such misbehavior in order to avoid those nodes, and some schemes also propose punishment as well [6, 7].

5.1. Watchdog and Pathrater

Two techniques were proposed by Marti, Giuli, and Baker [5], watchdog and pathrater, to be added on top of the standard routing protocol in ad hoc networks. The standard is Dynamic Source Routing protocol (DSR) [8]. A watchdog identifies the misbehaving nodes by eavesdropping on the transmission of the next hop. A pathrater then helps to find the routes that do not contain those nodes.

In DSR, the routing information is defined at the source node. This routing information is passed together with the message through intermediate nodes until it reaches the destination. Therefore, each intermediate node in the path should know who the next hop node is. In addition, listening to the next hop's transmission is possible because of the characteristic of wireless networks - if node A is within range of node B, A can overhear communication to and from B.

Figure 8 shows how the watchdog works. Assume that node S wants to send a packet to node D, which there exists a path from S to D through nodes A, B, and C. Consider now that A has already received a packet from S destined to D. The packet contains a message and routing information. When A forwards this packet to B, A also keeps a copy of the packet in its buffer. Then, it promiscuously listens to the transmission of B to make sure that B forwards to C. If the packet overheard from B (represented by a dashed line) matches that stored in the buffer, it means that B really forwards to the next hop (represented as a solid line). It then removes the packet from the buffer. However, if there's no matched packet after a certain time, the watchdog increments the failures counter for node B. If this counter exceeds the threshold, A concludes that B is misbehaving and reports to the source node S.

Pathrater performs the calculation of the"path metric" for each path. By keeping the rating of every node in the network that it knows, the path metric can be calculated by combining the node rating together with link reliability, which is collected from past experience. Obtaining the path metric for all available paths, the pathrater can choose the path with the highest metric. In addition, if there is no such link reliability information, the path metric enables the pathrater to select the shortest path too. As a result, paths containing misbehaving nodes will be avoided.

From the result of the simulation, the system with these two techniques is quite effective for choosing paths to avoid misbehaving nodes. However, those misbehaving nodes are not punished. In contrast, they even benefit from the network. In another word, they can use resources of the network - other nodes forward packets for them, while they forward packets for no one, which save their own resources. Therefore, misbehaving nodes are encouraged to continue their behaviors.

5.2. CONFIDANT

Buchegger and LeBoudec [6] proposed an extension to DSR protocol called CONFIDANT (Cooperation Of Nodes, Fairness In Dynamic Ad-hoc NeTworks), which is similar to Watchdog and Pathrater. Each node observes the behaviors of neighbor nodes within its radio range and learns from them. This system also solves the problem of Watchdog and Pathrater such that misbehavior nodes are punished by not including them in routing and not helping them on forwarding packets. Moreover, when a node experiences a misbehaving node, it will send a warning message to other nodes in the network, defined as friends, which is based on trusted relationship.

Figure 9 shows the components of the CONFIDANT protocol, which are the Monitor, the Trust Manager, the Reputation System, and the Path Manager. The process of how they work can be divided into two parts: the process to handle its own observations and the process to handle reports from trusted nodes.

- From observations: The monitor uses a "neighborhood watch" to detect any malicious behaviors with in its radio range, i.e., no forwarding, unusually frequent route update, etc. (This is similar to the watchdog in the previous scheme) If a suspicious event is detected, the monitor then reports to the reputation system. At this point, the reputation system performs several checks and updates the rating of the reported node in the reputation table. If the rating result is unacceptable, it passes the information to the path manager, which then removes all paths containing the misbehavior node. An ALARM message is also sent by the trust manager to warn other nodes that it considers as friends.
- From trusted nodes: When the monitor receives an ALARM message from its friends, the message will first be evaluated by the trust manager for the

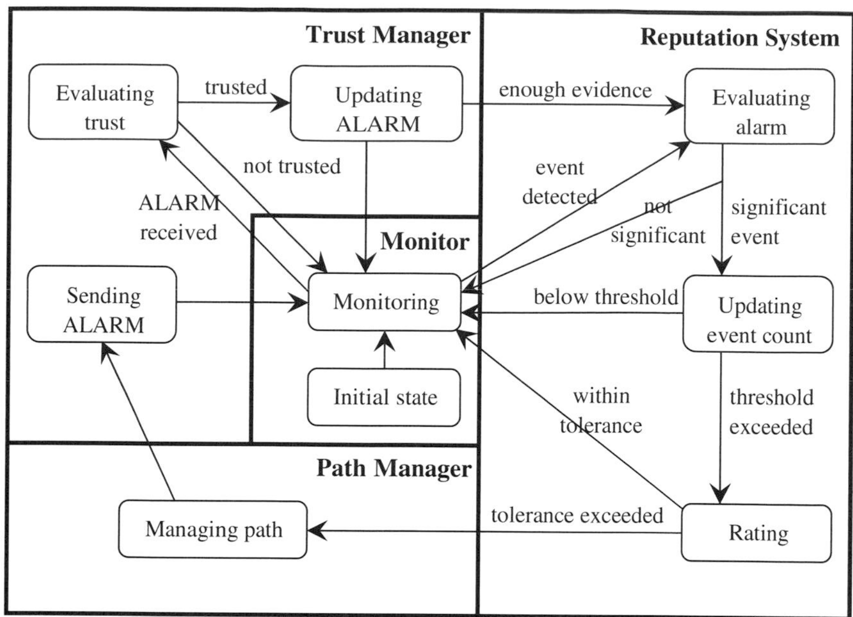

Figure 9. Components and State Diagram of CONFIDANT Protocol [6]

trustworthiness of the source node. If the message is trustworthy, this ALARM message, together with the level of trust, will be stored in the alarm table. All ALARM messages of the reported node will then be combined to see if there is enough evidence to identify that it is malicious. If so, the information will be sent to the reputation system, which then performs the same functions as described in the previous paragraph.

Since this protocol allows nodes in the network to send alarm messages to each other, it could give more opportunities for attackers to send false alarm messages that a node is misbehaving while it's actually not. This is one form of denial of service attacks.

5.3. CORE

Michiardi and Molva [7] presented a technique to detect a specific type of misbehaving nodes, which are selfish nodes, and also force them to cooperate. Similar to those in Section 5.1 and 5.2, this technique is based on a monitoring system and a reputation system, which includes both direct and indirect reputation from the system as will be described shortly.

As nodes sometimes do not intentionally misbehave, i.e., battery condition is low, these nodes should not be considered as misbehaving nodes and excluded from the network. To do this, the reputation should be rated based on past reputation, which is zero (neutral) at the beginning. In addition, participation in the network can be categorized into several functions such as routing discovery (in DSR) or forwarding packets. Each of these activities has different level of effects to the network; for example, forwarding packets has more effect on the performance of the system than that of routing discovery. Therefore, significance weight of functions should be used in the calculation of the reputation.

Like CONFIDANT, each node can receive a report from other nodes. However, the difference is CORE allows only positive reports to be passed while negative reports are passed in CONFIDANT. In another word, CORE prevents false accusation, thus, it also prevents a denial of service attack, which cannot be done in CONFIDANT. The negative rating is given to a node only from the direct observation when the node does not cooperate, which results in the decreased reputation for that node. The positive rating, in contrast, is given from both direct observation and positive reports from other nodes, which results in the increased reputation.

CORE can then be said to have two components, the watchdog system and the reputation system. The watchdog modules, one for each function, work the same way as in the previous two schemes above. For the reputation system, it maintains several reputation tables, one for each function and one for accumulated values for each node. Therefore, if there is a request from a bad reputation node (the overall reputation is negative), the node will be rejected and not be able to use the network.

5.4. OCEAN

Bansal and Baker [19] also proposed an extension on top of the DSR protocol called OCEAN (Observation-based Cooperation Enforcement in Ad hoc Networks). OCEAN also uses a monitoring system and a reputation system. However, in contrast to the previous approaches above, OCEAN relies only on its own observation to avoid the new vulnerability of false accusation from second-hand reputation exchanges. Therefore, OCEAN can be considered as a stand-alone architecture.

OCEAN categorizes routing misbehavior into two types: misleading and selfish. If a node has participated in the route discovery but not packet forwarding, this is considered to be misleading as it misleads other nodes to route packets through it. But if a node does not even participate in the route discovery, it is considered to be selfish.

In order to detect and mitigate the misleading routing behaviors, after a node forwards a packet to a neighbor, it buffers the packet checksum and monitors if the neighbor attempts to forward the packet within a given time. Then, a negative or positive event is given as the result of the monitoring to update the neighbor rating. If the rating falls below the faulty threshold, that neighbor node is added to a faulty list which will be added in the RREQ as an avoid-list. In addition, all traffic from the faulty neighbor node will be rejected. Nonetheless, the faulty timeout is used to allow the faulty node to join back to the network in case that it might be false accused or it behaves better.

Each node also has a mechanism of maintaining chipcounts for each neighbor to mitigate the selfish behavior. A neighbor node earns chips when forwarding a packet on behalf of the node and loses ships when asking the node to forward a packet. If the chipcount of the neighbor is below the threshold, packets coming from that neighbor will be denied.

5.5. Cooperative Intrusion Detection System

A cluster-based cooperative intrusion detection system, similar to Kachirski and Guha's system [4], has been presented by Huang and Lee [14]. In this approach, an IDS is not only able to detect an intrusion, but also to identify the attack type and the attacker, whenever possible, through statistical anomaly detection. Various types of statistics (or features), which are proposed in their previous work [15], are evaluated from a sampling period by capturing the basic view of network topology and routing operations, as well as traffic patterns and statistics, in the normal traffic. Hence, attacks could be identified if the statistics deviate from the pre-computed ones (anomaly detection).

Statistics can be categorized into two categories, non traffic-related and traffic-related. Non traffic-related statistics are calculated based on the mobility and the trace log files, which can be done separately on each node. Some of these statistics are route add count, route removal count, total route change, average route length, etc. Traffic-related statistics are involved in routing and packet forwarding and can be calculated by counting packets going in and out, e.g. the number of packet received, the number of packet forwarded, the number of route reply messages, etc. These statistics can be captured by the node itself or the neighboring nodes who overhear the transmission.

Several identification rules are pre-defined for known attacks by using relationships of the mentioned statistics. Once an anomaly is detected, the IDS will perform further investigation to determine the detailed information of the attack from a set of these identification rules. These rules enhance the system to identify the type of the attack and, in some cases, the attacking node. Some notations of statistics are presented as follows. Let M represent the monitoring node and m represent the monitored node.

- $\#(*, m)$: the number of incoming packets on the monitored node m.
- $\#(*, [m])$: the number of incoming packets of which the monitored node m is the destination.
- $\#(m, *)$: the number of outgoing packets from the monitored node m.
- $\#([m], *)$: the number of outgoing packets of which the monitored node m is the source.
- $\#(m, n)$: the number of outgoing packets from m of which n is the next hop.
- $\#([s], M, m)$: the number of packets that are originated from s and transmitted from M to m.
- $\#([s], [d])$: the number of packets received on m which is originated from s and destined to d.

- $\#(*, m)(TYPE = RREQ)$: the number of incoming RREQ packets on m.

These statistics are computed over a long period L. Let $FEATURE^L$ represents the aggregated $FEATURE$ over time L. Some identification rules are defined for well known attacks as follows.

- ***Unconditional Packet Dropping***:This rule uses Forward Percentage (FP) over a period L to define the attack.

$$FP_m = \frac{\text{packets actually forwarded}}{\text{packets to be forwarded}} = \frac{\#^L(m, M) \ - \ \#^L([m], M)}{\#^L(M, m) \ - \ \#^L(M, [m])}$$

If there are packets to be forwarded (denominator is not zero) and $FP_m = 0$, the unconditional packet dropping attack is detected and the attacker is m.

- ***Random Packet Dropping***: This rule also uses the same FP as unconditional packet dropping. However, the threshold ϵ_{FP} is defined ($\epsilon_{FP} < 1$). If $0 < FP_m < \epsilon_{FP}$, m is defined as an attacker using random packet dropping.

- ***Selective Packet Dropping***: This rule uses Local Forward Percentage (LFP) for each source s.

$$LFP_m^s = \frac{\text{packets from source } s \text{ actually forwarded}}{\text{packets from source } s \text{ to be forwarded}} = \frac{\#^L([s], m, M)}{\#^L([s], M, m) \ - \ \#^L([s], M, [m])}$$

If the denominator is not zero and $LFP_m^s = 0$, the attack is the unconditional packet dropping targeted at s. However, if LFP_m^s is less than the threshold ($\epsilon_{LFP} < 1$), the attack is detected as random packet dropping targeted at s.

- ***Blackhole***: This rule uses Global Forward Percentage (GFP) and it must be computed on M locally because the rule relies on information available only on the node. Let $N(M)$ denote M's 1-hop neighbors.

$$GFP_m^s = \frac{\text{packets to be forwarded}}{\text{packets from} N(M) \text{destined to other nodes than itself or another} N(M)} = \frac{\#^L(*, M) \ - \ \#^L(*, [M])}{\sum\limits_{i \epsilon N(M)} \#^L(i, M) \ - \ \sum\limits_{i,j \epsilon N(M)} \#^L(i, [j]) \ - \ \#^L(*, [M])}$$

If the denominator is not zero and $GFP = 1$, it means that the blackhole attack is detected and M is the attacker.

- ***Malicious Flooding on specific target***: This rule uses $\#^L([m], [d])$ for every destination d. If it is larger than the threshold the attack is Malicious Flooding. However, the attacker cannot be determined.

Table 1. Comparison among IDS for Node Cooperation

Techniques		Watchdog/ Pathrater	CONFIDANT	CORE	OCEAN	Cooperative IDS
Architecture		Distributed and cooperative			Stand-alone	Hierarchical
Type of data collection		Reputation				Statistics
Data distribution		negative to source node	negative to friends	positive from RREP	no	to clusterhead
Observation	self to neighbor	yes	yes	yes	yes	yes
	neighbor to neighbor	no	yes	no	yes	yes
Misbehavior detection	Selfish – routing	no	yes	yes	yes	yes
	Selfish – packet forwarding	yes	yes	yes	yes	yes
	Malicious – routing	no	yes	no	no	yes
	Malicious – packet forwarding	yes	yes	no	no	yes
Punishment		no	yes	yes	yes	n/a
Avoid misbehaving node in route discovery		no	no	no	yes	n/a

The authors also presented cluster formation algorithms and ensured that they are fair and secure. Each and every node has an equal chance of becoming a clusterhead and serves as a clusterhead for an equal service time. In addition, no node can manipulate the clusterhead selection process. Initially, each node forms a clique - a group of nodes where every pair of members can communicate via a direct wireless link. Then, members in the clique perform the selection of a clusterhead. The process of re-election, to enforce fairness, and the process of recovery from lost clusterheads are defined as well.

Monitoring is how data is obtained in order to analyze for possible intrusions, however it consumes power. Therefore, instead of every node capturing all features themselves, the clusterhead is solely responsible for computing traffic-related statistics. This can be done because the clusterhead overhears incoming and outgoing traffic on all members of the cluster as it is one hop away (a clique). As a result, the energy consumption of member nodes is lessened, whereas the detection accuracy is just a little worse than that of not implementing clusters. Besides, the performance of the overall network is noticeably better - decreases in CPU usage and network overhead.

5.6. Summary of IDS for Detecting Misbehaving Nodes

Although the watchdog is used in all of the above IDS, the authors in [5] have pointed out that there are several limitations. The watchdog cannot work properly in the presence of collisions, which could lead to false accusations. Moreover, when each node has different transmission ranges or implements directional antennas, the watchdog could not monitor the neighborhood accurately.

All of the above IDS's presented are common in detecting selfish nodes. However, CORE doesn't detect malicious misbehaviors while the others detect some of them, i.e., unusually frequent route update, modifying header or payload of packets, no report of failed attempts, etc. Table 1 shows the comparison among these IDS.

6. CONCLUSIONS AND FUTURE DIRECTIONS

As the use of mobile ad hoc networks (MANETs) has increased, the security in MANETs has also become more important accordingly. Historical events show that prevention alone, i.e., cryptography and authentication are not enough; therefore, the intrusion detection systems are brought into consideration. Since most of the current techniques were originally designed for wired networks, many researchers are engaged in improving old techniques or finding and developing new techniques that are suitable for MANETs.

With the nature of mobile ad hoc networks, almost all of the intrusion detection systems (IDSs) are structured to be distributed and have a cooperative architecture. The number of new attacks is likely to increase quickly and those attacks should be detected before they can do any harm to the systems or data. Hence, IDS's in MANETs prefer using anomaly detection to misuse detection [1, 3, 4, 14, 24]. Some techniques are proposed to implement on top of the existing protocols [5, 6, 7], others are proposed as independent modules to be added on mobile nodes [1, 3, 4, 14, 16, 24].

An intrusion detection system aims to detect attacks on mobile nodes or intrusions into the networks. However, attackers may try to attack the IDS system itself [5]. Accordingly, the study of the defense to such attacks should be explored as well.

Many researchers are currently occupied in applying game theory for cooperation of nodes in MANETs [20, 21, 22, 23] as nodes in the network represent some characteristics similar to social behavior of human in a community. That is, a node tries to maximize its benefit by choosing whether to cooperate in the network. There is not much work done in this area, therefore, it is an interesting topic for future research.

ACKNOWLEDGEMENTS

This work was supported in part by NSF grants CCR 0329741, CNS 0422762, CNS 0434533, ANI 0073736, EIA 0130806, and a grant from Motorola Inc.

7. REFERENCES

1. Y. Zhang, W. Lee, and Y. Huang, "Intrusion Detection Techniques for Mobile Wireless Networks," *ACM/Kluwer Wireless Networks Journal (ACM WINET)*, Vol. 9, No. 5, September 2003.

2. A. Mishra, K. Nadkarni, and A. Patcha, "Intrusion Detection in Wireless Ad Hoc Networks," *IEEE Wireless Communications*, Vol. 11, Issue 1, pp. 48-60, February 2004.

3. P. Albers, O. Camp, J. Percher, B. Jouga, L. Mé, and R. Puttini, "Security in Ad Hoc Networks: a General Intrusion Detection Architecture Enhancing Trust Based Approaches," *Proceedings of the 1st International Workshop on Wireless Information Systems (WIS-2002)*, pp. 1-12, April 2002.

4. O. Kachirski and R. Guha, "Effective Intrusion Detection Using Multiple Sensors in Wireless Ad Hoc Networks," *Proceedings of the 36th Annual Hawaii International Conference on System Sciences (HICSS'03)*, p. 57.1, January 2003.

5. S. Marti, T. J. Giuli, K. Lai, and M. Baker, "Mitigating Routing Misbehavior in Mobile Ad Hoc Networks," *Proceedings of the 6th Annual International Conference on Mobile Computing and Networking (MobiCom'00)*, pp. 255-265, August 2000.

6. S. Buchegger and J. Le Boudec, "Performance Analysis of the CONFIDANT Protocol (Cooperation Of Nodes - Fairness In Dynamic Ad-hoc NeTworks)," *Proceedings of the 3rd ACM International Symposium on Mobile Ad Hoc Networking and Computing (MobiHoc'02)*, pp. 226-336, June 2002.

7. P. Michiardi and R. Molva, "Core: A Collaborative Reputation mechanism to enforce node cooperation in Mobile Ad Hoc Networks," *Communication and Multimedia Security Conference (CMS'02)*, September 2002.

8. D. B. Johnson, and D. A. Maltz, "The Dynamic Source Routing Protocol for Mobile Ad Hoc Networks (Internet-Draft)," *Mobile Ad-hoc Network (MANET) Working Group, IETF*, October 1999.

9. P. Brutch and C. Ko, "Challenges in Intrusion Detection for Wireless Ad-hoc Networks," *Proceedings of 2003 Symposium on Applications and the Internet Workshop*, pp. 368-373, January 2003.

10. M. G. Zapata, "Secure Ad Hoc On-Demand Distance Vector (SAODV) Routing," *ACM Mobile Computing and Communication Review (MC2R)*, Vol. 6, No. 3, pp. 106-107, July 2002.

11. Y. Hu, D. B. Johnson, and A. Perrig, "SEAD: Secure Efficient Distance Vector Routing for Mobile Wireless Ad Hoc Networks," *Proceedings of the 4th IEEE Workshop on Mobile Computing Systems and Applications (WMCSA'02)*, pp. 3-13, June 2002.

12. Y. Hu, A. Perrig, and D. B. Johnson, "Ariadne: A secure On-Demand Routing Protocol for Ad hoc Networks," *Proceedings of the 8th Annual International Conference on Mobile Computing and Networking (MobiCom'02)*, pp. 12-23, September 2002.

13. A. Perrig, R. Canetti, D. Tygar and D. Song, "The TESLA Broadcast Authentication Protocol," *RSA CryptoBytes*, 5 (Summer), 2002.

14. Y. Huang and W. Lee, "A Cooperative Intrusion Detection System for Ad Hoc Networks," *Proceedings of the ACM Workshop on Security in Ad Hoc and Sensor Networks (SASN'03)*, pp. 135-147, October 2003.

15. Y. Huang, W. Fan, W. Lee, and P. Yu, "Cross-Feature Analysis for Detecting Ad-Hoc Routing Anomalies," *Proceedings of the 23rd IEEE International Conference on Distributed Computing Systems (ICDCS'03)*, May 2003.

16. D. Sterne, P. Balasubramanyam, D. Carman, B. Wilson, R. Talpade, C. Ko, R. Balupari, C.-Y. Tseng, T. Bowen, K. Levitt, and J. Rowe, "A General Cooperative Intrusion Detection Architecture for MANETs," *Proceedings of the 3rd IEEE International Workshop on Information Assurance (IWIA'05)*, pp. 57-70, March 2005.

17. Y. F. Jou, F. Gong, C. Sargor, X. Wu, S. Wu, H. Chang, and F. Wang, "Design and Implementation of a Scalable Intrusion Detection System for the Protection of Networks Infrastructure," *Proceedings of DARPA Information Survivability Conference and Exposition*, Vol. 2, pp. 69-83, January 2000.

18. E. Y. K. Chan *et al.*, "IDR: An Intrusion Detection Router for Defending against Distributed Denial-of-Service (DDoS) Attacks," *Proceedings of the 7th International Symposium on Parallel Architectures, Algorithms and Networks (ISPAN'04)*, pp. 581-586, May 2004.

19. S. Bansal and M. Baker, "Observation-Based Cooperation Enforcement in Ad hoc Networks," *Research Report cs.NI/0307012*, Stanford University, 2003.

20. P. Michiardi and R. Molva, "A Game Theoretical Approach to Evaluate Cooperation Enforcement Mechanisms in Mobile Ad Hoc Networks," *Modeling and Optimization in Mobile, Ad Hoc and Wireless Networks (WiOpt'03)*, March 2003.

21. A.Agah, S. K. Das, K. Basu, and M. Asadi, "Intrusion Detection in Sensor Networks: A Non-Cooperative Game Approach," *Proceedings of the 3rd IEEE International Symposium on Network Computing and Applications (NCA'04)*, pp. 343-346, 2004.

22. R. Mahajan, M. Rodrig, D. Wetherall and J. Zahorjan, "Experiences Applying Game Theory to System Design," *Proceedings of the ACM SIGCOMM Workshop on Practice and Theory of Incentives in Networked Systems (PIN'04)*, pp. 183-190, September 2004.

23. S. Zhong, L. Li, Y. G. Liu and Y. Yang, "On Designing Incentive-Compatible Routing and Forwarding Protocols in Wireless Ad-hoc Networks: An Integrated Approach Using Game Theoretical and Cryptographic Techniques," *Proceedings of the 11th Annual International Conference on Mobile Computing and Networking (MobiCom'05)*, pp. 117-131, 2005.

24. B. Sun, K. Wu, and U. W. Pooch, "Alert Aggregation in Mobile Ad Hoc Networks," *Proceedings of the 2003 ACM Workshop on Wireless Security (WiSe'03) in conjuction with the 9th Annual International Conference on Mobile Computing and Networking (MobiCom'03)*, pp. 69-78, 2003.

Part II

SECURITY IN MOBILE CELLULAR NETWORKS

8

INTRUSION DETECTION IN CELLULAR MOBILE NETWORKS

Bo Sun
Computer Science Department
Lamar University
Beaumont, TX 77710, USA
E-mail: bsun@cs.lamar.edu

Yang Xiao
Computer Science Department
University of Alabama
101 Houser Hall
Box 870290
Tuscaloosa, AL 35487-0290 USA
E-mail: yangxiao@ieee.org

Kui Wu
Computer Science Department
University of Victoria
BC, Canada V8W 3P6
E-mail: wkui@cs.uvic.ca

Security concerns have attracted a great deal of attentions for both service providers and end users in cellular mobile networks. As a second line of defense, Intrusion Detection Systems (IDSs) are indispensable for highly secure wireless networks. In this chapter, we first give a brief introduction to wired IDSs and wireless IDSs. Then we address the main challenges in designing IDSs for cellular mobile networks, including the topics of feature selection, detection techniques, and adaptability of IDSs. An anomaly-based IDS exploiting mobile users' location history is introduced to provide insights into the intricacy of building a concrete IDS for cellular mobile networks.

1. INTRODUCTION

The rapid development of cellular mobile services makes people rely heavily on cellular phones in their daily lives for important and sensitive tasks. While providing a great convenience, these booming new services have brought serious security concerns. The lack of security has become one of the main obstacles in preventing wireless communications carriers from providing business such as E-Banking or E-Shopping over wireless networks on a large scale basis. Although there are many security mechanisms in cellular mobile networks [36], the number of security incidents continues to increase. How to design a highly secure cellular mobile network is still a very challenging issue due to the open radio transmission environment and the physical vulnerability of mobile devices.

Generally, two complementary classes of approaches exist to protect the cellular mobile networks: *prevention*-based approaches and *detection*-based approaches. Prevention-based techniques, such as authentication and encryption, can effectively reduce attacks by ensuring that users conform to predefined security policies. They can keep most illegitimate users from entering the system. However, security research indicates that there are always some weak points in the system that is hard to predict, especially for a wireless network, in which open wireless transmission medium and low physical security protection of mobile devices pose additional challenges for prevention-based approaches. For example, although numerous security measures are taken into account in the design of second-generation and third-generation digital cellular systems, security flaws have been reported in literature [1] [2]. Security research indicates the necessity of multi-layer and multi-level protection because there are always some weak points in the system that attackers can exploit to break into the system. Currently, tamper-resistant hardware and software are still expensive or unrealistic for mobile devices. Therefore, if a device is compromised, all the secrets associated with the device become open to the attackers, rendering all prevention-based techniques helpless and resulting in great damage to service providers. For example, one of the basic threats is the illegitimate use of services, which leads to the serious problem of improper billing and masquerading. To solve these problems, Intrusion Detection Systems (IDSs), serving as the second wall of protection, could effectively help identifying malicious activities.

Although IDSs have been widely used in wired networks, not many research efforts have been dedicated to IDSs in cellular mobile networks. The communications paradigm in cellular mobile networks and traditional wired networks are fundamentally different. This makes attack scenarios in cellular mobile networks more complex. Moreover, it is challenging to model the normal and abnormal user behaviors because of the potential wide variety of users' activities. Feature selection, detection techniques, and adaptability of IDSs in the context of cellular mobile networks are still open research problems.

In this chapter, we provide a general introduction to IDSs in cellular mobile networks. Section 2 presents necessary background knowledge of cellular mobile networks. Section 3 focuses on the introduction of Intrusion Detection Systems, including IDSs for wired networks and IDSs for cellular mobile networks, respectively. Section 4 addresses one important and challenging topic regarding IDSs - Feature Selection. In Section 5, we discuss the adaptability issue of IDSs. Secion 6 presents the details of constructing a mobility-based anomaly detection system for cellular mobile networks. In Section 7, we conclude the chapter.

2. CELLULAR MOBILE NETWORKS

A mobile wireless network with a cellular infrastructure is illustrated in Figure 1. A typical network consists of a wired backbone and a number of Base Stations (BSs). Each BS controls a cell, and a group of BSs are managed by a Mobile Switching Center (MSC). When a mobile user moves into a different Location Area (LA), a location registration process happens. In cellular mobile networks, the Home Location Register (HLR) is a database used for storage and management of subscriptions. Usually, the HLR stores the permanent data about subscribers, while the Visitor Location Register (VLR) stores temporary information to serve visiting subscribers.

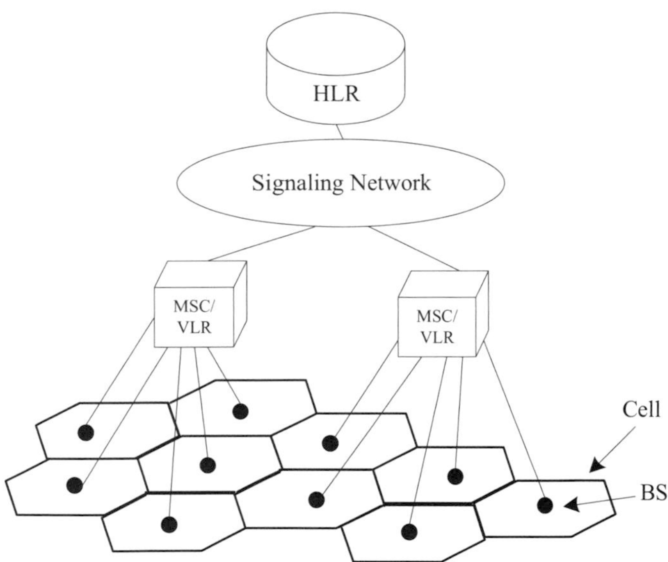

Figure 1. An Example of Cellular Mobile Network.

A mobile station communicates with another mobile station via a BS. To do so, the source mobile station needs to make a request through the BS of its current cell. If the request is granted by the MSC, a pair of voice channels is assigned. In cellular mobile networks, location updates often happen when the user traverses the border of an LA. When the user is inside an LA and is not making a phone call, the Mobile Switching Center, which is responsible for location and paging management, is not updated with the user's latest location information.

3. INTRUSION DETECTION SYSTEMS

Intrusions can be defined as any set of actions that compromise the confidentiality, availability, and integrity of the system. Intrusion detection is a security technology that attempts to identify individuals who are trying to break into and misuse a system without authorization and those who have legitimate access to the system but are abusing their privileges [3], [4]. An Intrusion Detection System (IDS) is a computer system that dynamically monitors the system and users' actions in order to detect intrusions. Because an information system can suffer from various kinds of security vulnerabilities, it is both technically difficult and economically costly to build and maintain a system that is not susceptible to attacks. IDSs, by analyzing the system and user operations in search of activity undesirable and suspicious, can effectively monitor and protect against threats.

Research on IDSs began with a report by Anderson [5] followed by Denning¡¯s seminal paper [6], which lays the foundation for most of the current intrusion detection prototypes. Since then, many research efforts have been devoted to wired IDSs. Numerous detection techniques and architectures for host machines and wired networks have been proposed. A good taxonomy of wired IDSs is presented in [4].

With the rapid proliferation of wireless networks and mobile computing applications, new vulnerabilities that do not exist in wired networks have appeared. Security poses a serious challenge in deploying wireless networks in reality. Moreover, the vast differences between wired and wireless networks make traditional intrusion detection techniques inapplicable. Wireless IDSs, emerging as a new research topic, aim at developing new architecture and mechanisms to protect wireless networks.

3.1. Intrusion Detection for Wired Networks

Focusing mainly on network traffic and computer audit data, there are two general approaches in wired IDSs to detect intrusions: misuse-based intrusion detection (also referred to as knowledge-based detection, or detection by appearance) and anomaly-based intrusion detection (also referred to as behavior-based detection or detection by behavior). They are complementary to each other for intrusion detection.

Misuse-Based Intrusion Detection Systems

Based on a database of known attack signatures and system vulnerabilities, misuse-based IDSs try to identify activities matching a signature that is stored in the database. An alarm is triggered whenever a match is found. The main advantage of misuse-based IDSs is that the false alarm rate is very low. The triggered alarms are meaningful because the attack signatures contain the diagnostic information about the cause of the alarm. The main disadvantage of misuse-based IDSs is that the attack signatures may not cover all attacks because new attacks are hard to predict. As such, the databases containing the attack signatures and system vulnerabilities need to be kept up-to-date. This is a tedious task because new attacks and system vulnerabilities are detected on a daily basis. Careful analysis of the vulnerabilities is also time-consuming. Misuse-based IDSs also face the generalization issues because most of the attack knowledge is focused on the different versions of operating systems and applications.

There are several approaches in misuse-based detection. They differ in the representation as well as the matching algorithm employed to detect intrusion patterns. Below are the mainly used approaches:

- *Expert System*: Expert systems provide strategies and mechanisms for processing facts regarding the state of a given environment, and derive logical inferences from these facts. Audit events and security policies are mapped to facts that are recorded and evaluated by the system. During the process of mapping, a semantic meaning is attached to increase the abstraction level of the audit data. The expert system contains a set of rules that describe the attacks. These rules are triggered when certain activities that meet their conditions happen. The execution speed of the expert system shell is usually poor because all of the audit data need to import into the shell as facts. Therefore, expert system based IDSs only exist in research prototypes, as performance is more important in commercial products.

  Event Monitoring Enabling Responses to Anomalous Live Disturbances (EMERALD) [7] is an extension of the Intrusion Detection Expert System (IDES) [8],[9] and Next Generation Intrusion Detection System (NIDES) [10] by SRI International. EMERALD uses a rule-based expert system component for misuse-based detection. A forward-chaining rule-based expert system development toolset called the Production Based Expert System Toolset (P-BEST) [11] is utilized to develop a modern generic signature-analysis engine. A chain of rules is established utilizing P-BEST to form the signature database.

- *Pattern Recognition*: In this approach, known intrusion signatures are encoded as patterns (e.g., strings, a sequence of events, etc.) and matched against audit data. An alarm is generated if a match can be found. This method allows a very efficient implementation. Therefore, they are commonly used in commercial tools, such as RealSecure of Internet Security Systems [12].

- *Colored Petri Nets*: In this method, signatures of known intrusions are modeled as a number of different states, which form Colored Petri Nets (CPNs). Compared with other approaches, CPNs have more generalities to represent signatures. This makes it easy to write complex intrusion scenarios. However, it is very computationally expensive to manifest misbehaviors in the audit trail. Intrusion Detection In Our Time (IDIOT) is the one example that uses CPNs [13].

- *State Transition Analysis*: In this approach, to represent an intrusion scenario, a sequence of actions is constructed starting from the initial state to the target compromised state. *State Transition Diagrams* identify the steps and the requirements of the penetration. The states that make up the intrusion form a simple chain that has to be traversed from the beginning to the end. It was a technique proposed by Porras and Kemmerer [14], which was implemented in Ustat - a real-time intrusion detection system for UNIX [15].

Anomaly-Based Intrusion Detection Systems

Anomaly-based IDSs assume that an intrusion can be detected by observing a deviation from normal or expected behaviors of systems or users. Normalcy is defined by the previously observed subject behavior, which is usually created during a training phase. The normal profile is later compared with the current activity. If a significant deviation is observed, IDSs flag the unusual activity and generate an alarm. The main advantage of anomaly-based IDSs is that they can detect attempts that try to exploit new and unforeseen vulnerabilities. They are also less system-dependent. Disadvantages include that they may have very high false alarm rate and are more difficult to configure because comprehensive knowledge of expected system behaviors is required. In order to build the up-to-date normal profiles, they also usually require a periodic online learning process. Anomaly-based detection techniques are harder to implement, making them inappropriate for commercial use.

Several anomaly-based detection techniques exist. They are different in the way of representing a normal profile and the method of inferring the difference between the normal profile and the observed activities. Below are the mainly used approaches:

- *Statistics*: Statistics-based anomaly detection techniques build a statistical profile (e.g., statistical distribution) of normal activities from historic data by measuring a number of variables over time. Examples of the variables are the login/logoff times, the time duration of one session, the number of packets transmitted in this session, and so on.

  In EMERALD [7], the statistical algorithms employ four classes of measures to track subject activities: categorical, continuous, intensity, and event distribution. The profile is subdivided into short and long-term elements. A short-term profile may characterize recent activities of the system, while a long-term profile is slowly adapted to the changes of system activities. Because of the popularity of the Internet, many traffic perspectives are used to profile TCP/IP streams [7]. For example, all ICMP exchanges can be parsed to analyze ICMP-specific transactions. The application-layer sessions from specific internal hosts to specific external hosts can also be analyzed for specific applications.

- *Neural Networks*: The use of neural networks in IDSs consists of three steps: learning the normal pattern of the system by collecting training data, training the neural networks to identify the subject, and applying the output of the neural networks to the observed activity to identify intrusions. Neural networks are computationally intensive. Therefore, they are not widely used in IDSs. Hyperview [16] is an example IDS that utilizes neural networks.

There are some other anomaly-based detection techniques. Detection techniques based on immunology [17] first capture a large set of event sequences from historic data to construct the normal profile. They then use either negative selection or positive selection algorithms to detect the difference of incoming event sequences from event sequences in the normal profile [18]. Expert systems can also be used to implement anomaly-based techniques [9]. To describe normal behaviors, these expert systems can study the activities of the target system to form a set of rules. Lee *et al.* proposed to use data mining approach to construct intrusion detection models [19]. Anomaly-based detection techniques utilizing *Chi-square Test* are introduced in [20] and [21]. There are also anomaly-based detection techniques that use a first-order or high-order Markov model of event transitions to represent a normal profile [22],[23],[24],[25]. In [22], utilizing a Markov Chain model, Jha *et al.* proposed a general framework to construct anomaly detectors.

Besides misuse-based detection and anomaly-based detection, there is a new class of detection algorithms: specification-based techniques [27]. They combine the advantages of both misuse-based detection and anomaly-based detection techniques. These approaches are based on manually developed specifications, thus avoiding the high rate of false alarms. IDSs detect deviations of observed program behaviors from these specifications, rather than detect the occurrence of specific attack patterns. Thus, attacks can be detected even though they have not previously been encountered.

3.2. Intrusion Detection for Cellular Mobile Networks

Most of the proposed work in the areas of wireless IDSs explores the regularity of users' behaviors (for example, mobility patterns, calling activities) to construct normal profiles. Regularity is one of the basic assumptions to develop realistic IDSs. For example, in terms of mobility patterns, a mobile user usually travels with a specific destination in mind and tends to follow the shortest path to it. A user's mobility pattern is a reflection of his/her daily routines and most mobile users have favorite routes and habitual movement patterns. In terms of calling activities, most mobile users have his/her regular calling activities. For example, because of the regular working rhythms like daily or weekly business telephone conference, most users demonstrate certain calling patterns. Although an attacker can compromise all the secrets associated with a mobile device, he/she could not follow the movement pattern of the authentic owner and mimic the authentic user's profile. By establishing an accurate normal profile that can reflect the normal pattern and comparing it with the current observed pattern, misbehaviors can be effectively identified.

Relatively few research efforts have been devoted to Intrusion Detection for Cellular Mobile Networks. Büschkes *et al.* [28] applied the Bayes decision rule to user's mobility patterns to increase the security in mobile networks. Through proper behavior predictions, they applied anomaly-based detection techniques to profile mobile users. Samfat *et al.* [29] proposed IDAMN (Intrusion Detection Architecture for Mobile Networks) that included two algorithms to model the behavior of users in terms of both telephony activity and migration patterns. IDAMN can perform intrusion detection in the visited location and within the duration of a typical call. Y. -B Lin [1] presented an excellent study to detect the potential fraudulent usage of cloned phones in cellular mobile networks. They showed how quickly the fraudulent usage can be detected under GSM/UMTS call setup procedures and how to reduce the possibility of fraudulent usage. Exploring mobility patterns of public transportation users, Hall *et al.* [30] utilized an Instance based Learning technique to classify different users' behaviors. There are also some research efforts dedicated to fraud detection systems in cellular mobile networks. Hollmén [31] presented fraud detection techniques in mobile communications networks by means of user profiling and classification. Call data is used to describe behavioral patterns of mobile users. Neural networks and probabilistic models were employed to learn their usage patterns. Based on these models, abrupt changes from established usage patterns can be detected.

It is worth mentioning that some of the above mentioned schemes require the tracking of uses' locations. This will cause location privacy issues because of the potential exposure of users' whereabouts. Fortunately, there is some work in the literature that are aimed to address the privacy issues. For example, He *et al.* [34] proposed to use *blind signature* to generate an authorized-anonymous-ID for the server to authorize the mobile device. Location-based IDSs should be properly integrated with these privacy-enhanced schemes in order to be readily deployed.

4. FEATURE SELECTION

One of the most important steps in constructing intrusion detection systems is to extract effective features. Features are security related measures that could be used to construct suitable detection algorithms. Desirable features must be selected to reflect the subject activities. Feature selection plays such a critical role in constructing effective features that its importance cannot be overemphasized.

Each intrusion detection approach is technically suited to identify a subset of security violations to which the system is subject. The selection of security measures should be based on good understanding about the system itself as well as all possible attacks that may influence the system's normal behaviors. Different attacks may be sensitive to different statistical features. Sometimes it requires domain expert knowledge to help selecting good features. In the history of IDSs, people have used various features to construct detection models. They tend to define the normal behavior of a user, a program, or a network element. Since the ground-breaking discovery of S. Forrect [32], people find that the short sequence of system calls of privileged programs is stable in characterizing system's behaviors. Therefore, many research efforts have focused on constructing different detection models using the short sequence of system calls since then.

Although there are some theoretical guidelines in optimal feature selection [33], it is still challenging to apply them in practice. In [26], Lee *et al.* utilized data mining algorithms to compute activity patterns from system audit data and extract temporal and statistical features from the pattern. They identified intrusion-only patterns from training data (a set of network connection records) and parsed these patterns to define features accordingly. Experiments based on test data were also needed to tell whether the selected features can be used to distinguish normal and abnormal activities. This process was repeated until a satisfactory set of features can be selected.

Today, features used in most anomaly-based IDSs are still selected empirically. It remains an open problem to decide the right set of features to construct IDSs in the context of cellular mobile networks. Some example features used include *call times and duration*, *roaming behavior*, *location coordinates*, *the list of traversed cells*, and so on.

5. ADAPTABILITY OF IDSS

It is necessary to integrate *adaptability* into the construction of IDSs. In reality, it is highly possible that a single user will demonstrate different mobility behaviors. Even if the user demonstrates the same mobility level, a user will have a set of mobility patterns during weekdays, while demonstrating a different set of mobility patterns during weekends. Therefore, established users' normal profiles need to be changed adaptively in order to reflect users' activities more accurately. Moreover, in constructing an anomaly-based IDS, a threshold-based scheme is often used. That is, the distance between observed activities and established normal profiles is compared with a threshold in order to decide whether the system needs to generate an alarm or not. It is also necessary to adjust the threshold adaptively in order to achieve desirable performance.

However, how to adaptively adjust the normal profile and the threshold of IDSs in the context of cellular mobile networks is a very challenging problem. Special mechanisms need to integrate with existing detection techniques to achieve adaptability. For example, an individual subject's activity may change over time. Therefore, it is necessary for the normal profile to be updated in order to reflect the recent activities. *Exponentially Weighted Moving Average* (EWMA) techniques [35] provide a suitable way to make activities in the recent past weigh more than activities long time ago. In this way, normal profiles can be adjusted accordingly. To adjust the threshold, usually an effective metric is needed to reflect the uncertainty of established normal profiles. *Entropy* may be a good choice here. We will see a more detailed example illustrating the integration of adaptability in Section 6.

6. CASE STUDY: AN EXEMPLARY IDS FOR CELLULAR MOBILE NETWORKS

6.1. Introduction

It is very difficult to design a once-for-all Intrusion Detection System for cellular mobile networks. Instead, an incrementally refined methodology is suitable. In this section, we introduce an exemplary IDS for cellular mobile networks [36],[37] that focuses on the exploitation of users' mobility patterns. Other important features like calling activities need to be integrated into the system to provide more comprehensive protections. In the sequel, we introduce system assumptions, models (threat model, network model, and mobility model), and detailed detection techniques.

6.2. Assumptions

First, we assume that most mobile users have favorite or regular itineraries. This makes it viable for us to establish each user¡¯s normal profile. This assumption is reasonable given that most users have regular daily lives. Studies in [38] conducted experiments over a period of six weeks to study the trajectories that users follow, and found out that users tend to follow regular trajectories more than 70% of time.

Actually, research on intrusion detection has two basic assumptions: 1) subject activities are observable via some system auditing mechanisms, and 2) normal and malicious activities should demonstrate distinct behaviors. Therefore, it is possible to reason about the evidence in the data to determine whether the system is currently under attack. If a user has totally random behavior, for example, the movement of a taxi driver, it will be very difficult, if not impossible, to create his normal movement profile. Our mobility-based detection algorithm alone is not suitable for such kind of users. Based on these considerations, our research is not motivated to build a system to accurately detect all intrusions. Instead, we aim at providing an optional service to end users as well as a useful administration tool to service providers. If the system observes some abnormal behaviors, other channels (e.g., email, phone calls to home) can be used to issue some warnings to the real users. Given the increasing number of security related incidents in wireless networks, these kinds of optional services can protect both the service providers and the end users from financial losses.

Second, we assume that there is a mobility database for each mobile user that describes his normal activities. This is a reasonable assumption in cellular mobile networks because this mobility database could be constructed by location tracking and prediction services. This mobility database could be stored together with the mobile user¡¯s personal information, such as billing information, in the Home Location Register (HLR). Note that in realistic networks, the locations of mobile users are actually tracked for the purpose of service provision and smooth handoff, even though the end users may be unaware of such monitoring. We assume that HLR is secure and the movement information is accurate. Usually, because of its importance, HLR is protected with highly secure measures, and thus it is extremely hard to be compromised. Also, the update and registration of the location are usually based on the device¡¯s current serving cell and the hardware registration such as the serial number of SIM card. Therefore, it will be hard for the attacker to hide or fabricate his location even if he has compromised all the secrets of the mobile device. Even if an attacker finds some magical ways to fabricate his location, he still has no idea about the normal movement profile of the real device owner.

Third, we assume that mobile devices can be compromised and all secrets associated with the compromised devices are open to attackers. Under this assumption, we do not need to assume or apply tamper-resistant hardware and software, which are still costly and impractical to handheld devices. This assumption justifies our research in anomaly detection, since all prevention techniques will be rendered helpless once the mobile device is captured and compromised. Actually, if we could assume the tamper-resistance of hardware/software, the whole security research could become much easier.

6.3. Model

Threat Model

The complex wireless mobile network system could incur software errors and design errors. This could make many attacks possible. One exemplary attack is *cell phone cloning*: a mobile phone card of an authenticate user A is cloned by an attacker B, which enables B to use the cloned phone card to make fraudulent telephone calls. If this kind of illegitimate use of service happens, the bills for the calls will go to the legitimate subscriber. Also, the *masquerader* can fake the International Mobile Equipment Identifier (IMEI) and the SIM (Subscriber Identity Module) card in order to obtain the service illegally. In the subscription fraud, fraudsters can also subscribe the service using the authentic user's name and obtain an account without intention to pay the bill. Our presented IDS can enhance system security to defend against these kinds of attacks.

Network Model

Different ways exist to model cellular mobile networks. For example, most previous work uses structured graph network topology models, such as hexagonal or square cell configurations. One disadvantage of this model is that it does not accurately represent a cellular network in practice, where the cell shape and size may vary depending on the antenna radiation pattern and propagation environment. In wireless cellular networks, each cell usually has a base station to serve it. Therefore, in our system, the wireless cellular network is modeled as a generalized graph $G = (V,E)$. The vertex set V represents all the base stations. If two cells are adjacent to each other, there is an edge between their two vertices. An example of this model is illustrated in Figure 2.(a) and Figure 2.(b). In this example, the vertex set is $V = \{a, b, c, d, e, f, g, h, i, j\}$, and the edge set is $E = \{(a, b), (a, c), ...(h, i)\}$.

There may exist other ways to model the networks in order to facilitate the intrusion detection tasks. For example, considering the fact that a mobile user usually drives along a road, cell-based models may not precisely locate a mobile user or model the trajectory of a user because they do not support fine granularity of the road network [39].

Usually, each user will follow a specific road for daily activities. Most users will follow the speed limit sign when driving. Also, each user has his own habit of traveling speed. Therefore, for a specific path, a user will take roughly the same amount of time to travel (if we do not consider the possible traffic jam). In reality, there exist a road network and the road network is overlapped with the Location Area, which consists of several cells. Considering all these factors, a network model as illustrated in Figure 3 could be adopted.

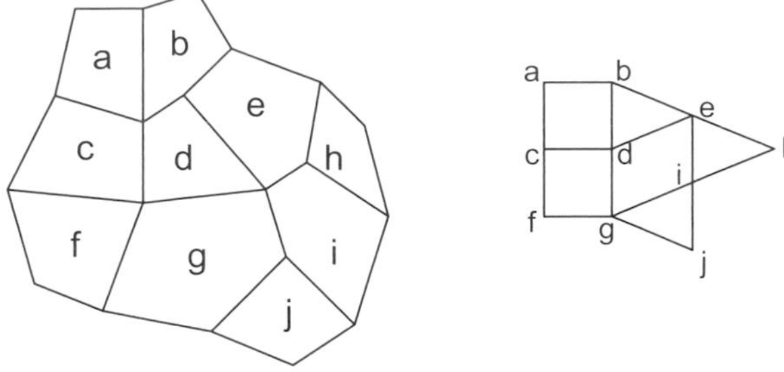

Figure 2. An Example Cellular Mobile Network and its Graph Model.

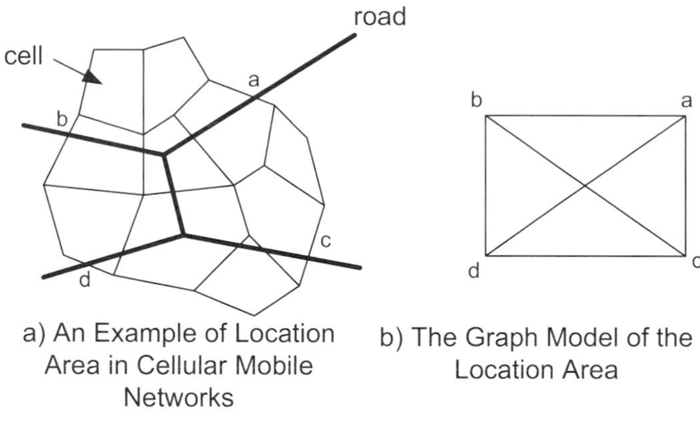

Figure 3. An Example Location Area in Cellular Mobile Networks and its Graph Model.

Figure 3.(a) illustrates the network topology in one LA, which consists of 10 cells. In Figure 3.(a), bold lines represent the road network. a, b, c, and d represent the intersection points of the road network and the boundary of the LA. For current mobile systems, location updates happen when the user enters or leaves one LA. This is the one of the most common ways to track the cellular mobile phones. This is true whenever a user is making a phone call or not. Considering this, we could adopt the corresponding network model as in Figure 3.(b).

In Figure 3.(b), each intersection between the road network and the LA is modeled as a vertex. In our example, we have four vertices, a, b, c, and d. These vertices form a fully connected graph, meaning that there is one path between any two vertices. In this way, we can ignore the complex internal road network inside one LA.

It is possible that in one LA, there are more than one possible path connecting two vertices. We assume that in one LA, one user prefers one specific path. This means that in Figure 3.(b), for a specific user, it will take him roughly the same amount of time to travel between any two vertices. If one user has variations in his traveling habit, i.e., if he takes two different paths between the same two vertices, we can have two entries for these two vertices in the user¡¯s mobility profile.

By integrating the current mechanisms that mobile networks use to track user¡¯s location information, this network model is more accurate than the model only considering the cell list traversed by each user. Furthermore, most routes have a speed limit and most users have a driving habit. For example, some users want to strictly follow the speed limit, while some others want to drive 10 miles/hour faster. This will cause the different time used by different users to traverse a specific route (edge). This network model is also more realistic because it ignores the potential different routes between two vertices. Therefore, it is more suitable for intrusion detection systems.

Different network models can be abstracted into different graphs. The vertices of the graph can be treated as the feature to construct different intrusion detection systems. In the following, we only use the cell list traversed by the user as the feature to illustrate a detailed detection technique. In this way, we denote each cell as a character. A string can be used to denote the cell list traversed by the user.

Mobility Model

The *random walk model* has been widely used in the literature, in which a mobile user will move to any one of the neighboring cells with equal probability after leaving a cell. This may not be realistic in practice, since mobile users normally travel with a destination in mind. Therefore, we adopt a m-th order Markov model. In such a model, the mobility of a user can be represented by a sequence of characters, $C_1, C_2, C_3, ..., C_i, ...$, where C_i denotes the identity of the cell visited by the mobile. Since the future locations of the mobile user are likely to be correlated with its movement history, the sequence of characters $C_1, C_2, C_3, ..., C_i, ...$ is assumed to be generated by an m-th order Markov source, where the states correspond to the context of the previous m characters. The probability that the user moves to a particular cell depends on the location of the current cell and a list of cells recently visited.

6.4. Mobility-Based Anomaly Detection Systems

In this section, we present two mobility-based anomaly detection schemes called LZ-based scheme and Markov-based scheme.

Figure 4 illustrates the LZ-based detection scheme. In the LZ-based detection scheme, based on users¡¯ regular itineraries, a mobility trie is constructed from the accumulative history of users¡¯ movement patterns. To integrate *adaptability* , the Exponential Weighted Moving Average (EWMA) [35] technique is applied to the mobility trie. This EWMA-based mobility trie serves as the normal profile of the user in the recent past, and reflects the stationary part of the user¡¯s regular mobility pattern. Based

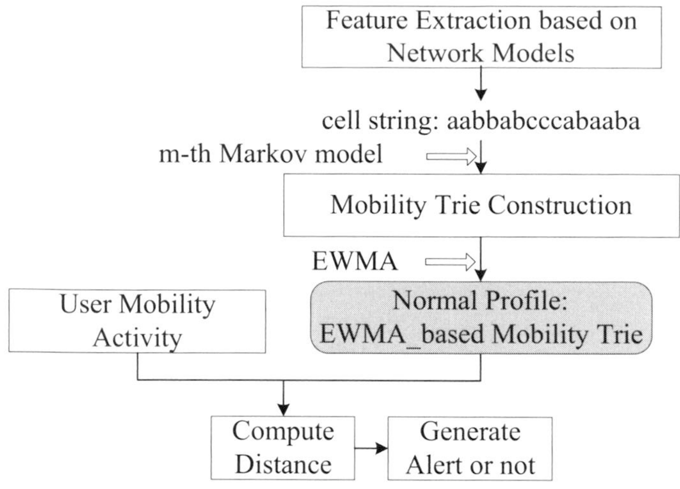

Figure 4. LZ-based Anomaly Detection Scheme.

on this, we use a *blending scheme* to calculate the probability of each user¡¯s activity in order to decide whether it is normal or not.

The second scheme, Markov-based scheme, is based on order-o Markov predictors. That is, given an order o, the probability of being the next cell given the previous o cells is constructed. In other words, the probability of the future activity can be calculated.

Both the LZ-based and the Markov-based schemes are online predictors, meaning that they examine the history so far, extract the current context, and predict the next cell location. Once the next location is known, the history is appended with one character (standing for one cell), and the predictor updates its history to prepare for the next prediction.

In the LZ-based scheme, we adopt Lempel-Ziv algorithms [40] [41]. In the rest of the chapter, when we discuss these algorithms, we use the word character. When we apply them to cellular mobile networks, we use the word cell. These two words have the same meaning in their respective contexts. Similarly, string is used in discussing Lempel-Ziv algorithms, while cell list is used in cellular mobile networks.

LZ-based Intrusion Detection

Data compression is a technique that encodes data in order to minimize its representation. Some of the most common lossless compression algorithms used in practice are dictionary-based schemes, where a dictionary $D = (M, C)$ is a finite set of phrases M and a function C that maps M onto a set of codes. In practice, when no *a priori* knowledge of the source characteristics is available, the problem of data compression becomes considerably complicated. Therefore, we often resort to universal coding schemes whereby the coding process is interlaced with a learning process for the varying source characteristics.

The family of Lempel-Ziv algorithms belongs to dictionary-based text compression and encoding techniques [42]. They are based on a popular incremental parsing algorithm by Ziv and Lempel [40],[41], and have been widely used in data compression. Since its invention, many variations have been developed. $LZ78$ is the most popular one.

The original $LZ78$ [40] is a word-based data compression algorithm. It parses the input string S of size n in a greedy manner into distinct substrings $x_1, x_2, \ldots, x_m$ with the following property: for $j > 1$, there exists a number $i < j$, which makes x_j equal to x_i concatenated by c, where c is one character in the alphabet. This is the so-called *prefix property* [42]. In the parsing process, if a phrase is the longest matching phrase seen previously concatenated by one character, the phrase, called a new phrase, is added to the *dictionary*. Substring x_j is encoded by the value i, using $\lceil \lg(j-1) \rceil$ bits, followed by the ASCII encoding of the last character of x_j, using $\lceil \lg \alpha \rceil$ bits, where α is the size of the input string's alphabet. Here the base of the logarithm is 2.

The Ziv-Lempel algorithm can be converted from a word-based method to a character-based algorithm by building a probabilistic model that feeds probability information to an arithmetic coder [43], which encodes a sequence of probability of p using $\lg(\frac{1}{p}) = -\lg p$ bits.

$LZ78$ is both theoretically optimal and good in practice. When the input text is generated by a stationary and ergodic source, $LZ78$ algorithms enjoy the property of being asymptotically optimal as the input size increases. That is, it encodes an indefinitely long string in the minimum size dictated by the entropy of the source. Here we omit the detailed proof. Being good in practice means that searching of $LZ78$ can be implemented efficiently by inserting each phrase in a *trie* data structure.

A *trie* is suitable to store the parsed phrases, and is a multiway tree with any path from the root to a unique node forming a string. In a trie, only the unique prefix of each string is stored because the suffix can be determined by searching the string. A longest match is found by following down the tree until no match is found, or the path ends at a leaf.

Here is an example of how to parse a string using $LZ78$ algorithm and construct a trie. Suppose the alphabet A is (a, b, c), and one possible string S over this alphabet is $aababccbababbabb\ldots$. Each element of the alphabet A could be one possible cell the user visits. S could be one possible cell list traversed by this user. Each substring in the parse is encoded as a pointer followed by an **ASCII** character. Based on the greedy parsing manner, this string will be parsed into phrases as follows: $(a)(ab)(abc)(c)(b)(aba)(bb)(abb)\ldots$.

In the character-based version of the Ziv-Lempel encoder, a trie is built when the previous substring ends. A trie at the start of the ninth substring is shown Figure 5.(a). The number associated with each node indicates the frequency in terms of number of times this node has been parsed in the construction of the mobility trie.

This trie characterizes the probability model of the string $aababccbababba$ $bb\ldots$. There are five previous substrings beginning with an a, two beginning with a b, and one beginning with a c. Therefore, the probability of a at the root is $\frac{5}{8}$. Similarly, the probability of b at the root is $\frac{2}{8} = \frac{1}{4}$ and the probability of c at the root is $\frac{1}{8}$. Of the 5 substrings that begin with an a, 4 begins with b. Therefore, the probability of b from a is $\frac{4}{5}$.

Probability Calculation

The probability calculation is based on the Prediction by Partial Matching (PPM) [44] scheme. Here, we use a context model to predict the next character based on the previous consecutive characters. Specifically, we use a *m-th Markov model* to model the sequence. That is, we use the consecutive previous m characters to predict the next character and calculate its probability. Here m is the *order* of the Markov model. For a first-order ($m = 1$) Markov model, it assumes that the next event only depends on the last event in the past. A high-order ($m > 1$ order) Markov model assumes that the next event depends on multiple (m) events in the past.

A trade-off exists here. If the order m is too small, the prediction will be poor in the long run because little audit data will be available to make a decision. However, if the order is too large, most contexts will seldom happen, and initially the probability estimation will have to solely rely on the resolve of zero-frequency problems [42]. Based on these considerations, we take a *blending* approach, where the predications of several contexts of different lengths are combined into a single overall probability. It uses a number of models with different orders to compute the probabilities respectively, assign a weight to each model, and calculate the weighted sum of the probabilities.

Let's denote the maximum order as m. The next character, denoted by α, is predicted on the basis of previous i characters. For each character α, let $p_i(\alpha)$ be the probability assigned to α by the finite-context model of order i. Note that when i is zero, the probability of each character is estimated independently of other characters. If the weight given to the model of order i is w_i and the blending weight vector is $[w_0, w_1, \ldots, w_m]$, the blended probability $p(\alpha)$ is computed as $p(\alpha) = \sum_{i=0}^{m} w_i * p_i(\alpha)$, where the sum of weights is normalized to 1. The larger the order, the larger the weight assigned to it, because context models with larger orders tend to be more accurate and should weight more in the current normal profile.

Anomaly Detection Algorithm

We adopt the character-based $LZ78$ to deal with the anomaly detection problem, and a classifier is trained with known "normal" data to distinguish normal behaviors from anomalous ones.

Integration of EWMA into Mobility Trie In anomaly detection, each subject (i.e., user in this application) has a normal profile. For an individual subject, its activity may change over time. Therefore, it is necessary for the normal profile to be updated in order to reflect the recent activities. In our situation, the normal profile of the user activity should be dynamic. Generally, activities in the recent past should weight more than activities long time ago. Adaptively modifying the normal profile correspondingly is a suitable mechanism.

Based on the above considerations, we integrate EWMA [35] to the mobility trie. The mobility trie is modified when a new phrase is formed during the string parsing. When a new phrase is inserted, we say an *event* happens. Note that this event corresponds to a sequence of characters. The insertion of the new phrase needs to modify the existing frequency of the mobility trie. We will call the modified frequency *EWMA-based frequency* hereafter. EWMA-based frequency measures how often the corresponding node appears in the recent past. Note that we do not need to do an extra trie search to modify the frequency. Instead, it is done at the same time with the update of the mobility trie to improve efficiency.

The EWMA-based frequency of each node in the mobility trie is updated as:

$$F(i) = \lambda * 1 + (1 - \lambda) * F(i), \tag{1}$$

where node i is one item of the corresponding events;

$$F(i) = \lambda * 0 + (1 - \lambda) * F(i), \tag{2}$$

where node i is not one item of the corresponding events.

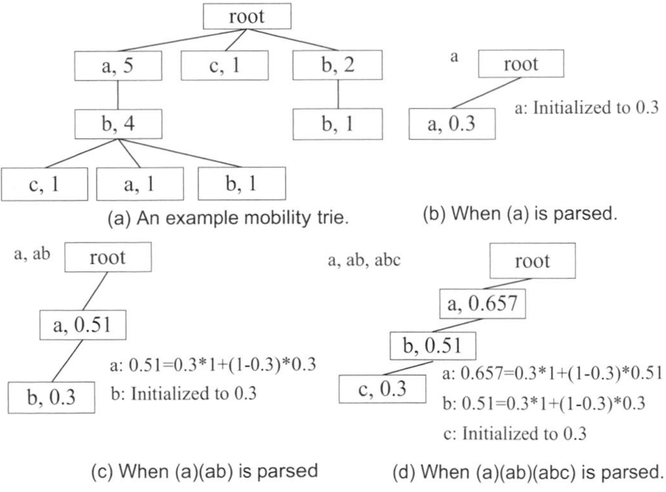

Figure 5. An Example of Mobility trie and an Example of Building Mobility Trie.

Here $F(i)$ is the EWMA-based frequency value stored in node i after a new phrase is inserted. For example, in Figure 5.c, the EWMA-based frequency associated with a is 0.51. The EWMA-based frequency associated with b is 0.3. Here λ is a smoothing constant that determines the decay rate. If a node i is not observed for continuous k

events (one *event* happens when a new phrase is inserted), the EWMA-based frequency of node i will be decayed to $(1-\lambda)^k$. In this way, the EWMA-based frequency of each node measures the intensity of this node over the recent past.

Continuing the example illustrated in Figure 5.(a), we illustrate how to integrate EWMA into the construction of the mobility trie. In this example, we let λ be 0.3. When the first character a is parsed, the corresponding mobility trie is illustrated in Figure 5.(c). When ab is parsed, the corresponding mobility trie is illustrated in Figure 5.(d). When abc is parsed, the corresponding mobility trie is illustrated in Figure 5.(d). As we can see, the EWMA-based frequency value associated with each node is exponentially faded.

The Similarity Measure EWMA-based mobility trie maintains the stationary part of each user's recent activities. Based on this, we could accurately predict whether the future activities are normal or not.

Let the sample space be all the possible cells traversed by a user. Because a user has his favorite routine of activity, this could lead to a small set of sample space. Let $S=(X_1, X_2, \ldots, X_n)$ denote the observed activities of the user, where X_i denotes a cell number. We want to identify whether or not it is normal based on our constructed mobility trie. We use a high-order Markov model to compute its blending transition probabilities.

Given an order o of the Markov model, we define the o-th order probability of S as:

$$P_o = \sum_{i=1}^{n-o} P(X_{i+o}|X_i, X_{i+1}, \ldots, X_{i+o-1}). \tag{3}$$

When it is order-0 model (o = 0), the probability of S is calculated as $P_o = P_0 = \sum_{i=1}^{n} P(X_i)$.

To calculate the probability of the transition $(X_i, X_{i+1}, \ldots, X_{i+o-1}) \longrightarrow X_{i+o}$ in equation 3, we need to search $(X_i, X_{i+1}, \ldots, X_{i+o-1})$ from the root. Let $F(X_{i+o})$ denote the EWMA-based frequency of node X_{i+o}. If $(X_i, X_{i+1}, \ldots, X_{i+o-1})$ is found, the probability $P(X_{i+o}|X_i, X_{i+1}, \ldots, X_{i+o-1})$ is defined as:

$$P(X_{i+o}|X_i, X_{i+1}, \ldots, X_{i+o-1}) = \frac{F(X_{i+o})}{F(X_{x+o-1})}. \tag{4}$$

If $(X_i, X_{i+1}, \ldots, X_{i+o-1})$ is not found, its probability is assigned 0.

To calculate $P(X_i)$, we compute the sum of the EWMA-based frequency of the root's children. $P(X_i)$ is then defined as $F(X_i)/\sum F(X_{\text{root's children}})$.

If X_i is not a child of the root, $P(X_i)$ is 0. That is, we only search from the root to decide the probability of each X_i.

Take the trie illustrated in Figure 5.d as an example, $P(b) = \frac{0.3}{0.357+0.3} = 0.4566$, $P(b|a) = \frac{0.51}{0.657} = 0.7763$, $P(c|ab) = \frac{0.3}{0.51} = 0.5882$.

Suppose that the blending weight vector is $[w_0, w_1, \ldots, w_m]$, where w_i is the weight value associated with the i-th order Markov model. $\sum_{i=0}^{m} w_i = 1$ and $w_i \geq 0, \forall i$. The probabilities of string S is defined as $P = \sum_{i=0}^{m} w_i * P_i$.

Intuitively, P increases with the increase of S's length because more transitions will be considered when S is longer. Therefore, P is not a good metric. We propose to use the following metric as our *similarity* measure $similarity(S) = \frac{P}{Length(S)}$, where $Length(S)$ is the length of string S.

Based on our definition, the similarity measure could be normalized by the length of the string and provides good criteria to evaluate its normalcy. Intuitively, $similarity$ indicates how good a mobile user follows its routines.

For the input string S, we calculate its $similarity(S)$. When a user follows one of its favorite itineraries, because this path is integrated into the mobility trie to construct the normal profile, many of its transitions illustrated in equation 4 at different order o will be found in the mobility trie, i.e., normal profile. Based on our definition, $similarity(S)$ will be a relatively large value. However, when the mobile is stolen, and the intruder takes an infrequent path, the similarity of this string tends to be a very small value, because many transitions cannot be found in the mobility trie.

We introduce a threshold, P_{thr}, which is a design parameter. When $similarity(S) \geq P_{thr}$, string S is evaluated as normal, otherwise string S is identified as anomalous.

Because our mobility trie records the most frequently used path of a user, it is very sensitive to anomalous paths, even if they are very short strings. This enables our detection algorithm to detect the abnormal very *quickly* - an important quality for reducing potential damage by a malicious user. At the same time, our detection algorithm has a very high detection rate. Also, when a frequently used path is taken, our detection algorithm can tolerate slight variations from the path and thus has small a false positive rate.

Implementation issues

In practice, an important issue is how to store the mobility information in a trie. A trie is actually a multiway tree with a path from the root to a unique node for each string represented in the tree. The fastest approach for processing is to create an array of pointers for each node in the trie with a pointer for each character of the input alphabet. Although this approach is easy for processing, it wastes memory space. Another approach is to use a linked list at each node, with one item for each possible branch. This method uses memory economically, but the processing is intensive. A trie can also be implemented as a single hash table with an entry for each node. For further details, the reader can consult books on algorithms and data structures.

6.5. Markov-based Anomaly Detection

Markov predictors are a very popular family of predictors. They have been widely used and studied in the literature. Let X_t be the cell visited by the user or the state of the user's activity at time t. The order-o Markov predictor assumes that the location can be predicted from the current *context*, which is the sequence of the previous o most recent characters in the location history $(X_{t-o+1}, X_{t-o}, \ldots, X_t)$. Under this Markov model, the transitions represent the possible cell locations that follow the context.

A Markov Chain with order-o of only one-step event transitions is a stochastic process with the following assumptions:

$$\begin{aligned}
P(X_{t+1} &= i_{t+1}|X_t = i_t, X_{t-1} = i_{t-1}, \ldots, X_0 = i_0) \\
&= P(X_{t+1} = i_{t+1}|X_t = i_t, X_{t-1} = i_{t-1}, \ldots, \\
&\quad X_{t-o+1} = i_{t-o+1}) \\
P(X_{t+1} &= i_{t+1}|X_t = i_t, X_{t-1} = i_{t-1}, \ldots, X_{t-o+1} = i_{t-o+1}) \\
&= P(X_{t+1} = j|X_t = i_o, X_{t-1} = i_{o-1}, \ldots, X_{t-o+1} = i_1) \\
&\equiv p_{\{i_1,\ldots,i_{o-1},i_o\}\to j}.
\end{aligned}$$

It describes the two important properties of the Markov Chain:

- Equation 5 states that the probability distribution of the user at time $t+1$ depends on the state at time $t, t-1, \ldots, t-o+1$, and does not depend on the previous states leading to the states at $t, t-1, \ldots, t-o+1$.
- Equation 5 states that the state transitions from time $t, t-1, \ldots, t-o+1$ to $t+1$ is independent of time.

If the system has a finite number of states $1, 2, \ldots, s$, these probabilities could be represented in a *transition probability matrix*, where each element in the matrix is $p_{\{i_1,\ldots,i_{o-1},i_o\}\to j}$, as illustrated in 5.

$$\begin{bmatrix}
p_{\{1,1,\ldots,1\}\to 1} & p_{\{1,1,\ldots,1\}\to 2} & \cdots & p_{\{1,1,\ldots,1\}\to s} \\
p_{\{1,1,\ldots,2\}\to 1} & p_{\{1,1,\ldots,2\}\to 2} & \cdots & p_{\{1,1,\ldots,2\}\to s} \\
\vdots & \vdots & \vdots & \vdots \\
p_{\{s,s,\ldots,s\}\to 1} & p_{\{s,s,\ldots,s\}\to 2} & \cdots & p_{\{s,s,\ldots,s\}\to s}
\end{bmatrix} \tag{5}$$

$p_{\{i_1,\ldots,i_{o-1},i_o\}\to j}$ could be learned from the observations of the user' locations in the past. When $o \geq 1$, $P(X_{t+1} = j|X_t = i_o, X_{t-1} = i_{o-1}, \ldots, X_{t-o+1} = i_1) = N(Lj)/N(L)$, where $L = \{i_1, \ldots, i_{o-1}, i_o\}$, $N(Lj)$ denotes the number of observation pairs of L and j. $N(L)$ denotes the number of observations of L.

When o is 0, the formula becomes:

$$P(X_{t+1} = j) = \frac{N(j)}{N}, \tag{6}$$

where N is the total number of observations (i.e., total number of cells). $N(j)$ is the number of observations of a.

Given this estimation, we can calculate the probability of the next location given the previous o locations for a specific user. The larger the probability, the more likely it is normal. We can then derive a *threshold* policy and use it to decide whether the current activity is normal or not.

That is, given a fixed order value o and an observed activity in terms of a cell list $S_{observed} = (X_1, X_2, \ldots, X_n)$, where each X_i denotes a cell number. For $o \geq 1$, we first calculate its o-order transition probabilities as $P_o = \sum_{i=1}^{n-o} P(X_{i+o} = j | X_i = i, X_{i+1} = i+1, \ldots, X_{i+o-1} = i+o-1) = \sum_{i=1}^{n-o} p_{\{i,i+1,\ldots,i+o-1\} \rightarrow j}$, where $p_{\{i,i+1,\ldots,i+o-1\} \rightarrow j}$ can be retrieved from the probability transition matrix whose element is obtained using Equation 5. If the transition does not exist in the transition matrix, we assign $P(X_{i+o} | X_i, X_{i+1}, ..., X_{i+o-1})$ to 0.

For $o = 0$, its probability could be calculated as $P_o = \sum_{i=1}^{n} P(X_i = j)$, where $P(X_i = j)$ can be obtained from Equation 6.

Similar to LZ-based mechanism, P_o increases with the increase of S's length. Therefore, for Markov-based prediction, we also define the following *similarity* metric: $similarity(S) = \frac{P_o}{Length(S)}$, where $Length(S)$ is the length of string S.

For the input string S, we calculate its $similarity(S)$. If most transitions can be found, $similarity(S)$ tends to be large. This indicates that S is more likely to be normal. However, if the mobile is stolen, and an infrequent or new path is taken, the $similarity$ of the string should be small.

When the mobile is at low mobility, the user usually travels one or two cells during the call. Given a fixed o, it is highly possible that the length of the transition $(o + 1)$ is larger than the length of the cell. The Markov-based prediction cannot make a decision under this situation. Therefore, high-order Markov-based prediction will become useless for low mobility data. We make a random guess when this situation happens. For example, with a probability of $1/2$, this cell list is identified as normal (abnormal).

For Markov-based prediction, we introduce a threshold P_{thr_markov}. If $similarity$ $(S) \geq P_{thr\_markov}$, string S is evaluated as normal. P_{thr_markov} should be tuned by taking into consideration both false alarm rate and detection rate.

6.6. Adaptive Anomaly Detection

In this section, we illustrate how to integrate adaptability into LZ-based detection schemes. EWMA-based mobility trie itself facilitates the differentiation between weekday and weekend routes because when the user changes its mobility patterns, for example, from weekday to weekend routes, the more recent the activities, the more weight they should have in the normal profile. The smoothing constant in EWMA techniques plays an important role in determining how much weight the more recent activities should have. Basically the larger the smoothing constant is, the more weight they should have. Therefore, intuitively, the shorter the recent activities last, the larger the smoothing constant should be.

The EWMA-based approach only partially addresses the adaptation of normal profiles. In the following, we detail our approach of how to tune the threshold for different users and different mobility levels.

Feedback-based Approach

One simple approach to adjust the threshold is to apply the feedback principle. That is, based on the output of the detection algorithm (for example, in terms of detection rate and false positive rate), the system administrator can adaptively adjust the detection threshold in order to achieve the required performance. If the false positive rate is a more important metric, for example, when the system has been detected raising too many false alarms, the system administrator could lower the detection threshold correspondingly. However, in this approach, the decrease of the false positive rate is achieved at the risk of a decreased detection rate.

Entropy-based Approach

We use Shannon's entropy measure to identify the uncertainness of the up-to-date normal profile. Based on this, we could adjust the detection threshold correspondingly.

Metric Selection

The first step we need is to identify a metric that can effectively reflect the location *uncertainty*. In our case, it is the EWMA-based mobility trie. Shannon's *entropy* measure [45] is an ideal candidate for quantifying this uncertainty. Our previous work showed that for the non-adaptive mechanism, given a mobility level, the more varied the mobility pattern is, the more dynamic the mobility trie is. This motivates us to use *entropy* as a measure to reflect the dynamic level of the normal profile. The lower the uncertainty under the movement pattern, the richer the movement pattern is.

Definition 1. Entropy: Suppose X is a dataset, $C_x = \{C_x[1], C_x[2], \ldots, C_x[m]\}$ is a class set. Each data item of X belongs to a class $x \in C_x[i]$. Then the entropy of X related to this $|C_x|$-wise classification is defined as $H(X) = \sum_{i=1}^{m} -P_i \log P_i$, where P_i is the probability of x belonging to class $C_x[i]$.

Entropy can be interpreted as the number of bits required to encode the classification of a data item. It measures the uncertainty of a collection of data items. The lower the entropy, the more uniform the class distribution. If all data items belong to one class, then its *entropy* is 0, which means that no bits needs to be transmitted because the receiver knows that there is one class. The more varied the class distribution is, the larger the *entropy* is. When all of the data items are equally distributed over the m classes, its *entropy* is $\log(m)$ (natural logarithm). In the context of anomaly detection, entropy is a measure of the regularity of audit data.

Definition 2. Conditional entropy: Suppose that X and Y are two datasets, and $C_x = \{C_x[1], C_x[2], \ldots, C_x[m]\}$ and $C_y = \{C_y[1], C_y[2], \ldots, C_y[n]\}$ are two class sets. Each data item of X belongs to a class $x \in C_x[i]$ and each data item of Y belongs to a class $y \in C_y[i]$. Then given Y and C_y, the entropy of X related to C_x is defined as $H(X|Y) = \sum_{i=1}^{m} \sum_{j=1}^{n} P_{ij} \log \frac{1}{P_{i|j}}$, where P_{ij} is the probability of $x \in C_x[i]$ and $y \in C_y[j]$, and $P_{i|j}$ is the probability of $x \in C_x[i]$ given $y \in C_y[j]$.

Conditional entropy describes the uncertainness of X given Y, i.e., it indicates the coefficiencies between X and Y. The smaller the *conditional entropy* is, the more correlated X and Y are. If X can be determined by Y, $H(X|Y)$ is 0. In the context of anomaly detection, *conditional entropy* can be used to explore the temporal sequential characteristics of audit data due to the temporal nature of the system activities.

Compute the Entropy of a Trie

When we compute the entropy of the EWMA-based mobility trie, we apply a weighted scheme at different orders. Specifically, based on the order of different finite contexts of the mobility trie, we calculate conditional entropies respectively and assign them different weights. The larger the order is, the larger the weight should be. The sum of these weighted entropies is used as the measurement for adjusting system detection threshold. Let's consider a more complex string $aaababbbbbaabccbaaaaaacabbbabcacb$. By applying $LZ78$ algorithm [42], we obtain a trie as illustrated in Figure 6.

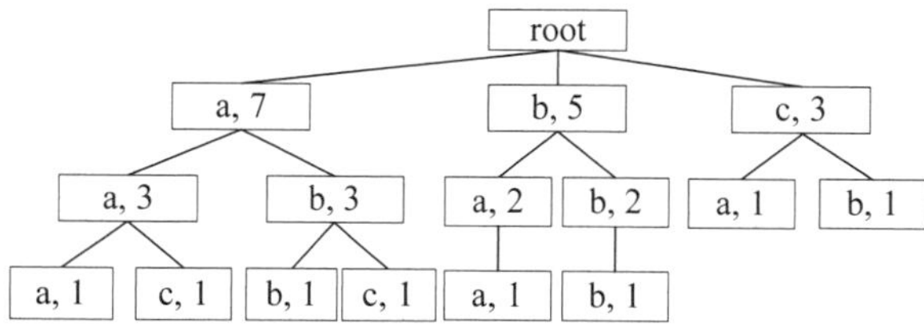

Figure 6. An Example of EWMA-based Mobility Trie.

The maximum order m and the corresponding weight w_i are design parameters. In this example, let's assign 2 to m.

- Order-0 Model

$$
\begin{aligned}
& H(V1) \\
= \; & \frac{7}{15} \log \frac{15}{7} + \frac{5}{15} \log \frac{15}{5} + \frac{3}{15} \log \frac{15}{3} \\
= \; & 1.0438.
\end{aligned}
$$

- Order-1 Model

$$\begin{aligned} & H(V2|V1) \\ = & \frac{7}{15}[(\frac{3}{6}\log\frac{6}{3}) \times 2] + \frac{5}{15}[(\frac{2}{4}\log\frac{4}{2}) \times 2] + \frac{3}{15}[(\frac{1}{2}\log\frac{2}{1}) \times 2] \\ = & 0.6931. \end{aligned}$$

- Order-2 Model

$$H(V3|V1V2) = \frac{3}{15}[(\frac{1}{2}\log\frac{2}{1}) \times 4] + 0 = 0.2773.$$

When the context of a specific length is not found in the trie, we assign its conditional probability to 0. Note that we treat $0 \log 0$ as 0.

Generally, the larger the order is, the larger the weight assigned to it should be, because context models with a larger order tend to be more accurate and should weight more in the current normal profile. If we assign 0.1, 0.2, and 0.7 to w_1, w_2, and w_3, respectively, the *weighted entropy* of the mobility trie in Figure 6 can be calculated as:

$$\begin{aligned} & weighted\_entropy \\ = & w_1 \times H(V1) + w_2 \times H(V2|V1) + w_3 \times H(V3|V1V2) \\ = & 0.4371. \end{aligned}$$

Adaptive Algorithm

The algorithm of constructing the adaptive normal profile is illustrated in Figure 7. It summarizes how to use EWMA to adaptively adjust the normal profile and how to use entropy to adaptively adjust the threshold.

7. SUMMARY

Significant security concerns exist in wireless networks. Although there are many prevention-based protocols in cellular mobile networks, how to design a highly secure cellular mobile network is still a very challenging issue due to the open radio transmission environment and physical vulnerability of mobile devices. Intrusion detection is indispensable to provide an enhanced protection for wireless networks.

This chapter presents the current status of major intrusion detection techniques developed for wired and wireless networks. We point out corresponding challenges that need to be addressed in the future. In the context of cellular mobile networks, we also present the detailed steps in developing one exemplary intrusion detection system. Our presented example mainly exploits users' location information to identify potential fraudsters and masqueraders. Future work may include the integration of users' calling activities. Because of the potential wide variety of users' behaviors, it is difficult to

```
INPUT: Observed user's mobility activities in terms of a cell list
OUTPUT: Adaptive normal profile

Initialize mobility database := null
LOOP
        Based on the LZ78 algorithm, wait for a sequence s
        IF (The mobility trie of the mobile exists)
          IF (A path p corresponding to s is found)
            Add s to the mobility trie
            Using EWMA to modify the frequencies of nodes
          ELSE
            Create new nodes, and initialize their frequencies to λ
        ELSE
            1) Create a mobility trie := single sequence s
            2) Initialize the frequencies for every node in sequence
               s to λ
        Compute the entropy e_1 of the EWMA-based mobility trie
        IF (e_1 > e)
        /* e is the entropy of the previous EWMA-based mobility trie */
            Decrease the detection threshold by Δ
        ELSE
            Increase the detection threshold by Δ
        e = e_1;
FOREVER
```

Figure 7. Adaptive Normal Profile.

accurately characterize users' activities. Moreover, considering the randomness of certain users' behaviors, not all users can be considered as potential candidates for the successful applications of anomaly detection techniques.

Intrusion detection in cellular mobile networks is a challenging problem. Not only will traditional challenges like feature selection continue to exist, but also new problems specific to cellular mobile networks keep appearing. All these deserve the further attention from the research community.

8. REFERENCES

1. Y.-B. Lin, M. Chen, and H. Rao, Potential fraudulent usage in Mobile Telecommunications Networks, *IEEE Transactions on Mobile Computing*, Vol.1 No.2 (2002) pp. 123-131.
2. M. Zhang, and Y. Fang, Security Analysis and Enhancements of 3GPP Authentication and Key Agreement Protocol, *IEEE Transactions on Wireless Communication*, Vol.4, No.2, (2005) pp. 734-742.
3. B. Mukherjee, T.L. Heberlein, and K.N. Levitt, Network Intrusion Detection, *IEEE Network*, Vol.8, No.3, (1994) pp. 26-41.
4. H. Debar, M. Dacier, and A. Wespi, A Revised Taxonomy for Intrusion-Detection Systems, *Annales des Telecommunications*, Vol. 55, (2000) pp. 361-378.
5. J. P. Anderson, Computer Security Threat Monitoring and Surveillance, *Technical Report, James P. Anderson Co.*, Fort Washington, PA, (April 1980).
6. D. E. Denning, An Intrusion-Detection Model, *IEEE Transactions on Software Engineering*, Vol.13, No.7, (Feb. 1987), pp. 222-232.
7. P. Porras and A. Valdes, Live Traffic Analysis of TCP/IP Gateways, *Proceedings of the 1998 ISOC Symposium on Network and Distributed System Security (NDSS'98),* San Diego, CA, March 1998.
8. T. F. Lunt, R. Jagannathan, R. Lee, S. Listgarten, D. L. Edwards, P. G. Neumann, H. S. Javitz, and A. Valdes, IDES: The Enhanced Prototype ¨C a Real-time Intrusion-Detection Expert System, Technical Report SRI-CSL-88-12, SRI International, Menlo Park, CA, Oct. 1988.
9. H.S. Javitz and A. Valdes, The SRI Statistical Anomaly Detector, *Proceedings of 1991 IEEE Symposium on Research in Security and Privacy,* pp. 316-326, May 1991.
10. R. Jagannathan, T. Lunt, D. Anderson, C. Dodd, F. Gilham, C. Jalali, H. Javitz, P. Neumann, A. Tamaru, and A. Valdes, System Design Document: Next-Generation Intrusion Detection Expert System (NIDES), Technical Report A007/A008/A009/A011/A012/ A014, SRI International, 333, Ravenswood Avenue, Menlo Park, CA, March 1993.
11. U. Lindqvist, and P.A. Porras, Detecting Computer and Network Misuse through the Production-Based Expert System Toolset (P-BEST), *Proceedings of the 1999 IEEE Symposium on Security and Privacy,* Oakland, CA, pp. 146-161, May 9-12, 1999.
12. Internet Security Systems, RealSecure Network Protection, Nov. 2003, Available at $http://www.iss.net/products\_services/enterprise\_protection/rsnetwork$.
13. S. Kumar and E. Spafford, A Pattern Matching Model for Misuse Intrusion Detection, *Proceedings of the 17th National Computer Security Conference,* pp. 11-21, Oct. 1994.
14. P.A. Porras and R. Kemmerer, Penetration State Transition Analysis ¨C a Rule-Based Intrusion Detection Approach, *Proceedings of the 8th Annual Computer Security Application Conference,* pp. 220-229, Nov. 1992.
15. K. Ilgun, Ustat: A Real-time Intrusion Detection System for Unix, *Proceedings of IEEE Symposium on Research in Security and Privacy,* Oakland, CA, pp. 16-28, May, 1993.
16. H. Debar, M. Becker and D. Siboni, A Neural Network Component for an Intrusion Detection System, *Proceedings of 1992 IEEE Symposium on Research in Security and Privacy,* Oakland, CA, pp. 240-250, May, 1992.
17. S. Forrest, S.A. Hofmeyr, and A. Somayaji, Computer Immunology, *Communications of the ACM,* vol. 40, no. 10, pp. 88-96, Oct. 1997.
18. C. Warrender, S. Forrest, and B. Pearlmutter, Detecting Intrusions Using System Calls: Alternative Data Models, *Proceedings of 1999 IEEE Symposium on Research in Security and Privacy,* Oakland, CA, pp. 133-145, May 1999.

19. W. Lee, S. J. Stolfo, and K. W. Mok, A Data Mining Framework for Building Intrusion Detection Models, *Proceedings of the 1999 IEEE Symposium on Security and Privacy,* Oakland, CA, pp. 120-132, May 1999.

20. N. Ye, X. Li, Q. Chen, S. M. Emran, and M. Xu, Probabilistic Techniques for Intrusion Detection Based on Computer Audit Data, *IEEE Transactions on Systems, Man, and Cybernetics,* vol. 31, no. 4, pp. 266-274, 2001.

21. N. Ye, S. M. Emran, Q. Chen, and S. Vilbert, Multivariate Statistical Analysis of Audit Trails for Host-based Intrusion Detection, *IEEE Transactions on Computers,* vol. 51. no. 7, pp. 810-820, 2002.

22. S. Jha, K. Tan, and R.A. Maxion, Markov Chains, Classifiers, and Intrusion Detection, *Proceedings of the 14th IEEE Computer Security Foundations Workshop,* Cape Breton, Nova Scotia, Canada, pp. 206-219, 2001.

23. N. Ye, A Markov Chain Model of Temporal Behavior for Anomaly Detection, *Proceedings of 2000 IEEE Workshop on Information Assurance and Security,* United States Military Academy, West Point, NY, pp. 171-174, June 6-7, 2000.

24. N. Ye, T. Ehiabor, and Y. Zhang, First-order versus High-order Stochastic Models for Computer Intrusion Detection, *Quality and Reliability Engineering International,* vol. 18, no. 3, pp. 243-250, 2002.

25. M. Nassehi, Anomaly Detection for Markov Models, Technical Report Tech Report RZ 3011(#93057), IBM Research Division, Zurich Research Laboratory, March 1998.

26. W. Lee, and S. J. Stolfo, A framework for constructing features and models for intrusion detection systems, *ACM Transactions on Information and System Security (TISSEC)*, Vol. 3 Issue 4, (2000).

27. C. Ko, G. Fink, and K. Levitt, Execution Monitoring of Security-Critical Programs in Distributed Systems: A Specification-Based Approach, *Proceedings of 1997 IEEE Symposium on Security and Privacy,* Oakland, CA, pp. 134-144, May 1997.

28. R. Buschkes, D. Kesdogan, and P. Reichl, How to Increase Security in Mobile Networks by Anomaly Detection,¡± *Proceedings of the Computer Security Applications Conference*, Phoenix, AZ, Dec. 1998, pp. 3-12.

29. D. Samfat, and R. Molva, IDAMN: An intrusion detection architecture for mobile networks, *IEEE Journal on Selected Areas in Communications,* Vol.15, No.7, (Sept. 1997), pp. 1373-1380.

30. J. Hall, M. Barbeau, and E. Kranakis, Anomaly-based Intrusion Detection Using Mobility Profiles of Public Transportation Users, *IEEE Wireless and Mobile Computing, Networking and Communications (WiMob'2005)*, (2005), pp. 17-24.

31. J. Hollmén, User profiling and classification for fraud detection in mobile communications networks, *PhD Thesis*, Helsinki University of Technology, (2000).

32. C. Warrender, S. Forrest, and B. Pearlmutter, Detecting Intrusions Using System Calls: Alternative Data Models, *Proceedings of 1999 IEEE Symposium on Research in Security and Privacy*, Oakland, CA, pp. 133-145, May 1999.

33. D. Koller, and M. Sahami, Toward Optimal Feature Selection, *Proceedings of the 13th International Conference on Machine Learning*, Bari, Italy, July 1996, pp. 284-292.

34. Q. He, D. Wu, and P. Khosla, Quest for personal control over mobile location privacy, *IEEE Communications Magazine,* Vol. 42, No. 5, May 2004, pp. 130-136.

35. R.A. Johnson and D.W. Wichern, Applied Multivariate Statistical Analysis, Upper Saddle River, NJ: Prentice Hall, 1998.

36. B. Sun, F. Yu, K. Wu, Y. Xiao, and V.C.M. Leung, Enhancing Security using Mobility-Based Anomaly Detection in Cellular Mobile Networks, IEEE Transactions on Vehicular Technology, 2005, in press.

37. B. Sun, Z. Chen, R. Wang, F. Yu, and V.C.M. Leung, Towards Adaptive Anomaly Detection in Cellular Mobile Networks, *IEEE Consumer Communications and Networking Conference (CCNC'06)*, Las Vegas, NV, 2006.

38. S. Schonfelder, Some notes on space, location and travel behaviour, *Swiss Transport Research Conference*, Monte Verita, Ascona, 2001.

39. H. A. Karimi, X. Liu, A predictive location model for location-based services, *Proceedings of the 11th ACM international symposium on Advances in geographic information systems*, New Orleans, LA, Nov. 2003, pp. 126 - 133.

40. J. Ziv, and A. Lempel, Compression of individual sequences via variablerate coding, *IEEE Transactions on Information Theory*, Vol. 24, Noo. 5, Sept. 1978, pp. 530-536.

41. J. Ziv and A. Lempel, A Universal Algorithm for Sequential Data Compression, *IEEE Transactions on Information Theory*, Vol. 23, pp. 337-342, 1977.

42. T. C. Bell, J. G. Cleary, and I. H. Witten, Text Compression, Prentice-Hall Advanced Reference Series, Prentice-Hall, Englewood Cleffs, NJ, 1990.

43. J. S. Vitter and P. Krishnan, Optimal prefetching via data compression, *Journal of ACM*, vol. 43, no. 5, Sept. 1996, pp. 771-793.

44. J. G. Cleary and I. H. Witten, Data compression using adaptive coding and partial string matching, *IEEE Transactions on Communications*, vol. 32, no. 4, Apr. 1983, pp. 396-402.

45. T. Cover and J. Thomas, *Elements of Information Theory,* John Wiley & Sons, 1991.

9

THE SPREAD OF EPIDEMICS ON SMARTPHONES

Bo Zheng
Dept. of Computer Science and Technology
Tsinghua University
Beijing 100084, P. R. China
E-mail: bzheng@csnet1.cs.tsinghua.edu.cn

Yongqiang Xiong
Microsoft Research Asia
5F, Beijing Sigma Center, No. 49, ZhiChun Road,
Haidian District, Beijing 100080, P. R. China
E-mail: yqx@microsoft.com

Qian Zhang
Dept. of Computer Science
Hong Kong University of Science and Technology
Clear Water Bay, Kowloon, Hong Kong
E-mail: qianzh@cs.ust.hk

Chuang Lin
Dept. of Computer Science and Technology
Tsinghua University
Beijing 100084, P. R. China
E-mail: chlin@tsinghua.edu.cn

The emergence of epidemics such as worms and viruses on smartphones severely threaten the Internet and telecom networks. Two important features of smartphone, i.e., static shortcuts and mobile shortcuts, bring great challenge for traditional epidemic spread model. In this paper, we propose a novel epidemics spread model (ESS) for smartphone which is an SIR model based on the analysis of the unique features of smartphones. With this ESS model, we study the "static shortcuts" and "mobile shortcuts" effects brought by smartphones and consider the influence of the epidemic spread rate, network topology, patching and death rate as well as the initial pre-patch to the propagation of the smartphone epidemics. Critical condition of epidemic fast die out is derived from the ESS model, and the detailed analysis is given to the individual parameters in the model to study their effects to the epidemics spread. Extensive simulations in typical network topologies (small-world network, power law graph, and Waxman network) have been performed to verify the ESS model and demonstrate the effectiveness and accuracy. The guidance to prevent the epidemics of smartphones is also given based on our theoretical analysis and the simulations.

1. INTRODUCTION

As of 2004 smartphones are an increasingly large part of the mobile phone market. At the same time, epidemics[1] begin to appear in the smartphone. In this section, we introduce the features of the smartphone and the attacks on the smartphone. And then describe the difference between the smartphone epidemics and the PC epidemics.

1.1. Smartphones

The Wikipedia [1] defines the smartphone as the following: A smartphone is generally considered any handheld device that integrates personal information management and mobile phone capabilities in the same device. Often, this includes adding phone functions to already capable PDAs or putting "smart" capabilities, such as PDA functions, into a mobile phone.

In recent years, the global market for smartphones takes on a meteoric rise. According to analyst house Canalys [2] smartphone shipments increased over 100% from 2004Q2 to 2005Q2, with over twelve million devices shipped in the latter period. And according to market research from IDC [3], 50 million smartphones will be shipped in 2005, and more than 110 million smart-phones will be shipped by 2008. In a couple years, it is likely that most phones sold will be considered "smart", except for disposable phones.

Most Smartphones connect the Internet and telecom networks together. Smartphones tend to unify communications which integrate telecom and Internet services onto a single device because it has combined the portability of cell-phones with the computing and networking power of PCs.

The key feature of smartphones is that they has common operating systems (OSes), and one can install additional applications to the device. The applications can be developed by the manufacturer of the handheld device, by the OS vendor, by the operator or by any other third-party software developer.

Most common operating systems are Symbian [4, 5] (developed by a group of renowned mobile phone solution providers), Windows CE / Mobile [6, 7] (developed by Microsoft), Palm OS [8] (developed by PalmSource), BREW [9] (technically a platform developed by Qualcomm), and Linux (such as Montavista [10]). Although the detailed design and functionality vary among these OS vendors, all share the following features [11].

[1] In this chapter, we use the term "epidemic" to denote the epidemic-like phenomena in the computer and smartphone networks, including worms, viruses and Trojans that can spread from one device to another.

- Access to cellular network with various cellular standards such as GSM /CDMA and UMTS.
- Access to the Internet with various network interfaces such as infrared, Bluetooth, GPRS/CDMA1X, and 802.11; and use standard TCP/IP protocol stack to connect to the Internet.
- Multi-tasking for running multiple applications simultaneously.
- Data synchronization with desktop PCs.
- Open APIs for application development.

While common OSes, open APIs, and sophisticated capabilities enable powerful services, they also create common ground and opportunities for security breaches and increase worm or virus spreading potentials. Given the PC-like nature of smart-phones and the trend of full-fledged OSes, software vulnerabilities seem inevitable for their OSes and applications. Moreover, with the Internet exposure, smartphones become ideal targets for Internet worms or viruses since smart-phones are always on, and their user population will likely exceed that of PCs, observing from the prevalence of cell phone usage today.

1.2. The Smartphone Attacks

Smartphones get a rapid growth in 2004, but this rapid growth also draws the attacker's attention. In June 2004, Cabir was developed. This worm, capable of spreading via Bluetooth was the first notable piece of malware seen on mobile phones and the Symbian OS. It was not released to the public in order to infect phones, however, but was instead sent to security experts as a proof of concept of a "wireless worm". Up to now, more than 60 attacks [12, 13, 14] (virus, worms, Trojan horse, malware, etc.) are found in the smartphones.

The following things make the attacks on the smartphones may have many new features compared with attacks on PC. First of all, smartphones connect with many other things, they connect to Internet and telecom network togather, and they often contact with another bluetooth devices and sync with host PC. Moreover, when smartphones got infected, they may infect other Smart-phones/PCs, more seriously, the unauthorized outgoing may be phone, SMS, MMS, even the user's private information, and these are not free! After a extensive survey, we summarize all the attacks we found in [12, 13, 14] and some attacks which we thought may appear in the near future. Most attack ways on PC appeared on smartphone, we list some special attack way on smartphone in Table 1.

The security of smartphone has drawn great attention recently. On the one hand, the open mobile operating system, flexible programmability, and powerful computational/network capabilities of smartphones inevitably create opportunities for software vulnerabilities. On the other hand, as mentioned before, with the fast growth of the smart-phone customer base, smart-phones have become ideal targets because, with a

Table 1. Attacks which may appear on smartphones

Attack/Spread Ways	**Explanation**
Hot synced	Infect PC/Smartphone while hot synced
SPAM	Spread SPAM via SMS, MMS, or email
Malformed SMS	Send a certain malformed SMS and make the victim smart-phone shutdown
Limit some functions	Limit most functions of the Smart-phone, such as restrict the Smart-phone to only receive phone call, or set random password to media card and make it unaccessible
Overwrite system ROM	Overwrite system ROM and make the system crashed
Worms	Not only through the internet, smartphones can spread worms through MMS, Bluetooth, WiFi, etc.
Sleep deprivation torture attack	Vastly shortened battery life caused by the constant scanning. Because Smart-phone is a resource restraint device, energy is a very important resource to it.
Unapproved dial (DoS)	Some applications (usually Trojan or worm) dial a certain phone number to make phone-DoS attack
Unapproved dial (Theft)	Hacker uses the victim's Smart-phone to make phone call through some backdoors. They can make phone call paid by the victim and receive the voice by a VoIP connection to the Smart-phone
Unapproved SMS/ MMS (DoS/ Spam/ Worm)	Like unapproved dial, hacker uses the victim's Smart-phone to send spam SMS/MMS through some backdoors or Trojan keeps sending SMS/MMS to some Smart-phone to make DoS attacks, or worm sends MMS message includes a copy of itself as an attachment.
SIM Card cloned	SIM Card is cloned, another person use the cloned SIM Card (STK) to make phone call or something else.
Dial/SMS/MMS redirection	Some malware redirects the dial/SMS/MMS number just before the user press the send key.
Remote wiretapping	Hacker wiretaps the Smart-phone through a VoIP connection or some Trojan send the phone record via email as an attachment
Remote watcher/ Private information theft	Hackers use the DC/DV in the Smart-phone to watch the owner, or steal the private information such as pictures or videos in the Smart-phone

large cohort of subverted smart-phones, attackers can cause damage not only to the Internet but also to the telecom infrastructure [15]. Moreover, Smartphones are often used to store the private or confidential information, which also attracts crackers launching attacks to them. Consequently, attacks to smartphones including worms, email viruses, MMS virus, and Trojan horses have emerged recently with growing frequency.

1.3. The Spread of Epidemic on Smartphones

In order to deal better with the smartphone security problem and provide some guidance for building security scheme for smartphone in the near future, we need study the propagation behavior of the smartphone worms, viruses, and Trojan horses, and identify which factors will influence the propagation. In the literature, these attacks can be modeled as spread of epidemic through the network. So we also leverage it to investigate the spread of epidemics in mobile network. However traditional epidemic models can not be applied to smartphones, because they only consider the static network topology, but the new effects of mobile nodes that bring to the network are not modeled. Firstly, the smartphone is movable between networks. This handheld device may carry epidemics and spread them to devices encountered while moving in different networks, which are not physically connected. Secondly, the smartphones often have multiple networking interfaces, such as GPRS or CDMA for wide area networks, and WiFi or Bluetooth for local area networks. So smartphone can connect to both telecom networks and Internet/enterprise/home networks simultaneously, which would speed up the epidemics spread in both networks.

We use Figure 1 to illustrate the smartphone's new effects on spread of epidemics. In this figure, the enterprise LAN is well shielded by firewalls or security gateways, so it is difficult for worms, viruses and spams to infect the company's computers from outside. With smartphone, two cases of epidemics can happen, For case A, the smartphone connects to both telecom network and the enterprise WLAN, so it becomes a static shortcut between the two networks. An MMS worm from the telecom network may compromise the smartphones and then spread to the enterprise network. In case B, even though the smartphone has only single interface, if it is compromised outside and being carried in the company by an employee, it may infect the computers and smartphones in the LAN, and we call this smartphone creating a mobile shortcut between internal corporate LAN and outside networks. When numerous compromised smartphones (with multiple interfaces) move to other places, and infect more and more smartphones and computers, we can conclude the "static shortcuts" and "mobile shortcuts" may cause the epidemic spread faster than that in normal static network, which are not studied in traditional epidemic models in the literature.

In this chapter, we propose a novel epidemics spread model on mobile network called ESS (Epidemic Spread on Smartphones) model. To the best of our knowledge, this is the first work to model smartphones epidemics. The major contributions of our work are as follows.

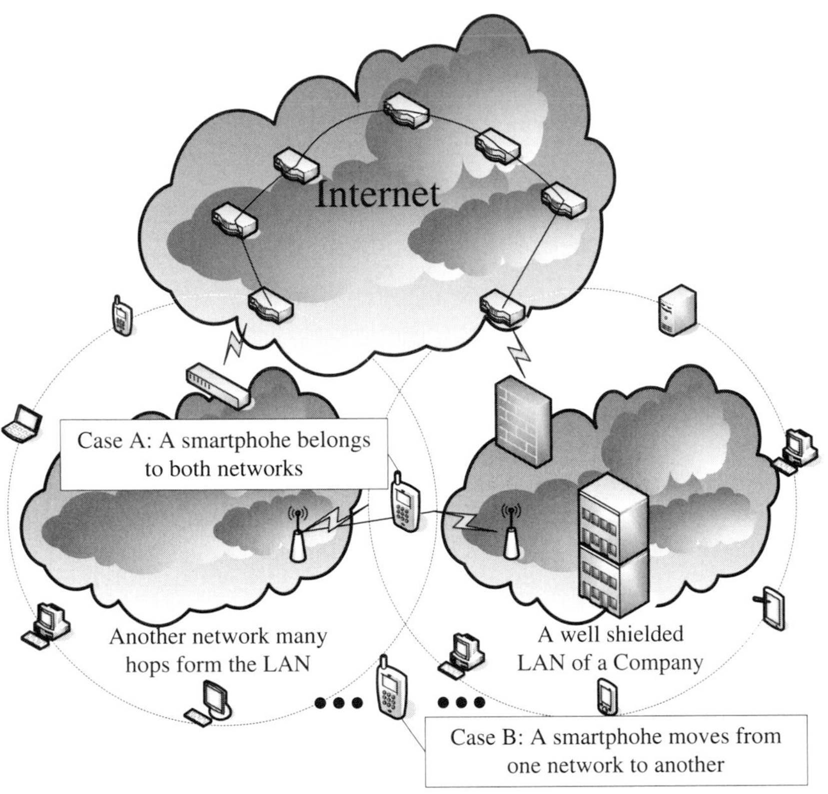

Figure 1. The mobility of smartphone makes itself belongs to both networks. Smartphones may become the "Mobile shortcut" between two networks, thus speed up the propagation of epidemic.

1) We model epidemics spread in mobile networks while taking the aforementioned two unique characteristics of smartphones into account. When smartphone moves, these mobile and static shortcuts change accordingly, and we integrate this dynamics into the factor of network topology in our model. Based on the analysis of the smartphone features and spread of epidemics, the ESS model is a comprehensive SIR (Susceptible-Infected-Removed) model which considers the influence of the epidemic spread rate, the effect of the network topology, the influence of the nomadism of smartphones. And we also model the patching and death rate as well as the initial pre-patch in our proposal, which is also often ignored in previous models[2].

[2] Smartphones are computationally powerful and flexibly programmable, thus patches or shields [16] technology can be applied to smartphones to fix the software vulnerabilities. Pre-patch means some nodes can be patched between time of the software vulnerability release and the time of the corresponding epidemic

2) With this ESS model, we solve the critical conditions problem for the fast die out of an epidemic on smartphones. We give the theoretical analysis on the effect of different parameters and summarize the importance of each parameter in the spread of epidemic in smartphones.

3) Extensive simulations have been performed to verify the ESS model and the critical condition. We have also performed some experiments to study the effect to the spread speed of nomadism and the nodes density. Based on these theoretical analysis and experimental simulations, we give some guidance for prevention attacks on smart-phones.

The remainder of this chapter is structured as follows. The related work is introduced in Section 2. Then we present how epidemics spread on smartphones in section 3. Based on the study in section 3, the ESS model is proposed, and the effect of topology on epidemic spread and the critical conditions for the fast die out of an epidemic are derived in section 4. Section 5 analyzes the proposed ESS model and the individual parameters that affect the propagation behavior of the epidemics. Simulations results in different network topologies are found in Section 6. Section 7 summarizes the chapter and describes further directions to pursue.

2. RELATED WORK

Much work about epidemic or virus spread model has been done in both physics field [17, 20] and computer science field [22, 23, 24] The researchers in physics usually model the general case. Based on these physical models, the researchers of computer science further study the spread of worm [22, 24], or viruses using the contact list [23] (e.g. email). They consider many specific parameters of computer viruses and computer networks. The two most well studied classes of epidemic models are SIS and SIR model [27]. In SIS model, individuals can only exist in two discrete states, namely, susceptible and infected. When an infected individual is cured, it changes back to susceptible one, just like the way of many diseases in the world. While in SIR model, individuals can exist in three discrete states, namely, susceptible, infected and removed. When an infected individual is cured or dies, it becomes remove one. SIR model is similar to the condition in the Internet and smartphones, when a device is patched, it immunes to the certain attack.

Watts and Strogatz presented a simple infectious disease model in [28]. In their model, the contact infection rate is always 1; the nodes of the infection withdraw the system after a unit time. The study of its spread time shows that in the regular network, small world network and random network, the spread time is in direct proportion to the shortest path. This explained the function of the shortest path. With this model, the infectious disease will break out in the whole network for any network. So they can not derive the critical outbreak condition, which would be helpful for prevention.

based upon such vulnerabilities spread. The patch, shield and pre-patch rate have different impact on the propagation of the epidemics

In [18], Newman studies the percolation and epidemic model in small world model. This proposal mapped the epidemic problem into a percolation problem, and found out the threshold of the break out of the epidemic. If the probability of susceptible nodes is greater than the threshold, the epidemic will break out. The study shows that the threshold in small world networks is much smaller than that in regular networks. In [21], the author gives the threshold of arbitrary distribution of vertices degree. However, in these models, there're no special considerations on the mobile nodes which will affect the propagation as we illustrated.

In [19], R. Pastor-Satorras and A. Vespignani study the case on the scale-free networks and point out that there is no similar threshold exists in infinite scale-free networks for the SIS or SIR model. In other words, once infectious disease occurs, it will spread out in big scope. Therefore, only curing the inflected nodes is not enough, changing the structure of the network is also needed. The typical method that breaks the network structure is to quarantine or cut down some connections forcedly. The model for finite scale-free networks is studied in [20, 25, 26]. In such models, they often ignore the difference between patch, remove, death and pre-patch which leads to inaccuracy of the results.

In [22], the authors study the spread of active worm in the Internet. They consider the characteristic of Internet worms such as hit-list, scanning rate, death rate and patching rate, but they assume the worms randomly scan the Internet to find the victims and do not consider the influence of network topology.

In summary, all the previous models treat the network as a static one, focusing on the influence of the distribution of the vertices degree, or network topology. They don't study the effect of mobile nodes in the network as we discussed in the section I, which motivate our work described in the following sections.

3. HOW EPIDEMIC SPREAD ON SMARTPHONES

In this section, in order to model the propagation of the epidemic, we first describe how the epidemic spread on smartphones and then study which parameters will influence the spread speed of the epidemics.

When an epidemic is fired into the mobile Internet connected with many computers and movable smartphones and laptops, it attempts to send itself to vulnerable machines to infect them. The epidemics may spread from one infected device to its neighbors through the following ways.

- Some epidemics may disguise as some interesting game or useful software and be published to the Internet waiting for some smartphone users to download and play;
- Some spread from PC to smartphones by sending epidemic-contained files through email as an attachment, or propagating while the smartphones synchronizing with PC;
- Similarly, smartphons can infect the PCs using the same way;

- Between smartphones, they can infect each other using Bluetooth, WiFi or other wireless connection to scan and spread epidemic to all its neighbors during its movement;

- Smartphones can also send epidemic-contained files through email or MMS as an attachment to infect other smartphones in its contact list;

- The epidemics can also use the neighbors as "hosts" and infect the devices multi-hops from it. For example, a PC-target epidemic may spread from a PC to a smartphone without damage it, and then infect other PCs when they contact with the smartphone;

- Moreover, some epidemics which may infect both smartphone and PC will appear in the near future, because both smartphones and PCs use the similar common operating systems.

After any of the aforementioned propagation successes, a copy of this epidemic is transferred to the new device (smartphone or PC). This newly infected device then tries to infect other devices using the same way at a certain probability, which is influenced by some factors described in the remaining of this section. Hence, the epidemics can spread from PC to smartphone and then infect other PCs, and vice versa. Compromised smartphones may also start attacks to the Internet, and then infect more smartphones.

The spread speed of the epidemic is influenced by many factors. It's mostly determined by its propagating attempt (or in other word determined by its codes), e.g., a worm spread much faster than a download Trojan horse. The topology of the network also influence the epidemic spread greatly, as mentioned in the Section II, epidemic spread much faster in the scale-free network than in the regular network. The mobile nodes also influence the spread speed. If the infected node is a mobile device, such as a smartphone or laptop, acting as mobile shortcuts, it may move to another place with the user and infect the computers and smartphones in other networks.

Correspondingly, there are some prevention approaches which can slow down the spread speed of the epidemic. Nowadays, the epidemics often come after the announcement of vulnerability [30]. After the announcement of vulnerability and related patch available, some cautious people pre-patch their smartphone or PC to make their machine immune to this epidemic. The more devices are patched between the announcement of vulnerability and the appearance of associated exploit code, the harder the epidemics spread. When the attack is detected, more people will try to slow it down or stop it. The patch or shield, which repairs the security vulnerability of the devices, is widely used to defend against the epidemic. When an infected or vulnerable node is patched, it becomes an invulnerable node. During the process of epidemic spreading, some nodes might stop functioning properly, crashed, or be shutdown or at least made offline by the users; all these make the infected nodes eliminated in the network.

4. EPIDEMIC SPREAD MODEL ON SMARTPHONES

In this section, we describe a comprehensive model of epidemic spread on smartphones which considers the influence of the various factors mentioned in Section III. For convenience we introduce the basic ESS model which considers the death rate, patch rate and the nomadism feature of smartphones at first. After that, we enhance the basic ESS model with topology effect and mobile shortcuts effect, and present the final ESS model.

Table 2 lists the parameters used in the spread of epidemic on the mobile network.

4.1. The Basic ESS Model

The ESS model is a comprehensive model which combines the influence of the epidemic spread rate, patching and death rate, the effect of the network topology, as well as the influence of the nomadism of smartphones.

In this chapter we focus on SIR (susceptible-infective-removed) model. In SIR models, a population of N individuals is divided into three states: susceptible (S), infective (I), and removed (R). In this context "removed" means individuals who are either recovered from the disease and immune to further infection, or dead.

However, the traditional SIR model, such as Kermack- McKendrick model [31], doesn't consider the uniqueness of epidemics on the computers and smartphones such as patch rate. According to the characteristics of the spread of epidemic on computers and smartphones, "removed" means either vulnerable nodes (includes the infected nodes) are patched and immune to further infection, or infected nodes die and eliminated from the network. We use d to denote the *death rate* and p to denote *patch rate*.

If an infected node moves to other network clusters, it may infect nodes in those clusters. The increase of inflected node should plus the nodes that are infected because of the movement. We use ϕ to denote the *density of mobile shortcuts*, and use $m(t)$ to denote the *average move speed* (move times in unit time) of mobile nodes at time t.

Let $S(t), I(t)$ and $R(t)$ denote the *proportion of vulnerable nodes*, the *proportion of infected nodes* and the *proportion of removed nodes* at time t ($t \geq 0$) respectively, and use S', I', R' to denote the increment of $S(t), I(t), R(t)$ (i.e. $S'(t), I'(t), R'(t)$, the derivative of $S(t), I(t), R(t)$). The infective nodes contact with randomly chosen nodes of all states at an average rate α per unit time. At the beginning of the epidemic spread $I(0) = I_0$, and $0 < I_0 \ll 1$ is a very small proportion of the total number of vulnerable nodes.

Assumes that the nodes are fully mixed, meaning that the individuals with whom a susceptible individual has contact are chosen at random from the whole nodes (the effect of network topology will be considered in the following subsection), all individuals have approximately the same number of contacts at the same time, and that all contacts transmit the disease with the same probability. In the time t the newly infected nodes because of normal contact is $\alpha S(t)I(t)$. At the same time, some mobile node move to other places and infect some more nodes, the newly infected nodes because of the nomadic nodes is $\alpha S(t)I(t)\phi m(t)$. Because of patch and death $(d+p)I(t)$ infected nodes and $pS(t)$ susceptible nodes are removed. Then we get the basic ESS model:

Table 2. Notions Used in The Model

Notion	**Explanation**
t	Time
$S, S(t)$	Proportion of susceptible nodes
$I, I(t)$	Proportion of infected nodes
$R, R(t)$	Proportion of removed nodes (patched or dead)
S', I', R'	The derivative of $S(t), I(t), R(t)$, i.e. $S'(t), I'(t), R'(t)$
S_0, I_0, R_0	Initial value of $S(t), I(t), R(t)$, i.e. $S(0), I(0), R(0)$
$S_k(t)$	Proportion of susceptible nodes in the group of vertices degree k
$I_k(t)$	Proportion of infected nodes in the group of vertices degree k
$R_k(t)$	Proportion of removed nodes in the group of vertices degree k.
N	Total vulnerable nodes
λ	Infected probability of each link
α	Epidemic spread speed (i.e. contacted rate)
d	Death rate
p	Patching rate
t_0	The average time between the announcement of vulnerability and the appearance of associated exploit code
p_0	Patching rate during t_0
ϕ	The density of "mobile shortcut"
$m, m(t)$	The average move speed (move times in unit time) of mobile nodes
$P(k)$	Vertices degree distribution of the whole network
$Q(k)$	Vertices degree distribution of the mobile vertices

$$\begin{cases} S' = -\alpha S(t)I(t) - \alpha S(t)I(t)\phi m(t) - pS(t) \\ I' = \alpha S(t)I(t) + \alpha S(t)I(t)\phi m(t) - (d+p)I(t) \\ R' = pS(t) + (d+p)I(t) \end{cases} \tag{1}$$
$$0 < S(t), I(t), R(t) < 1, S(t) + I(t) + R(t) = 1, \alpha, \phi, d, p > 0.$$

When $I' = \alpha S(t)I(t) + \alpha S(t)I(t)\phi m(t) - (d+p)I(t) < 0$, the epidemic will die out, assume the mobile nodes move at uniform velocity m, and now the sufficient condition of epidemic dies out is $\frac{\alpha(1+\phi m)S(t)}{d+p} < 1$, or $\frac{\alpha(1+\phi m)(1-I(t)-R(t))}{d+p} < 1$.

As mentioned before, some nodes are pre-patched between the announcement of vulnerability and the appearance of associated exploit code. We denote this period of time as t_0, and the pre-patch rate in this period is p_0, and $R_0 = p_0 t_0$. If we want to restrict the spread of epidemic from the beginning, $\frac{\alpha(1+\phi m)(1-I_0-R_0)}{d+p} < 1$ should be satisfied.

The epidemiological threshold is now:

$$\frac{\alpha(1+\phi m)(1-I_0-p_0 t_0)}{d+p} < 1. \tag{2}$$

If the sufficient condition (2) is satisfied, the epidemic will die out and not spread all over the network. Usually, at the beginning of virus spread, the infected nodes are a very small set, i.e. I_0 is very small and can be ignored in (2).

4.2. The Extended ESS Model for Smartphones

Both the topology of network and the nomadism of mobile nodes can influence the spread of the epidemic. We will enhance the basic ESS model (1) by adding the effect of topology and the effect of nomadism in this subsection.

Effect of Topology on the Spread of Epidemic

In our model, we assume a connected network $G = (N; E)$, where N is the number of nodes in the network and E is the set of edges. The edges of a node are the set of links to nodes with whom the node may have contact during the time it is infective, such as the devices that in the same subnet, in the email or phone call contact list, next to the node and can build up a Bluetooth connection, and so forth.

So we can vary the number of connections of each node by choosing a particular degree distribution for the network. We use λ to denote the infected probability of each link (assumes that λ is a universal infection rate for each edge connected to an infected node and independent with the vertex degree of the node).

Let us assume initially that the vertices degree distribution is $P(k)$. For the group of vertices that have the same vertex degree k, in time t the proportion of newly infected nodes because of normal contact is now: $\lambda k S_k(t)\Theta(t)$, where $I_k(t)$ and $S_k(t)$ denote the probability of infected nodes and the probability of susceptible nodes in the group of vertex degree k, and $\Theta(t)$ is the probability of a randomly chosen link has an infected node and a susceptible node in each side, since the epidemic only spreads through such type of links. Because the node has higher vertex degree is more possible to have an infected node connect to it, we get:

$$\Theta(t) = \frac{1}{\overline{k}} \sum_{k} kP(k)I_k(t) \sum_{k} P(k)S_k(t), \tag{3}$$

where $\overline{k}$ is the mean degree of the network, $\overline{k} = \sum_k kP(k)$.

The number of infected nodes in the group of vertices degree k been removed in time t because of death and patch is: $(d + p)I_k(t)$

Effect of Nomadism on the Spread of Epidemic

As mentioned before, when an infected node moves to other network clusters, it may infect other nodes in the clusters. It becomes a mobile shortcut between the two network clusters (although it may not connect to the first cluster now, but the epidemic has been taken to the second cluster by it). The increase of inflected node should plus the nodes that are infected because of the movement.

$$\begin{aligned} \Delta I_{m,k}(t) &= \lambda k S_k(t)\Theta_m(t), \\ \Theta_m(t) &= \tfrac{1}{\overline{k}} \textstyle\sum_k kP(k)P_m(k)I_k(t) \sum_k P(k)S_k(t), \end{aligned} \tag{4}$$

where $\Delta I_{m,k}(t)$ is the incensement of infected nodes in the group of vertex degree k in time t because of the movement of the mobile nodes, and $\Theta_m(t)$ is the probability of a randomly chosen link connecting an infected mobile node and a susceptible node after movement. Moreover, $P_m(k) = \frac{\phi m(t)Q(k)}{P(k)}$, where $Q(k)$ is the distribution of the degree of the movable nodes (The probabilities of mobile nodes are different in different group of degree. The leaf nodes of a network which has a small vertex degree may have a higher probability to be a mobile node. But the kernel nodes of the network which connect a lot of nodes are more likely to be static nodes. There are no such previous researches on the distribution of vertices degree of mobile networks; we use some hypothetic distribution in this chapter).

Now, we get the ESS model consider both the effect of topology and the effect of mobile shortcuts on the spread of epidemic:

$$I'_k = \lambda k S_k(t)(\Theta(t) + \Theta_m(t)) - (d + p)I_k(t). \tag{5}$$

Redefine Θ as

$$\Theta = \frac{1}{\overline{k}} \sum_{k} kP(k)[1 + P_m(k)]I_k(t) \sum_{k} P(k)S_k(t). \tag{6}$$

Then the ESS model (5) becomes

$$I'_k = \lambda k S_k(t)\Theta(t) - (d + p)I_k(t).$$

Eventually, we get the final ESS model

$$\begin{cases} S'_k = -\lambda k S_k(t)\Theta(t) - pS_k(t) \\ I'_k = \lambda k S_k(t)\Theta(t) - (d+p)I_k(t) \\ R'_k = dI_k(t) + p(S_k(t) + I_k(t)) \end{cases} \tag{7}$$
$$0 < S_k(t), I_k(t), R_k(t) < 1, S_k(t) + I_k(t) + R_k(t) = 1, \lambda, d, p > 0.$$

5. ANALYSIS OF THE ESS MODEL

In this section, we give the critical conditions for the fast die out of an epidemic from our model. We also give the analysis on the effect of different parameters and summarize the importance of each parameter in the spread of epidemic in smartphones.

5.1. A Critical Condition for Epidemic Fast Die Out

In this section, we want to find out the epidemiological threshold from (7).

Because $S_k(t) + I_k(t) + R_k(t) = 1$ applying it to formulation (7), we get the following equations:

$$I'_k = \lambda k S_k(t)\Theta_m(t) - (d+p)(1 - S_k(t) - R_k(t)), \tag{8}$$
$$I'_k = \lambda k(1 - I_k(t) - R_k(t))\Theta_m(t) - (d+p)I_k(t). \tag{9}$$

To find out the critical condition, let (8)=0 and (9)=0, then

$$S_k(t) = \frac{(d+p)(1-R_k(t))}{d+p+\lambda k\Theta}, \tag{10}$$
$$I_k(t) = \frac{\lambda k(1-R_k(t))\Theta}{d+p+\lambda k\Theta}. \tag{11}$$

Apply (10) and (11) to (6) we get the following equation:

$$\Theta = \frac{1}{\bar{k}}\sum_k P(k)[1+P_m(k)]\frac{\lambda k(1-R_k(t))\Theta}{d+p+\lambda k\Theta}\sum_k P(k)\frac{(d+p)(1-R_k(t))}{d+p+\lambda k\Theta},$$
$$0 < \Theta \leq 1.$$

We can find that when the epidemic just appears ($t \to 0$) or a very long period after the epidemic begin to spread ($t \to \infty$), $\Theta \to 0$. Using Taylor expansion, when $\Theta \to 0$

$$\Theta = \{\frac{1}{\bar{k}(d+p)}\sum_k \lambda k^2 P(k)[1+P_m(k)][1-R_k(t)]\sum_k P(k)[1-R_k(t)]\}\Theta + A\Theta^2 + \cdots$$

Then, the critical condition can be got as follows.

$$\frac{\lambda}{\overline{k}(d+p)}\sum_k k^2P(k)[1+P_m(k)][1-R_k(t)]\sum_k P(k)[1-R_k(t)] = 1. \quad (12)$$

When the epidemic just appears ($t \longrightarrow 0$), the proportions of removed nodes in every group are almost equal, we can get $\frac{\lambda(1-R_0)^2}{(d+p)\overline{k}}\sum_k k^2P(k)[1+P_m(k)] = 1$, since we have $R_0 = p_0t_0$, then

$$\frac{\lambda(1-p_0t_0)^2}{(d+p)\overline{k}}\sum_k k^2P(k)[1+P_m(k)] = 1. \quad (13)$$

Because $P_m(k) = \frac{\phi m(t)Q(k)}{P(k)}$, assume the mobile nodes move at uniform velocity m, then the critical condition (13) becomes

$$\frac{\lambda(1-p_0t_0)^2}{(d+p)\overline{k}}\sum_k k^2[P(k)+\phi mQ(k)] = 1. \quad (14)$$

And $\sum_k k^2P(k) = \overline{k^2} = \overline{k}^2 + Dev(k)$, we get

$$\frac{\lambda(1-p_0t_0)^2}{(d+p)\overline{k}}[\overline{k}^2 + Dev(k) + \sum_k \phi mQ(k)] = 1. \quad (15)$$

This critical condition shows that if two networks have the same mean vertices degree, the network which has larger deviation $Dev(k)$ is more vulnerable to virus spread.

5.2. Analysis on the Effect of Different Parameters

After getting the critical condition of the spread of epidemic in mobile networks, we would like to analyze the model and find out the influence of each parameter, and find efficient defensive way further.

In SIR model, the state of "Removed" includes the dead nodes and the patched nodes. Although both of them are removed from the flow of propagation, they have totally different features. The patched nodes are healthy nodes, but the dead ones mean that the damage already taken. Hence, the prevention of epidemic should not only prevent the epidemic from spread but also make the patched ratio as higher as possible.

From the ESS model (7), the increment of the proportion of the patched at time t is $p(S_k(t) + I_k(t))$. It's hard to give the exact solution of the final patched ratio when the epidemic levels off, in the performance evaluation section, we do experiments to give some numerical results.

In the viewpoint of epidemiology, the defense ways of disease can be divided into three kinds: prophylaxis, quarantine and cure. In the following part, we'll analyze the parameters in (14) and present relative defense ways. For convenience, the analyses of all parameters are listed in Table 3.

In the critical condition (14), we can see that it's affected by network topology, patch rate and death rate, as well as pre-patch ratio. In the equation, $\sum_k k^2[P(k) + \phi m Q(k)]/\overline{k}$ reflects the influence of topology. We call this factor *Topology Factor* and denote it as T.

Inside the Topology Factor there are following parameters. The first one is the distribution of the vertices degree k in the whole network $(P(k))$, which is determined by the topology of the whole network. Network topology can be changed to restrict the spread of epidemic. The effective way includes changing the routing table, setup firewall, quarantining infected subnet, or setting black list in the gateways, etc. The second parameter is the distribution of the "mobile vertices" $(Q(k))$. The number of mobile device increase rapidly in recent years, it makes the density of mobile device in a certain area increase rapidly too. And with the development of wireless technology, the smartphones will connect more and more mobile devices within a single hop. All these will increase the mean value of the degree distribution of the mobile vertex. It's hard to restrict the spread of epidemic by changing $Q(k)$, unless we limit the access right of mobile node to the Internet. The third parameter inside topology factor is the density of mobile shortcuts in the whole network. The more nodes are mobile nodes the faster the virus may spread. The mobile nodes increase the mix degree of the whole vertices and cause the virus spread faster. Therefore, the frequent movements make the infective mobile nodes spread the epidemic widely. According to the trend of smartphone and computer market, the influence of mobility will become more and more significant.

In the equation (14), there are some other parameters affecting the critical condition including the death rate and patch rate. The higher death rate (d) is, the more hardly the worm spread. But death means that the infected nodes may be crashed or unable to access the Internet, which is we unwilling to see (And most worms do not crash the computer, Witty was the first widely propagated Internet worm to carry a destructive payload, it tries to destroy the system after sending 20,000 packets). Patch rate influences the denominator of the critical condition, the higher the better. We can also see that if we only patch the vertices after the virus appear and take no other prophylactic treatment , the virus will spread out unless $(d + p)$ is greater than λT. It means that to avoid the virus spread, $(d + p)$ should at least be the same magnitude as the virus spread rate λ, while this would be very difficult when we suffer from fast-spreading worms where $(d + p)$ will not catch up with the worm's spread speed. Moreover, even if we have a very high speed patch method to satisfy the condition $((d + p) >= \lambda T)$, they may still cause congestion in the network just like there are two kind of worm spreading in the same period (Thinking about the way using AntiBlaster to remove the Blaster).

As we can see in the formulation of the critical condition, there is $(1 - p_0 t_0)^2$ in it. It's not a linear change when we increase this pre-patch ratio $(p_0 t_0)$. Hence, pre-patch is very important for preventing the spread of epidemic. However, we can not explicitly control the time (t_0) between the announcement of vulnerability and the appearance of associated exploit code since it is determined by the exploit coding difficulty and

Table 3. The Effect of Different Parameters

Parameters	Descriptions	Actions needed to defense epidemics	Effectiveness [a]
λ	The spread rate of the epidemic in each link.	Decrease	Moderate, but hard to adjust[b]
T	The total effect of topology, including the effect of static and mobile topology	Decrease	Moderate, inconvenience to adjust, quarantine is the common way
$P(k)$	The distribution of the vertices degree k of the whole network	Decrease mean value and standard deviation	Moderate. Some topologies may cause the T tending to infinite
$Q(k)$	The degree distribution of the "mobile vertex".	Decrease mean value and standard deviation	Moderate. Some topology may cause the T tending to infinite
ϕ	The percentage of "mobile shortcuts" in the whole network.	Decrease	Moderate. Hard to manually control it, and it would become larger since number of smartphones will increase
m	The move speed (average move times in unit time).	Decrease	Moderate
d	The death rate. Determined by the function way of epidemic and the action of people.	Increase	Moderate, but increasing death rate takes much damage
p	Patch rate after the virus appeared	Increase	Moderate, There're some approaches to increase it, while having difficulties[c].
t_0	The time between the announcement of vulnerability and the appearance of associated exploit code.	Increase	Great, but hard to increase, since it 's determined by the coding difficulty, the virus maker's interest, etc.
p_0	Patching rate during t_0	Increase	Great, we have effective ways to increase p_0

[a] The effectiveness of each parameter to prevent the epidemic from spread out.

[b] The spread rate of the epidemic is mostly determined by the exploit code and the capability of victim devices, little can be done to reduce it.

[c] Some factors slow down the increase of patch rate after the virus appears. A) Patch needs more tests and evaluation before it's installed. B) Users may not patch their computer timely due to lack of professional skills or poor network conditions, but viruses exhaust the system to spread itself very fast. C) The patch size is usually larger than the size of virus and there often exists bottleneck at the patch servers, they would also slow down the patch speed.

the virus maker's interest, etc. However, we can increase the pre-patch speed (p_0) and make most of the devices immune the epidemic from the beginning for smartphones, e.g, mobile network operators can enforce patch to their subscribers.

In summary, as analyzed in Table 3, changing the topology of network can influence the spread of epidemic, but it's hard and brings inconvenience to the quarantined devices. Reducing spread rate λ and increasing death rate d can have moderate effect to slow down the spread of epidemic, but it's hard to adjust them explicitly and dependant on the user behavior. Patch is a good way, but after the epidemics start propagation, the patch rate might not be higher enough to stop the spread. Pre-patch takes great effect on preventing the spread of epidemic and there're some effective ways to increase the pre-patch speed.

6. PERFORMANCE EVALUATION

We have performed extensive simulation of epidemic spread to validate the ESS model and check our analytic results, and to investigate further the behavior of the models under typical network topologies including small world network, Waxman random network, and power-law network. We have also conducted some experiments to analyze the effect of individual parameters in the proposed ESS model.

The simulations are performed using Matlab. In order to compare the simulation results with the analysis results, firstly, we educe the discrete-time ESS model. And then we use Matlab to generate some network topologies and simulate the spread of epidemic in these networks. Finally, we compare the simulation results with the analysis results in the same topology or the critical condition we derived from the equation (14). The process of simulations will be described amply in the following subsections.

6.1. Comparing ESS Model with Simulation Result

In order to perform numerical calculations, we transform the continuous-time ESS model (7) into discrete-time ESS model. Let $S_{k,t}$, $I_{k,t}$ and $R_{k,t}$ denote the number of vulnerable nodes, the number of infected nodes and the number of removed nodes in the group of vertex degree k at time tick $t(t \geq 0)$ respectively. For the group of vertices that have the same vertex degree k, from time tick t to time tick $t+1$ the newly infected nodes because of normal contact is $\lambda k S_{k,t}\Theta$, where $\Theta = \frac{1}{\bar{k}} \sum_k k[P(k) + \phi m Q(k)] I_{k,t} \sum_k P(k) S_{k,t}$. And the number of infected nodes in the group of vertex degree k been removed after time stick t because of death and patch is $(d+p)I_{k,t}$. Then we get the discrete-time ESS model (For convenience we only list the infected part):

$$I_{k,t+1} = I_{k,t} + \lambda k S_{k,t}\Theta - (d+p)I_{k,t} \tag{16}$$

We begin each simulation with a set of randomly chosen infected nodes and a set of randomly chosen pre-patched nodes on a given network topology (the number of initially-infected nodes and pre-patched nodes does not affect the equilibrium of the propagation). And a set of randomly chosen mobile nodes are also initialized according to the "density of mobile shortcut" ϕ and the "degree distribution of mobile nodes" $Q(k)$. Simulation proceeds in steps of one time unit. During each step, an infected node attempts to infect each of its neighbors with probability λ, and a mobile node moves to another network cluster with speed (probability) m. In addition, every node is patched with probability p, and every infected node is dead with probability d.

Small World Network

The small world phenomenon is the theory that everyone in the world can be reached through a short chain of social acquaintances. Previous study of many researchers discovered that many real networks have such small world phenomenon. Watz and Strogatz define the following properties of a small world graph[28]:

1. The clustering coefficient C is much larger than that of a random graph with the same number of vertices and average number of edges per vertex.

2. The characteristic path length L is almost as small as L for the corresponding random graph. C is defined as follows: If a vertex v has k_v neighbors, then at most $k_v * (k_v - 1)$ directed edges can exist between them. Let C_v denote the fraction of these allowable edges that actually exist. Then C is the average over all v.

Figure 2 shows the simulation result compared with the analysis result derived from our model. The results are rather satisfied: our model yields precisely to the simulation result which demonstrates the effectiveness of the ESS model.

Power-law Network

Power-law networks [29, 33, 34], including current Internet, are characterized by an uneven distribution of connectedness. The nodes in these networks do not have a random pattern of connections, instead, some nodes act as "very connected" hubs, which dramatically influences the way the network operates.

The degree distributions of power-law networks have power-law tails, i.e. $P(k) \sim k^{-\tau}$, typically $2 < \tau \leq 3$.

Figure 3 shows the time evolution of epidemic in a power-law network (generated using Inet model [35]). Our model conforms very close to the simulation results.

Waxman Random Network

In the random network, a (fixed) set of nodes is distributed in a plane uniformly at random. A link is added between each pair of nodes with a certain probability. The Waxman method [36] is an instantiation of this method where the probability of adding a link is given by:

$P(u, v) = \alpha e^{-d/\beta L}$, where $0 < \alpha, \beta \leq 1$, d is the Euclidean distance from node u to v, and L is the maximum distance between any two nodes.

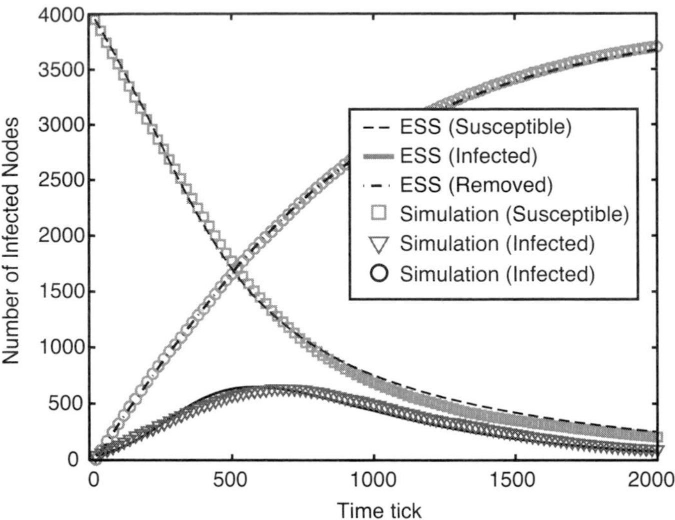

Figure 2. Simulation result on small world network with vertices number $N = 4000$, and average vertices degree is 6.0. $\lambda = 0.0025$, $m = 0.1$, $d = 0.001$, $p = 0.001$, $\phi = 0.02$ and $p_0 t_0 = 0$.

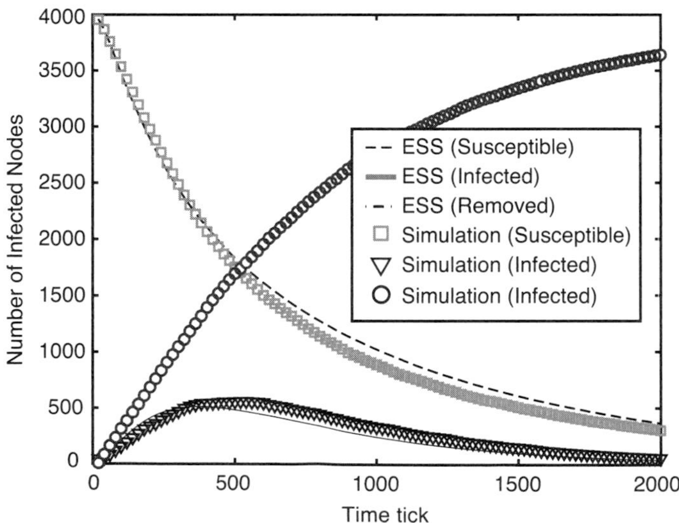

Figure 3. Simulations on power-law network with vertices number $N = 4000$, and average vertices degree is 3.3218. $\lambda = 0.0015$, $m = 0.0005$, $d = 0.001$, $p = 0.001$, $\phi = 0.02$ and $p_0 t_0 = 0$.

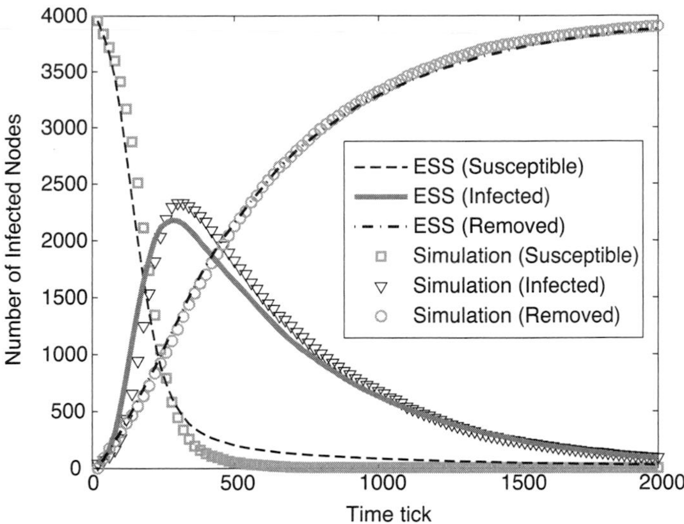

Figure 4. Comparison of ESS Model and simulation result on Waxman network with vertices number $N = 4000$, and average vertices degree is 15.5520. $\lambda = 0.002$, $m = 0.001$, $d = 0.001$, $p = 0.001$, $\phi = 0.02$ and $p_0 t_0 = 0$.

The Waxman random network is much like some wireless network, such as Bluetooth and WiFi network, where a mobile device only connects to its neighbors within a certain distance using their wireless connection. Studies in this kind of network topologies are helpful to learn the spread of epidemic on smartphones, because nowadays a lot of smartphone worms spread through Bluetooth connections.

Figure 4 shows the simulation result compared with the analysis result derived from our model. Our model conforms very close to the simulation results.

6.2. Critical Condition Verification

In this subsection, we measure the point of epidemic threshold for the fast die out of epidemic for comparison with our analytic results. We generate the network topology and simulate the spread of epidemic in these networks with Matlab. The procedure of spread of epidemic is simulated as follows. Each infected node scans its neighbors one by one, and randomly infects them. Mobile nodes will move to another cluster, and all nodes will be patched, or die if infected in some random way. We then calculate at each step the size of infected nodes, susceptible nodes and removed nodes. The position of the percolation threshold can then be estimated from the point at which the derivative of this size with respect to the number of infected nodes takes its maximum value. Since there are N nodes on the network in total and the action of infecting, moving, patching and death takes time $O(N)$, such simulation runs in time $O(N^2)$.

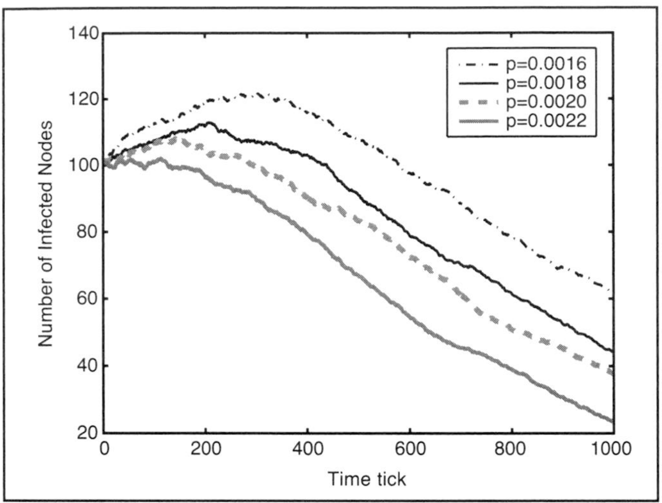

Figure 5. Simulations on the small world network. All case for $N = 10000$ vertices network, with the "small world" short cut density of 0.01. The average vertices degree $k = 8$, and the Topology Factor $T = 8.0037$, $\lambda = 0.0004$, $p_0 t_0 = 0$, $\phi = 0.01$, $m = 0.01$, and $p = 0.0016, 0.0018, 0.002$ and 0.0022 (form top to bottom).

In order to make the figures clear, we only change one parameter in one group of simulations. According to the simulation setup, we can calculate the critical condition of the chosen parameter in advance using equation (14). And then, we choose a set of value near the critical condition to perform the simulations, and the figure of simulation results can reflect the real critical condition. All the experiments validate the critical condition (14) we presented in section 5.

Small World Network

In Figure 5 we show the simulation result on a small world network which has $N = 10000$ vertices and the static shortcut density is 0.005. The average vertices degree $k = 8$, and the Topology Factor $T = 8.0037$. In these simulations we randomly choose 1% of nodes as mobile nodes i.e. $\phi = 0.01$, and fix $\lambda = 0.0004$, $p_0 t_0 = 0$, $m = 0.01$, $d = 0.001$, and do the simulation in condition of $p = 0.0016$, 0.0018, 0.002 and 0.0022. Following the critical condition (14), when the patch rate p is equal to 0.00220148 the critical condition equals to 1. The bottom one is very close to the epidemic threshold. As we can see, the number of infected nodes of the epidemic for $p = 0.0022$ does not increase from the beginning and then peter out because of patch and death. Once we get above the epidemic threshold a large number of cases appear and then peter out slowly because of death and patch.

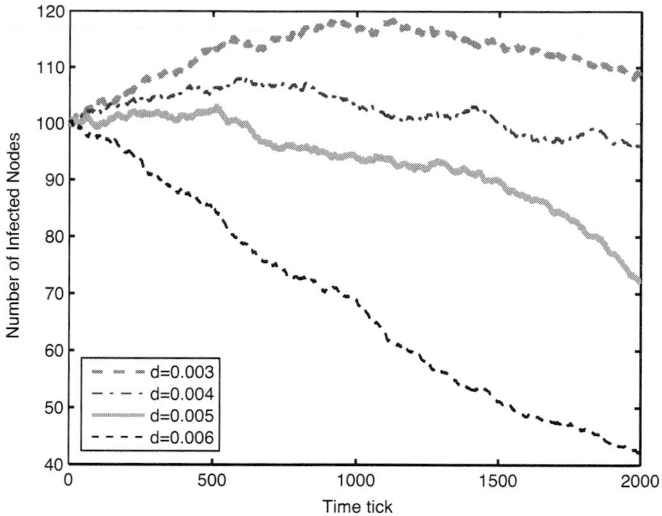

Figure 6. Simulation on the power-law network with vertices number $N = 10000$, average vertices degree k=4.1151, and Topology Factor $T = 19.8087$. $\lambda = 0.0004$,$\phi = 0.02$,$m = 0.001$, $p = 0.003$, $p_0t_0 = 0$, and $d = 0.003$, 0.004, 0.005 and 0.006 (form top to bottom).

Power-law Network

We generate a power-law network using Inet model. The network has $N = 10000$ nodes, the average vertices degree $k = 4.1151$, and the Topology Factor $T = 19.8087$. In these simulations we fix $\lambda = 0.0004$, $m = 0.001, p = 0.003$, $\phi = 0.02, p_0t_0 = 0\%$, and then simulate in the condition of death rate $d = 0.003$, 0.004, 0.005 and 0.006. Following the critical condition (14), if the death rate d is lower than 0.00049235 the epidemic will break out. The last one is below the epidemic threshold and the third one is very close to the epidemic threshold. As Figure 6 shows, the number of infected nodes of the epidemic for $d = 0.006$ die out fast, and while others increase first then decrease because patch and death.

Waxman Random Network

Figure 7 shows the simulation result of the infected nodes number as a function of time on a Waxman random network which has $N = 10000$ vertices, average vertices degree $k = 6.5195$, and Topology Factor $T = 10.009$. In these simulations we take the density of mobile shortcut $\phi = 0.02$, death rate $d = 0.0002$, patch rate $p = 0.003$, mobile nodes move speed $m = 0.001$, and no pre-patch applied (i.e. $p_0t_0 = 0$). The curves in the figure have (from bottom to top) the epidemic spread rate per link $\lambda = 0.0004, 0.0005, 0.0006$ and 0.0007, which implies, following (14), that the

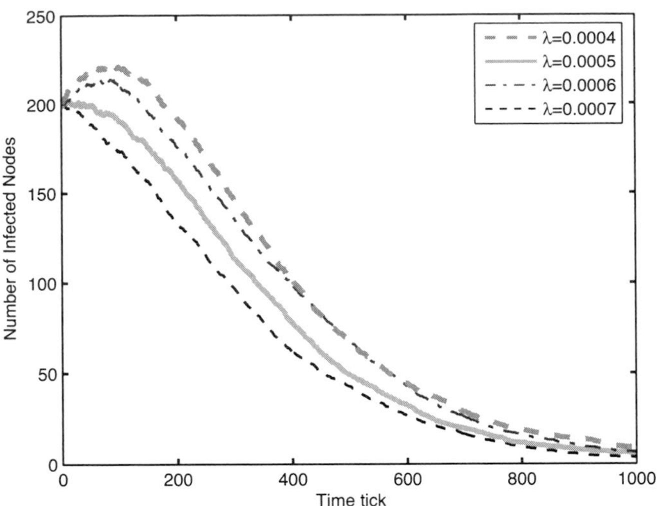

Figure 7. Simulation on the random Waxman network with vertices number $N = 10000$, average vertices degree $k = 6.5195$, and Topology Factor $T = 10.009$. $\phi = 0.02$, $m = 0.001$, $d = 0.002$, $p = 0.003$, $p_0 t_0 = 0$, and $\lambda = 0.0004, 0.0005, 0.0006$ and 0.0007 (form bottom to top).

epidemic will break while the epidemic spread rate per link λ is greater than 0.00049955. Only the bottom one is below the epidemic threshold. As we can see, the number of infected nodes of the epidemic for $\lambda = 0.0004$ shows that the epidemic die out directly without getting more nodes infected. And the second curve with $\lambda = 0.0005$, which is almost equal to the epidemic threshold, maintains its infected nodes number for a little while and then peter out because death and patch. Once we get above the epidemic threshold, the number of infected nodes increases, which indicating the onset of epidemic behavior.

6.3. The Effect of Mobile Shortcut Density and Move Speed

The previous researches consider either influence of the distribution of the vertices degree or the density of shortcut. They don't study the mobility of the nodes in the network. But the density of mobile shortcut influence greatly on the spread of epidemic. Hence, the previous models not accurate enough on the network of smartphones.

Figure 8 illustrates the simulation result of the spread of epidemic in a small world network with the total nodes of $N = 1000$, the density of static shortcut is 0.01, average vertices degree is 6, and the move speed of mobile nodes $m = 0.3$. We start the simulation at an initial state of zero mobile nodes in the network and then increase the mobile shortcut density step by step. As we can see, the traditional SIR model (i.e. $\phi = 0$, the bottom curve) is inaccurate in mobile condition: the number of infected

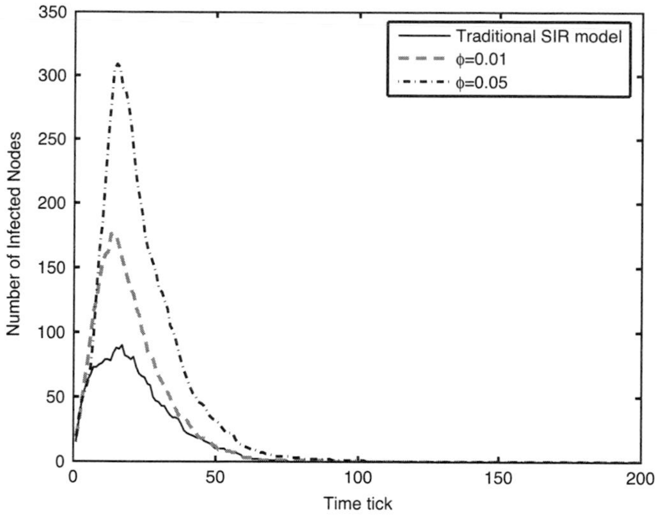

Figure 8. The effect of mobile shortcut on a small world network with total nodes number $N = 1000$, the density of static shortcut is 0.01, the average vertices degree $k = 6$, move speed of mobile nodes is $m = 3$, and mobile shortcut density $\phi = 0.01$, and 0.05, compared with traditional SIR model.

nodes increase slower than real condition in mobile network, even a very few (1%) mobile shortcuts exist in the network.

Figure 9 illustrates the simulation result of the same network topology with mobile shortcut density $\phi = 0.05$. We speed up the move speed of mobile nodes step by step. The simulations show that with the increase of m from 0 to 0.9, at first, the spread speed of the epidemic increase very fast and then slow down. It's because that with the movement of mobile nodes, they act as shortcut between deferent network clusters, this greatly decrease the diameter of the network; when the nodes move faster there are some mobile shortcuts duplicated, and this makes the increase of epidemic spread speed slow down.

6.4. The Effect of the Uptrend of Peering Spread of Smartphone

In these years, more and more smartphones connect to the Internet. The number of smartphones may even exceed the number of computers in the foreseeable future. As mentioned before the epidemics can spread from one smartphone to another locally with Bluetooth or WiFi connection, what will happen with the rapid growth of smartphone? This uptrend firstly increase the density of mobile shortcut density ϕ directly. And with more and more smartphone in the world, the density of smartphone in a certain area will increase too. Hence, the uptrend of smartphones will also increase the connectivity degree of most nodes.

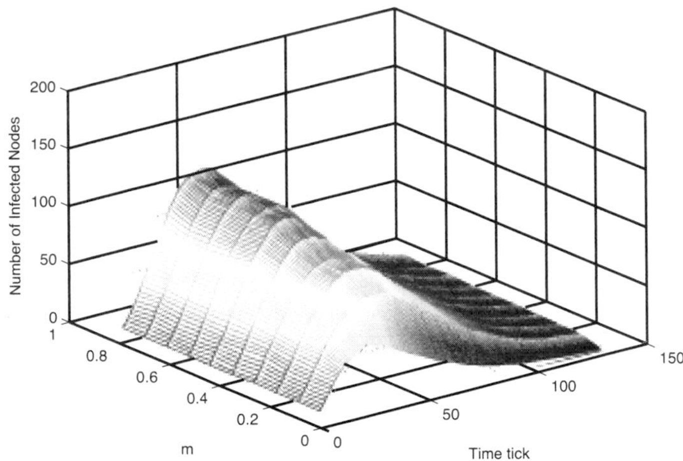

Figure 9. The effect of mobile shortcut on a regular network with nodes number $N = 1000$, degree of each node $k = 6$, mobile shortcut density $\phi = 0.05$, and move speed of mobile nodes $m = 0, 0.1, 0.2 \ldots 0.9$.

We simulate this uptrend in the Waxman random networks with fixed maximum distance L and fixed connection probability α, β between any two nodes. We assume that the initial static nodes and mobile nodes are 90% and 10% and the total number of the nodes is initialized as 2000. The increase of static nodes and mobile nodes are 10% and 30% and all these nodes are placed in a 10x10 area. The configuration of each experiment is listed in Table 4.

In Figure 10 we show the results of the mean degree and the Topology Factor of the networks as a function of the nodes density on Waxman Random networks. We can see that they are two straight lines in the logarithmic axes. It means that the mean degree and the Topology Factor are power-law functions of the nodes density (and nodes density is an exponential function of time).

Figure 11 shows the simulation results of the relationship of nodes density and the spread of epidemic. The spread speed increase very fast when the nodes density increases. Hence, the threat of epidemic will be more and more serious with the uptrend of smartphones in the future.

6.5. Protection of Epidemic

The purpose of the studies of epidemic is to provide some guides to prevent the spread of epidemic in the future. As mentioned in Section "Analysis of the ESS Model", in the "Removed" nodes, only the patched nodes are healthy. Hence, the defense of epidemic should not only prevent the epidemic from spread but also make the patched ratio as higher as possible.

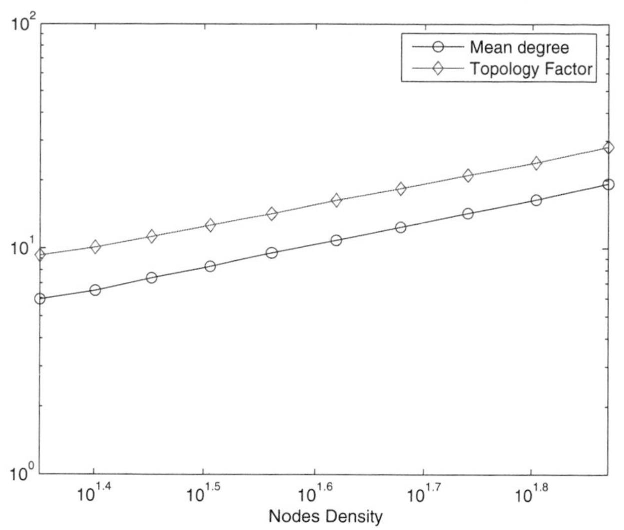

Figure 10. Mean degree and Topology Factor are power-law functions of nodes density.

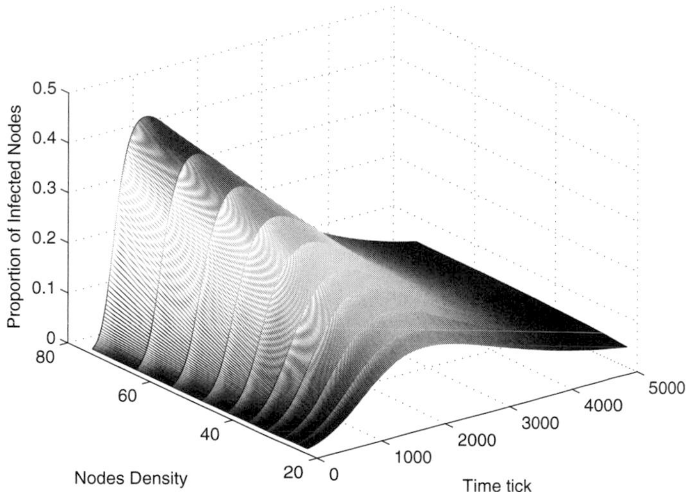

Figure 11. The effect of nodes density on Waxman random network with the configurations listed in Table 4.

Table 4. Configuration of Epidemic Spread Parameters on Waxman Random Network

#	# Static nodes	# Mobile nodes	Mean degree k	Topology Factor
1	1980	260	5.9375	9.2820
2	2178	338	6.5008	10.0760
3	2395	439	7.3888	11.2417
4	2635	571	8.3054	12.6225
5	2898	742	9.5467	14.2051
6	3188	965	10.8392	16.2766
7	3507	1254	12.3896	18.3849
8	3858	1631	14.2911	21.0538
9	4244	2120	16.3974	23.9066
10	4668	2757	19.3006	28.1408

The configurations of the spread of epidemic on Waxman Random Network with $\lambda = 0.0005$, $d = 0.0003$, $p = 0.0003$ and $m = 0.05$. And all the Waxman Random Network has $\alpha = 0.2$, $\beta = 0.05$.

Figure 12 shows the simulation results of healthy nodes at the end of an epidemic in a small world network with $N = 10000$, static shortcut density 0.005, $m = 0.01$, $d = 0.005$ and no pre-patch. We plot the proportion of final patched nodes as a function of patch rate p and epidemic spread rate per link λ. We can find that both increasing p and decreasing λ can heighten the final patched ratio, but the effect of increasing p is better than decreasing λ.

Figure 13 illustrates the experimental results in the same network topology with the fixed the spread rate per link $\lambda = 0.01$, which is the worst case in the pervious experiment. We plot the final patched ratio as a function of patch rate p and pre-patched ratio $p_0 t_0$. We can see that increasing p and pre-patched ratio can both rapidly increase the healthy ratio. Smartphones are different from the Internet; they are well managed by the operators. Hence, the operator can push the patch using GPRS, MMS, etc. to the subscriber. This service of pre-patch and patch will be very helpful to prevent the epidemic on the smartphones.

7. CONCLUSION

In this chapter, we propose a novel epidemics spread model (ESS) for smartphone which is based on the analysis of the unique features of smartphones and SIR model. With the ESS model, we study the "static shortcuts" and "mobile shortcuts" effects brought by smartphones and consider the influence of the epidemic spread rate, network topology, patching and death rate as well as the initial pre-patch to the propagation of the smartphone epidemics. Critical condition of epidemic fast die out is derived from the

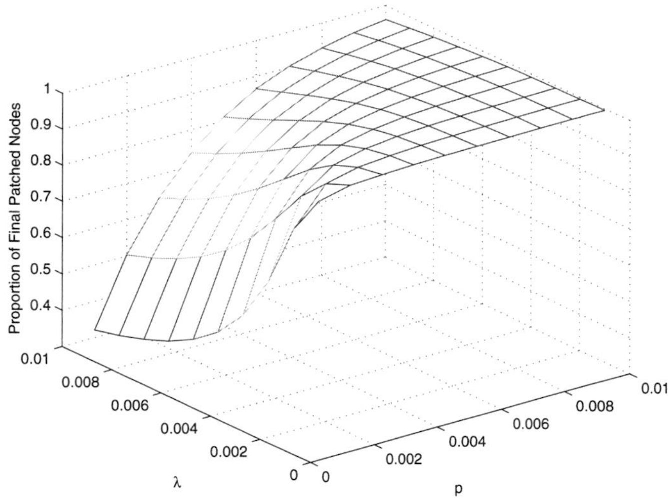

Figure 12. Influence to the proportion of patched nodes at the end of epidemic, patch rate p vs. epidemic spread rate per link λ.

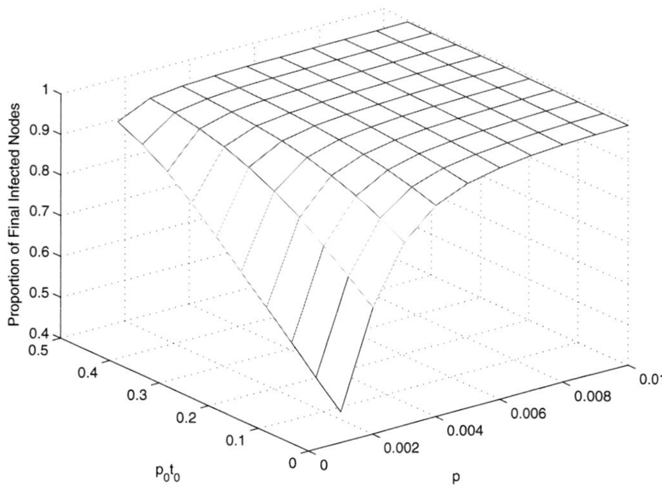

Figure 13. Influence to the proportion of patched nodes at the end of epidemic, patch rate p vs. pre-patched ratio $p_0 t_0$.

ESS model, and the detailed analysis is given to the individual parameters in the model. We demonstrate the effectiveness and accuracy of the ESS model using simulations with the typical network topologies.

From the theoretical analysis and experimental simulations, we give some guidance to defend attacks on smart-phones. We find that the pre-patch before epidemics spread is very important for prevention, and shield is especially useful because of its non-interruptive nature and small size of shield filter. Moreover, some intrusion prevention system can also be used to help reduce the epidemics spread rate to slow down the propagation. This also motivates the future research work on smartphone security.

8. REFERENCES

1. http://en.wikipedia.org/wiki/Smartphone
2. http://www.canalys.com/pr/2005/r2005071.htm
3. http://www.idc.com/getdoc.jsp?containerId=31554
4. Sander Siezen, Product Manager, Symbian Ltd, Symbian OS Version 9.1 Product description, Revision 1.1, February 2005
5. http://www.symbian.com/
6. Windows CE Home Page on MSDN. http://msdn.microsoft.com/embedded/windowsce/default.aspx
7. Microsoft Corporation. Windows Mobile-based Smartphones. http://www.microsoft.com/windowsmobile/smartphone/default.mspx.
8. http://www.palmsource.com/
9. http://brew.qualcomm.com/brew/en/about/about_brew.html
10. http://www.mvista.com/
11. S.J. Vaughan-Nichols. OSs battle in the smart-phone market. IEEE Computer, 36(6), 2003.
12. http://www.symantec.com/avcenter/
13. http://www.kaspersky.com/cyberthreats
14. http://www.trendmicro.com/vinfo/
15. C. Guo, H. J. Wang, and W. Zhu, Smartphone Attacks and Defenses, in Proc. ACM HotNets 2004.
16. H. Wang, C. Guo, D. Simon, and A. Zugenmaier, Shield: Vulnerability- driven network filters for preventing known vulnerability exploits, In ACM Sigcomm'04, Portland, OR, Aug. 30 - Sep. 3 2004.
17. C. Moore and M. E. J. Newman, Epidemics and percolation in small-world networks, Physical Review E 61, 5678-5682.
18. C. Moore and M. E. J. Newman, Exact solution of the site and bond percolation on small-world networks, Phys. Rev. E, 62(2000), 7059-7064.
19. R. Pastor-Satorras and A. Vespignani, Epidemic spreading in scale-free networks, Physical Review Letters, 86 (2001), 3200-3203.
20. R. Pastor-Satorras and A. Vespignani, Epidemic dynamics in finite scale-free networks, Physical Review E, 65 (2002).
21. M. E. J. Newman, Spread of epidemic disease on networks, Phys Rev E, 2002, 66, 016128

22. Zesheng Chen, Lixin Gao, and Kevin Kwiat, Modeling the Spread of Active Worms, In Proceedings of IEEE INFOCOM 2003, San Francisco, CA, April 2003.

23. Michele Garetto, Weibo Gong, and Don Towsley, Modeling Malware Spreading Dynamics, In Proceedings of IEEE INFOCOM 2003, San Francisco, CA, April 2003.

24. Cliff Changchun Zou, Weibo Gong, and Don Towsley, Code Red Worm Propagation Modeling and Analysis, CCS'02, November 18-22, 2002, Washington, DC, USA.

25. Y. Wang, D. Chakrabarti, C. Wang and C. Faloutsos, Epidemic spreading in real networks: An eigenvalue viewpoint, Proc. IEEE SRDS, 2003.

26. A. Ganesh, L. Massoulié, D. Towsley, The Effect of Network Topology on the Spread of Epidemics, IEEE Infocom, 2005.

27. O. Diekmann and J. A. P. Heesterbeek, Mathematical epidemiology of infectious diseases: model building, analysis and interpretation, (JohnWiley & Sons, New York, 2000).

28. D. J. Watts and S. H. Strogatz, Collective dynamics of 'small-world' Networks, Nature, 393(1998), 440-442.

29. Réka Albert and Albert-László Barabási, Statistical mechanics of complex networks, Rev. Mod. Phys. 74(2002), 47-97.

30. Symantec Internet Security Threat Report, Trends for January 1, 2004 - June 30, 2004, Volume VI, Published September 2004.

31. W. O. Kermack, and A. G. McKendrick, A Contribution to the Mathematical Theory of Epidemics, Proc. Roy. Soc. Lond. A 115, 700-721, 1927.

32. A. Rowstron and P. Druschel, Pastry: Scalable, distributed object location and routing for large-scale peer-to-peer systems, Proc. Middleware 2001, Germany, November 2001.

33. M. Faloutsos, P. Faloutsos and C. Faloutsos, On power-law relationships of the Internet topology, Proc. ACM Sigcomm, 1999.

34. A. L. Barabási and R. Albert, Emergence of scaling in random networks, Science 286,509-511 (1999).

35. C. Jin, Q. Chen and S. Jamin, Inet: Internet Topology Generator, Technique Report CSE-TR-433-00, University of Michigan, EECS dept. 2000, http://topology.eecs.umich.edu/inet.

36. B. Waxman, Routing of Multipoint Connections, IEEE J. Select. Areas Commun., SAC-6(9):1617-1622, December 1988.

Part III

SECURITY IN WIRELESS LANS

CROSS-DOMAIN MOBILITY-ADAPTIVE AUTHENTICATION

Hahnsang Kim
INRIA, Sophia Antipolis, France
E-mail: hahnsang.kim@inria.fr

Kang G. Shin
Department of Electrical Engineering and Computer Science
University of Michigan, Ann Arbor, U.S.A.
E-mail: kgshin@eecs.umich.edu

When mobile users with on-going sessions cross the domain boundary, their re-authentication affects significantly the inter-domain handoff latency as each inter-domain handoff requires remote contact with the home authentication server across domains, making it difficult to employ existing authentication protocols as they are. This chapter focuses on cross-domain authentication over wireless local area networks (WLANs) that minimizes the need for remote contact/access. We analyze the security requirements suggested by the IEEE 802.11i authentication standard, and consider additional requirements to help reduce the authentication latency without compromising the level of security. We propose an enhanced protocol called the *Mobility-adjusted Authentication Protocol* (MAP) that performs mutual authentication and hierarchical key derivation with minimal handshakes, relying on symmetric cryptographic functions. We also introduce *security context routers* (SCRs) that handle security context in conjunction with MAP, eliminating the need for continual remote contact with the home authentication server. In contrast to Kerberos that favors inter-domain authentication, MAP achieves a 26% reduction of authentication latency without degrading the level of security.

1. INTRODUCTION

Time-sensitive applications, such as Voice over IP (VoIP) or video streaming, are now possible over wireless local area networks (WLANs), such as those based on the IEEE 802.11 Standard [4], thanks to their high bandwidth. WLAN technologies also allow the mobiles to roam within public/corporate buildings or university campuses. Furthermore, we anticipate that mobile users might cross the domain boundary without their on-going application sessions disrupted. However, VoIP requires a handoff to be completed in less than 50*ms* for acceptable Quality-of-Service (QoS) [33], including the execution of the IEEE 802.11i authentication [6] as part of a secure handoff mechanism.

Minimizing the number of messages to be exchanged is important as cross-domain authentication needs to contact the remote home server. Moreover, the authentication latency increases in proportion to the round-trip time between two points involved in inter-domain message exchanges. Optimization of the authentication protocol is of utmost importance since an existing redundant combination of authentication and key negotiation functions incurs more rounds of message exchange than necessary.

We propose an enhanced protocol for cross-domain authentication, Mobility-adjusted Authentication Protocol (MAP) that relies on far less costly symmetric cryptography. (1) MAP reduces the cross-domain authentication latency by reducing the number of message exchanges. MAP requires less message exchanges without degrading security or the re-authentication mechanism, reducing the authentication latency significantly. (2) MAP replaces the 4-way handshake of the IEEE 802.11i authentication. In coordination with the authenticator within an access point, MAP defines hierarchical key derivation and generates consecutive keys during authentication operations. This leads to optimizing the 802.11i authentication mechanism by removing the need for the 4-way handshake. (3) MAP leverages the concept of security context to mostly avoid remote contact. With the mobile moving along, its security context is transferred via security context routers (SCRs) we present in this chapter. An SCR also plays a role of an authentication server in a foreign domain; it provides security context for MAP operating as if in the home server. Via a prototype implementation, our evaluation results show that the cross-domain authentication latency of MAP accounts for 74% and 85% that of Kerberos [17] and Needham-Schroeder symmetric-key protocol (NS) [26, 27], respectively. It makes up to 53% improvement in the authentication latency which is proportional to the end-to-end domain distance until the round-trip time counts up to 100*ms*.

The remainder of this chapter is organized as follows. Section 2 gives an overview of the 802.11i authentication mechanism, the related cross-domain protocols, design requirements, and prerequisites of BAN logic. Section 3 first describes MAP including its architecture and a relevant interaction between SCRs. Subsequently we details defined keys and types of messages, an example of message exchanges for a successful authentication, and the corresponding pseudo code of each module. Section 4 considers possible threats and analyzes the security of MAP. Section 5 examines the performance via measurements and simulation. Finally, we discuss related work in Section 6 and conclude the chapter in Section 7.

2. OVERVIEW OF AUTHENTICATION MECHANISM AND REQUIREMENTS

In this section, we first introduce the 802.11i authentication scheme and protocols applicable to the cross-domain authentication, and then describe the design requirements of authentication protocols. Finally, we explore prerequisites to BAN logic.

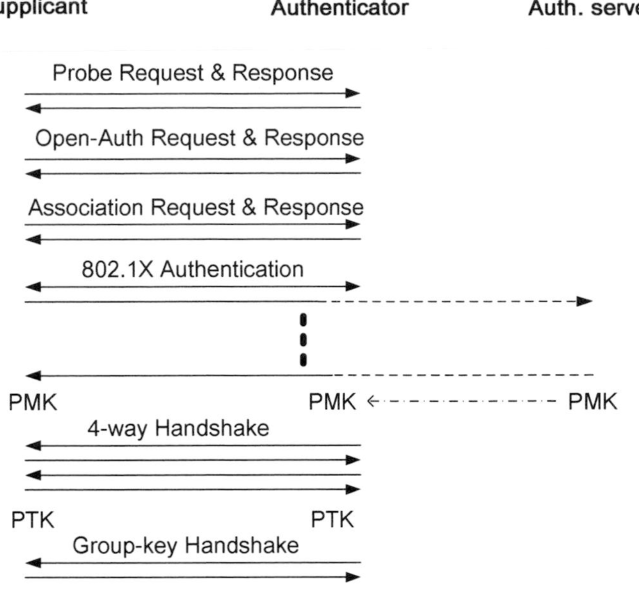

Figure 1. Message exchanges in the IEEE 802.11 and .11i systems

2.1. The IEEE 802.11i Authentication

The IEEE 802.11i authentication is responsible for mutual authentication and key derivation for securing WLANs via the 802.1X and 4-way handshake [6]. Figure 1 shows a typical scenario of message exchanges in the context of the IEEE 802.11 and 11i. Our focus is on two main steps after (re)association. First, the 802.1X authentication, where an authentication protocol like TLS [13] operates, is to verify the authenticity of end-to-end principals: the mobile (STA) and the authentication server (AS) via the authenticator (AUTH) (which operates in an AP). In particular, the AUTHs and AS construct an *authentication authorization and accounting* (AAA) architecture [1]. Successful mutual verification of each identity leads to the derivation of a pair-wise master key (PMK). This key is transferred to the AUTH via a secure tunnel. Second, the STA and AUTH perform the 4-way handshake, exchanging their nonces, so that a pair-wise temporary key (PTK) with which the wireless link will be secured is produced using the PMK as a seed.

The performance of the IEEE 802.11i authentication depends on the efficiency of this authentication protocol. Recent efforts on security associations have been limited to distribution of keys to access points within a domain [24]. For inter-domain handoffs, however, the authentication latency is critical to the application QoS.

2.2. Cross-domain-related Protocols

There are two protocols: Kerberos that supports the cross-domain authentication and NS that can be effectively extended to do so. We will use the two protocols to

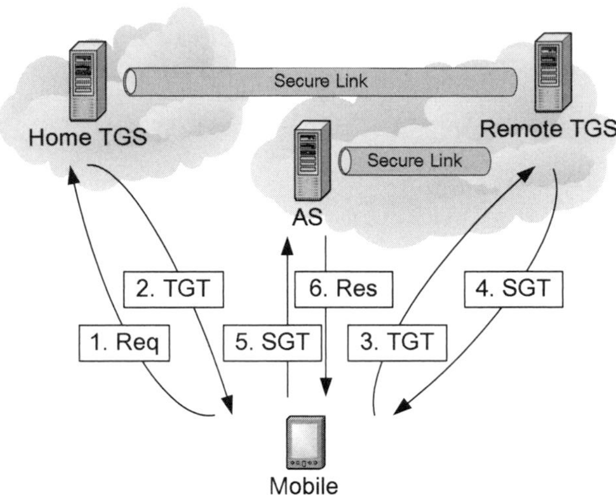

Figure 2. Message exchanges for a remote access grant in Kerberos. This sequence is repeated each time the mobile is bound to a remote TGS.

comparatively evaluate the throughout of our protocol via simulation. The following are the descriptions of the message exchanges of each protocol.

The Kerberos protocol provides cross-domain operations. By establishing inter-domain keys, the administrators of two domains allow the mobile to receive services in a remote domain. It receives a remote ticket granting ticket (TGT) from the ticket granting server (TGS) in the local domain. It then obtains a service granting ticket (SGT) from the remote TGS in the other domain by using the issued remote TGT. With the SGT containing a secret key, the mobile and AS can authenticate each other. Figure 2 illustrates a sequence of message exchanges for a remote authentication in Kerberos. The link among TGSs is assumed to be secure; a secret key of each TGS is shared to identify itself. In addition to the secure link, the AS has security association with its TGS. The remote TGT issued earlier can be reused to get TGTs in the current domain within a given period of time. However, each time the mobile moves into a foreign domain, the mobile needs to get a remote TGT again by contacting its home TGS.

The NS protocol on which Kerberos is based is not intended to operate over cross-organization boundaries. However, it can support cross-domain authentication with minor modifications, which we call a modified NS protocol (MNS). At first, the original protocol operates, in principle, as can be seen in Figure 3. The initiator A and its correspondent B share secret keys AK and BK with the AS, respectively. In the beginning, A obtains two copies of a pair-wise key encrypted with BK and AK by the AS, respectively, during their communication. Then, A sends B the BK-encrypted pair-wise key along with SK-encrypted AN. AN will be returned in the next message in order for A to ensure that B with which A is communicating is legitimate. B also adds

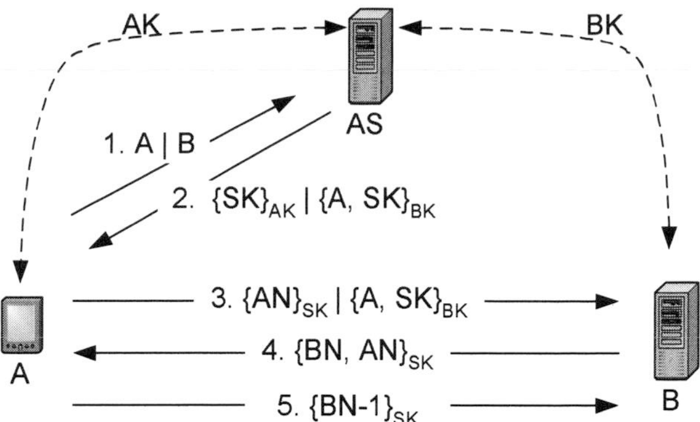

Figure 3. An example of message exchanges in symmetry-key-based NS protocol. AK and BK are pre-shared between A and AS, and B and AS, respectively. Elements used to authenticate A or AS in messages 1 and 2 are omitted. AN and BN are A's and B's nonces, respectively.

BN to the message encrypted by SK, and verifies the decremented BN that A sends eventually; those exchanged nonces may be used for key generation for need. We can view A as STA, and B and AS as foreign and home ASs, respectively. When the foreign AS requires a set of pair-wise keys, the home AS generates and sends a set of multiple different keys. Once receiving them, the foreign AS has no need for contacting the home AS, which may lead message exchanges to be reduced into the 3-way handshake.

2.3. Design Requirements

The IEEE 802.11i authentication suggests several requirements that must be preserved to secure WLANs.

- **Requirement 1**: The STA and AS must be able to authenticate each other. Since the STA establishes a wireless link to the AS via anonymous APs, it should be able to identify the AS, so should the AS.
- **Requirement 2**: A successful mutual authentication leads to the derivation of a fresh key for the AS and STA. After the successful mutual authentication, a 256-bit key (i.e., PMK) is generated by the AS and STA, and is eventually used by the STA and AP. This key must not be reused, and becomes obsolete whenever the STA binds with a new AP.
- **Requirement 3**: Mutual authentication should be strong enough to be protected from any unauthorized reception. It is uneasy to demonstrate the safety of the authentication protocol, but there are theoretical approaches for this purpose. For example, formal verification methods based on model checking and theorem proving, modal logic, and modular approach are widely used. We will show

a logical proof of MAP using BAN logic in Section 4.1. In addition to these requirements, we present the following recommendations for the authentication protocol design to help achieve fast handoffs in WLANs.

- **Recommendation 1**: Minimizing message exchanges during the authentication process helps improve the performance of cross-domain handoffs. We evaluate the effects of the number of message exchanges.
- **Recommendation 2**: The use of lightweight cryptographic algorithms helps low-power mobile terminals, like personal data assistants, mitigate the performance overhead of computation-intensive cryptographic algorithms.

Based on the above requirements, we will design a protocol supporting cross-domain mobility.

2.4. BAN Logic

BAN logic [11] is a modal logic developed for authentication protocol analysis. It presents the proof that a simple logic could be used to describe the beliefs of trustworthy communicating parties. It found redundancies or security flaws in authentication protocols in the literature [10]. BAN logic reasons that the protocol operates as correctly as expected. It is effective to prove the correctness of the authentication mechanisms with logical reasoning.

We introduce the several constructs and logical postulates in BAN logic that will be used for the proof of MAP. Full details of its rules are given in [11]. First, the following are the constructs that we use:

- P *believes* X: P believes X. In particular, the principal P may act as if X is true. This construct is essential to the logic.
- P *sees* X: P sees X. Someone has sent a message containing X to P, who can read and repeat X possibly after doing some decryption.
- P *said* X: P once said X. The principal P at some time sent a message including the statement X. It is unknown when the message was sent, but it is known that P believed X then.
- P *controls* X: the principal P is an authority on X and should be trusted on this matter, e.g., a server is often trusted to generate encryption keys properly. This may be expressed by the assumption that the principals believe that the server has jurisdiction over statements about the generated keys.
- *fresh(X)*: the formula X is *fresh*, i.e., X has not been sent in a message at any time before the current run of the protocol. This is usually true for *nonces* that is randomly generated for use only once.
- $P \overset{K}{\longleftrightarrow} Q$: P and Q may use the shared key K to communicate. It is never disclosed by any principal except for P and Q.

- $P \overset{X}{\Longleftrightarrow} Q$: the formula X is a *secret* known only to P and Q, and possibly to principals trusted by them. Only P and Q may use X to prove their identities to one another.

- $\{X\}_K$: this represents the formula X encrypted under the key K.

- $< X >_Y$: this represents X combined with the formula Y; it is intended that Y be a secret and that its presence prove the identity of whoever utters $< X >_Y$.

Then, we use the following logical postulates in proof.

- The *message-meaning* rules are applied to the interpretation of messages for shared keys

$$\frac{P\ believes\ Q \overset{K}{\longleftrightarrow} P,\ P\ sees\ \{X\}_K}{P\ believes\ Q\ said\ X}$$

and for shared secrets,

$$\frac{P\ believes\ Q \overset{Y}{\Longleftrightarrow} P,\ P\ sees\ < X >_Y}{P\ believes\ Q\ said\ X}.$$

- The *nonce-verification* rule represents the check that a message is recent and that the sender still believes in:

$$\frac{P\ believes\ fresh(X),\ P\ believes\ Q\ said\ X}{P\ believes\ Q\ believes\ X}.$$

- The *jurisdiction* rule states that if P believes that Q has jurisdiction over X then P trusts Q on the truth of X:

$$\frac{P\ believes\ Q\ controls\ X,\ P\ believes\ Q\ believes\ X}{P\ believes\ X}.$$

In addition, the HMAC (Hash Message Authentication Code) represented as MAC (m, K), where m and K denote a message and a pair-wise secret key, respectively, is used to verify whether or not the verifiee possesses the same K as the verifier. In other words, only if the generated codes are different, the applied Ks are different. Therefore, $MAC(m, K)$ is interpreted as a unit of the secret $< X >_Y$.

3. MAP

In this section, we describe an authentication architecture that extends the AAA architecture to SCR communications, and design MAP. The description of MAP includes the definition of keys and messages, message exchanges, and detailed operations in each functional module.

3.1. Architecture

Authentication operations work basically with three entities: STA, AS, and AUTH. An STA represents the end user with a WLAN-interface-equipped device. An AS verifies the STA's authenticity and provides each key to secure their wireless link. An AUTH relays authentication traffic between the STA and AS. In addition to dealing with these entities, our protocol solves the cross-domain authentication problem by introducing so-called *security context routers* (SCRs). An SCR is usually placed between multiple AUTHs and an AS. The SCR is logically distinct from the AS in terms of enforcing authentication policy, although both may reside on the same physical machine or the SCR can be integrated into the AS. The SCR functions as follows. After receiving a security context[1] issued by MAP on the AS, it can perform re-authentication on behalf of the AS. The SCRs are distributed in each domain so that they can reduce the authentication latency while the STA roams around the domain. It is assumed that in case of the communication of inter-administration domains they have a security association agreement on roaming and are securely connected to one another by sharing inter-domain keys. This combination is adaptable to the security architecture of the IEEE 802.11i authentication and Wi-Fi Protected Access 2 (WPA2) [2]. The protocol describing how messages are exchanged between the SCRs is part of our future work. In this chapter, we will give a rough idea of how to exchange messages between SCRs shortly.

Figure 4 depicts the MAP architecture. The MAP server module on the AS, which is described in Section 3.6, is an end-point authentication protocol that is assumed to securely be connected to the AUTHs via the SCR. The AS used in the architecture is functionally equivalent to the AAA server. The MAP security context module (SC module) in the SCR, which is described in Section 3.6, helps the AS communicate with the other MAP-support AS for cross-domain authentication. The AUTH is an authentication client as a pass-through authenticator. It relays authentication traffic from the STA to the AS, and vice versa. The MAP client module in the STA, which is described in Section 3.6, is an end-point authentication party that requests authentication and eventually establishes a secure link with the attached AP.

3.2. Communication between SCRs

The SCRs communicate with each other, based on a peer-to-peer manner. There are two ways of transferring security context among the SCRs involved. In case of no security context cached in an SCR with which an STA has just associated, the targeted SCR fetches security context from the original SCR with which the STA associated previously; *reactive transfer* introduces the latency of fetching security context. On the other hand, the original SCR may somehow forward the targeted SCR(s) security context before the mobile is handed off; *proactive transfer* emphasizes the availability

[1] Its contents vary with individual protocols. MAP is expected to have a set of authentication value pairs, identity (= mobile Id), validity time, time stamp, mean handoff time, counter and other security information.

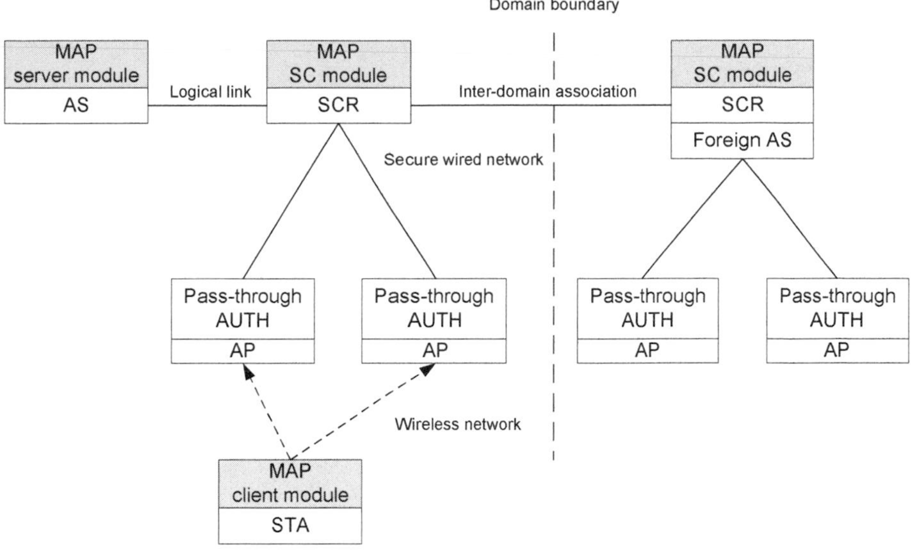

Figure 4. Authentication architecture

of the context ahead of time. On the other hand, estimation of the STA's direction and management of security context can be emphasized, which is referred to as *predictive forwarding* of security context. Their combination yields a tradeoff between storage overhead and latency performance. Elaboration on such issue is part of our future work.

3.3. Authentication

The MAP's authentication relies on Message Authentication Code (MAC) algorithms [18]. The MAC values rely on shared symmetric keys, the management of which is uneasy to scale in that two communication parties must somehow exchange the key in a secure way, compared to that of asymmetric-key pairs. However, on the other hand, signing and verifying public keys are very time-consuming; the MAC values are preferred to digital signatures because the MAC computation is two to three orders of magnitude faster. There is a tradeoff between scalability and CPU usage; we chose cost efficiency since it matches our design goal.

3.4. Defined Keys

We define three types of keys for different purposes: primary key (PK), domain key (DK) and temporary key (TK). PK is a long-term symmetric key which may be periodically updated and deployed, e.g., online subscription to a service provider or off-line set-up with a purchased card. PK is assumed to have guaranteed protection against disclosure for a sufficiently long period of time. DK is a quasi-primary key in a (sub)domain, which is derived from PK and the previous DK. The STA generates a

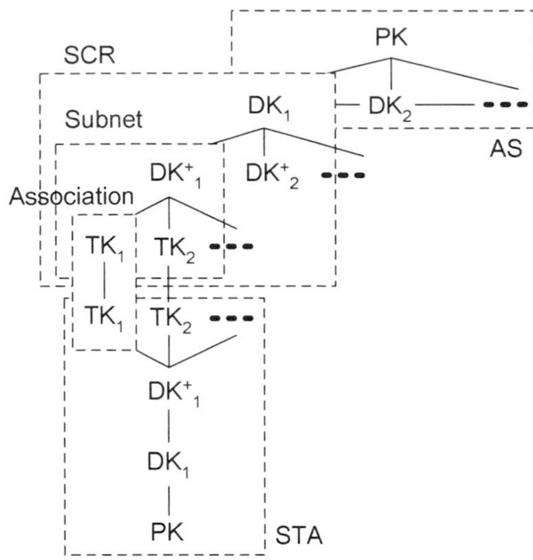

Figure 5. Defined keys hierarchy and boundary. An SCR controls a DK derived from the PK. A subnet uses a DK^+ hashed from the DK to generate TK that will be used for each association.

new DK as it changes a domain; an old DK must be revoked. In addition, DK^+, an n-time-hashed DK, is defined for use in a subnet within a domain—if no concept of such subnet is applicable DK^+ is generated from each DK; it plays a role of *loose coupling* of DK and TK. TK that is derived from DK^+ is a link key affiliated with securing a wireless link established between the STA and AP. TK binds with the addresses of two involved physical devices. Therefore, in case of re-associating or changing a binding address, TK is also changed. Figure 5 shows a hierarchical derivation and boundary of the defined keys. An association is made of each TK; the disclosure of any TK has no effect on other (re)associations. DK^+ also provides a key-disclosure barrier for relation between TK and DK. AS affects only the generation of DKs.

3.5. Defined Messages

We define six types of messages exchanged, with the server, SC, authenticator, and client modules interacting with each other during initial and re- authentication in MAP. The first four messages are used during the initial authentication and the last two are used during re-authentication.

- *Auth-req* message, sent by the client module in the STA, triggers a negotiation on authentication and key agreement from scratch.
- *Auth-chal* message, sent by the server module, as a return message, is used for the purpose of challenging the STA, with an encrypted code used for verifying the AS's authenticity to the STA.

- *Chal-res* message, sent by the client module, as a response message, contains a nonce-response encrypted code so that the AS verifies the STA's authenticity.
- *Auth-res* message, sent by the server module, is a reply to a challenge-response message.
- *Reauth-req* message is sent by the client module in the authenticated STA. The SCR captures this message and verifies if the authentication code is legitimate.
- *Reauth-res* message is a reply to the reauthentication-request message including the authentication result.

3.6. Message Exchanges

The following is an example of exchanging messages in case of a successful authentication. Only authentication-related information is highlighted in the messages.

M1.STA $\rightarrow$ AS: *Auth-req*(STAId, SN_i)
The STA sends the AS an authentication request containing its identity (Id) and a fresh nonce. On receipt of the message, the AS fetches credential corresponding to the Id and extracts its key from it.

M2.AS $\rightarrow$ STA: *Auth-chal*(MAC_{PK}[STAId, SN_i, ASN_j, 'authch'])
The AS uses the STA's nonce to compute a MAC, which protects from a reply attack.

M3.STA $\rightarrow$ AS: *Chal-res*(MAC_{PK}[STAId, SN_{i+1}, ASN_j, 'authres'])
If the received MAC is matched with the one that the STA generates, the AS is authenticated to the STA. Subsequently, the STA responds to the challenge. Otherwise, the message is silently ignored.

M4.AS $\rightarrow$ STA: *Auth-res*(ENC_{DK^+}[SN_{i+1}, ASN_{j+1}, AUN_k])
If the STA is successfully authenticated as well, the AS adds a *secret* value, i.e., ASN_{j+1} to a response message. During transfer of the message, the authenticator inserts a newly generated nonce AUN_k that is used to compute a temporary key (TK). Meanwhile, the AS computes and sends a set of authentication value pairs (AVPs) to the SCR.

When the STA re-associates with another AP in (another) subnet, the following messages are exchanged.

M5.STA $\rightarrow$ SCR: *Reauth-req*(MAC_{PK}[STAId, ASN_{j+1}, DK_i, 'reauth'])
The STA computes a MAC, using the secret value obtained in the previous round of authentication. The SCR can verify if the STA holds the same nonce.

M6.SCR $\rightarrow$ STA: *Reauth-res*(ENC_{DK^+}[SN_{i+2}, ASN_{j+2}, AUN_{k+1}])
If the STA is authenticated successfully, the SCR adds another nonce for the next challenge in the message. The STA can authenticate the SCR by verifying if the nonce is identical of the one that it sent previously.

In the subsequent section, we describe in details how MACs and hierarchical keys are computed and used in each module.

MAP Server Module

The server module handles two types of incoming messages (i.e., *auth-req* and *chal-res*) that are related only to authentication from scratch. The following is the description of the pseudo code of the server module.

```
var: sn_{1..n}, cn_{1..n} := 0; %Server and client nonce queues are initialized.
for all i: auth-req of Id_i in buffer do
    sn_i=refresh(sn_i); %A fresh nonce is generated.
    send auth-chal: sn_i | MAC_{PK_i}(Id_i, cn', sn_i, "authch");
    cn_i=cn'; %Client nonce from the message is buffered.
end for
for all i: chal-res of Id_i in buffer do
    DK_{i,j-1} = PRF(PK_i, cn_i, sn_i); %cn_i is obtained from auth-req.
    if MAC_{PK_i}(Id_i, cn', sn_i, "authres") && MIC_{DK_{i,j-1}} verified
    %cn' is obtained from chal-res.
        sn_i=refresh(sn_i);
        DK^+=H^{α_i}(DK_{i,j-1})
        send auth-res: sn_i | cn' | DK^+; %DK^+ is transferred to the authenticator.
        make SC_i:
        for e = 1..n do
            MAC_{PK_i}(Id_i, sn_i, DK_{i,j-1}, "reauth");
            DK_{i,j}=PRF(PK_i, DK_{i,j-1}, sn_i);
            AVP_e:(Id_i, sn_i, MAC, DK_{i,j}) ∈ ∪_{1..e-1} AVP;
            sn_i=refresh(sn_i);
        end for
    end if
end for
```

A MAC, including client nonce cn' from the received message and server nonce sn_i, is computed and sent to the STA of Id_i. The MAC allows the STA to verify the AS's authenticity. DK is computed by calculating an n-bit key generating pseudo random function (PRF)—in most cases n=128 is sufficient—with PK and the previously-exchanged nonces. An MIC provides a means of verifying authenticity once the associated MAC is verified successfully. A hashed domain key, DK^+, is generated by applying α times a cryptographic one-way function H, equivalently $H^{\alpha}(x) = H^{\alpha-1}H(x)$ and $H^0(x) = x$. The α value is a sync-one shared between the STA and the AS/SCR. DK^+ allows DK to be hidden from the authenticators. After the message exchanges, the server module creates the STA's security context that is composed primarily of the set of AVPs. It is then transferred to the corresponding SCR. The AVPs enable the SCR to conduct the re-authentication and re-keying process on behalf of the AS.

MAP SC Module

The SC module handles an incoming message (i.e., *reauth-req*) and an outgoing message (i.e., *reauth-res*) which are related to re-authentication. In particular, this module can be implemented, combined with the server module. The following is the description of the pseudo code of the SC module.

for all i: *reauth-req* of Id$_i$ in buffer **do**
  AVP$_l$=(Id$_i$, sn_i, MAC, $DK_{i,j}$) $\leftarrow \bigcup_{l..n}$ AVP; `%Select one of AVPs.`
  if MAC && MIC$_{DK_{i,j}}$ verified **do** `%The integrity of the message is checked.`
    send *reauth-res*: $cn'|sn_i|H^{\alpha_i}(DK_{i,j})$; `%DK+ is derived by α-time hashing.`
  end if
end for

The SC module first retrieves one of AVPs from the security context corresponding to Id$_i$ and then verifies MAC$\neq$MAC$'$ or MIC$_{DK_{i,j}}$(*reauth-req*)$\neq$MIC$'$. If they are matched correctly, it computes DK$^+$ and sends the authenticator it along with the exchanged and retrieved nonces. If DK is not allowed to be reused, the AVP is dethroned when it is notified somehow that the STA of Id$_i$ de-associates with the current AP. If no more AVP exists, the re-authentication request is forwarded to the AS which will, in turn, handle the request from scratch. Note that the SC module does not possess any PK.

Authenticator

A primary role of this module is to relay incoming messages. It also computes an TK with which the STA and AP establish a secure link after a successful authentication.

var: an; `%This is an authenticator nonce.`
if *auth-req* | *auth-chal* | *chal-res* | *reauth-req* received
  relay it;
end if
if *auth-res* | *reauth-res* received
  if success in authentication `%This is determined by AS/SCR.`
    an=refresh(an); `% A new an is used to generate TK.`
    send *auth-res*: $ENC_{DK^+}[sn' \mid an \mid cn']$;
    `%DK+ and cn' are obtained from the message.`
    TK=PRF(DK^+, $Addr_{STA} \mid Addr_{AP}$, $an \mid cn'$);
  end if
end if

Authenticator is beyond access to DK; DK$^+$ received from the AS/SCR is used to compute TK by calculating a PRF—the key-size varies with cryptographic protocols to be used for securing a wireless link, yet it is either 256 or 512 bits. TK binds with media access control addresses of the STA and AP; de-association revokes TK and a new TK must be recomputed.

MAP Client Module

The client module incurs an authentication request message (e.g., *auth-req* or *reauth-req*) when the STA (re)associates with an AP. It also handles incoming messages (i.e., *auth-chal*, *auth-res*, and *reauth-res*) and outgoing messages (i.e., *chal-res*).

```
var: secret := 0, cn; %cn is a client nonce.
if (re)associated
    cn=Refresh(cn);
    if !secret %In case of authentication from scratch
        send auth-req: Id | cn;
    else %In case of the previous successful authentication
        DK_i=PRF(PK, DK_{i-1}, secret);
        send reauth-req: Id|MAC_PK(Id, secret, DK_{i-1}, "reauth")|cn|MIC_{DK_i};
    end if
end if
if auth-chal received
    if MAC_PK(Id, cn, sn', "authch") verified %It authenticates AS.
        DK_{i-1}=PRF(PK, cn, sn'); %sn' is obtained from auth-chal
        cn=Refresh(cn);
        send chal-res: Id|cn|MAC_PK(Id, cn, sn', "authres")|MIC_{DK_{i-1}};
    end if
end if
if auth-res | reauth-res received
    DK+=H^α(DK_{i-1} or DK_i);
    DEC_{DK+}[ENC[sn' | an' | cn']];
    if cn==cn' %It authenticates AS.
        secret=sn'; %sn' is stored as secret
        TK = PRF(DK+, Addr_STA | Addr_AP, an' | cn);
        %cn is obtained from the previous message.
    end if
end if
```

It retains the *secret* value provided by the AS after completion of the previous successful authentication. Confidentiality of the secret is guaranteed since it is transferred in ciphertext. The secret determines whether the authentication process is conducted from scratch. The α value is matched to that of the AS/SCR.

4. SECURITY CONSIDERATIONS

In this section, first, using BAN logic, we show the logical proof that MAP performs its authentication mechanism correctly as it is expected, and then examine security threats to our protocol.

4.1. Protocol Analysis

The analysis procedure works as follows. First, we translate the original protocol into the idealized one and then make assumptions about the initial state. Finally, we make logical formulas as assertions and apply the logical postulates to the assumptions and assertions to arrive at conclusions.

Translation; we extract the encrypted forms of messages from MAP communications as follows:

M1.B $\rightarrow$ A: $< N_a, N_b >_{PK}$

M2.A $\rightarrow$ B: $< N_b, N_a >_{PK}$

M3.B $\rightarrow$ A: $\{N_{b'}, N_{a'}\}_{DK}$

M4.A $\rightarrow$ B: $< N_{b'}, A \overset{DK}{\longleftrightarrow} B >_{PK}$

M5.B $\rightarrow$ A: $\{N_{b''}, N_{a''}\}_{K_{ab'}}$

We have STA and SCR, referred to as A and B, respectively—the functionality of AS and AUTH is integrated into SCR for simplicity; DK$^+$ is identical of DK. We also omit communication in clear-text. There is a slight difference by representing $(N_a \oplus N_b)$ as (N_a, N_b), which is acceptable since this means that N_a and N_b were uttered at the same time and their XOR-ed value is straightforwardly obtained.

For authentication, each party verifies the MAC which requires the nonces generated by itself and the other. That is, the correct MAC can only be generated with the fresh nonces from the two. Thus, authentication between A and B might be deemed complete if each of the two believes that the other has recently sent the nonce, and proving sound mutual authentication is sufficiently satisfied by deriving the facts:

$$A \textit{ believes } B \textit{ believes } N_a \text{ and } B \textit{ believes } A \textit{ believes } N_b$$

for initial authentication and

$$A \textit{ believes } B \textit{ believes} N_{a''} \text{ and } B \textit{ believes } A \textit{ believes } N_{b'}$$

for re-authentication.

Making assumptions; we then write the following assumptions:

$(1) A \; believes \; A \overset{PK}{\Longleftrightarrow} B$, $(2) \; B \; believes \; A \overset{PK}{\Longleftrightarrow} B$,

$(3) A \; believes \; A \overset{DK}{\longleftrightarrow} B$, $(4) \; B \; believes \; A \overset{DK}{\longleftrightarrow} B$,

$(5) A \; believes \; A \overset{DK'}{\longleftrightarrow} B$, $(6) \; B \; believes \; A \overset{DK'}{\longleftrightarrow} B$,

$(7) A \; believes \; fresh(N_a)$, $(8) \; B \; believes \; fresh(N_b)$,

$(9) A \; believes \; fresh(N_{a'})$, $(10) \; B \; believes \; fresh(N_{b'})$,

$(11) A \; believes \; fresh(N_{a''})$, $(12) \; B \; believes \; fresh(N_{b''})$,

$(13) A \; believes \; fresh(N_{b'})$, $(14) \; A \; believes \; fresh(N_{b''})$,

$(15) A \; believes \; B \; controls \; N_{b'}$,

(16)A *believes* B *controls* $N_{b''}$.

Assumptions (1) and (2) are made from the fact that A and B initially share a secret, PK. Assumptions (3), (4), (5) and (6) are derived from the fact that only A and B can generate a shared key only if the sound authentication is achieved. Assumptions (7) to (12) state that A and B believe that the nonces generated by themselves are fresh; freshness of nonces holds by verification of MAC and MIC associated with the nonces. The nonces, $N_{b'}$ and $N_{b''}$, also play a role of secrets since they are transferred with proper encryption. Thus, A can believe that B has generated the nonces that was not used in the past,which leads to Assumptions (13) and (14), and also (15) and (16), indicating that A trusts B to generate the secret.

Reasoning; we analyze the idealized version of MAP by applying the logical postulates presented in Section 2.4 to the assumptions.

A receives Message M1. The annotation rule yields that A *sees* $< N_a, N_b >_{PK}$ holds afterward. With the hypothesis of (1), the message-meaning rule for shared secrets applies and yields A *believes* B *said* (N_a, N_b). Breaking conjunctions produces A *believes* B *said* N_a. With the hypothesis of (7), we apply the nonce-verification rule and yield A *believes* B *believes* N_a. On the other hand, B receives Message M2 and the following result is obtained in the same way as that of Message M1, via the message-meaning and nonce-verification rules with hypotheses (2) and (8), respectively, B *sees* $< N_b, N_a >_{PK}$ and B *believes* A *believes* N_b. This concludes the analysis of Message M2. The analysis of Messages M1 and M2 confirms that MAP performs mutual authentication successfully.

A receives Message M3 and the annotation rule yields that A *sees* $\{N_{b'}, N_{a'}\}_{DK}$ holds thereafter. The message-meaning rule for shared keys with the hypothesis of (3) via breaking conjunctions yields: A *believes* B *said* $N_{b'}$, and A *believes* B *said* $N_{a'}$. Taking the former, with hypotheses (13) and (15), the nonce-verification and jurisdiction rules apply and yield A *believes* B *believes* $N_{b'}$, and A *believes* $N_{b'}$, respectively. Taking the latter, the nonce-verification rule with hypothesis (9) yields A *believes* B *believes* $N_{a'}$. This concludes the analysis of Message M3. This message may appear redundant since authentication is completed from Message M1, but it is essential not because it is for authentication, but because it is for transmission of a secret, nonce $N_{b'}$.

B receives Message M4 and the annotation rule yields that B *sees* $< N_{b'}, A$ and $\overset{DK}{\longleftrightarrow} B >_{PK}$ holds thereafter. By applying the message-meaning rule for the secrets with (2) via breaking conjunctions, we obtain: B *believes* A *said* $(N_{b'}, A \overset{DK}{\longleftrightarrow} B)$, and B *believes* A *said* $N_{b'}$. The nonce-verification rule with hypothesis (10) yields that B *believes* A *believes* $N_{b'}$. On the other hand, A receives Message M5 and the annotation rule yields that B *sees* $\{N_{b''}, N_{a''}\}_{DK'}$ holds thereafter. By applying the message-meaning rule for the shared keys with hypothesis (6) via breaking conjunctions, we obtain A *believes* B *said* $N_{b''}$ and A *believes* B *said* $N_{a''}$. Taking the former, the nonce-verification and jurisdiction rules with (14) and (16) yield A *believes* B *believes* $N_{b''}$, and A *believes* $N_{b''}$, respectively. Taking the latter, nonce-verification with (11) yields that $A\ believes\ B\ believes\ N_{a''}$. The analysis of Messages M4 and M5 confirms that MAP also achieves mutual re-authentication.

4.2. Possible Attacks

Key recovery attack: This relies on finding the key K itself from a number of message–MAC pairs. Ideally, any attack allowing key recovery requires about 2^k operations where k is the length of K. The adversary tries all possible keys with a small number of message–MAC pairs available. Choosing a sufficiently long key is a simple way to thwart a key search. Another possible attack is to choose an arbitrary fraudulent message and append a randomly-chosen MAC value. Ideally, the probability that this MAC value is correct is equal to $1/2^m$, where m is the number of bits in the MAC value. Repeated trials can increase the corresponding expected value, but a good implementation will be alert to repeated MAC verification errors.

Forgery attack: This attack relies on prediction of $\text{MAC}_K(x)$ for a message x without initial knowledge of K. For an input pair (x, x') with $\text{MAC}_K(x)$=$g(H)$ and $\text{MAC}_K(x')$=$g(H')$, where g denotes the output transformation and H is a chaining variable, a collision occurs if $\text{MAC}_K(x)$ = $\text{MAC}_K(x')$. Its feasibility depends on an n-bit chaining variable and the MAC result. Given g that is a permutation, a collision can be found using an expected number of $\sqrt{2} \cdot 2^{n/2}$ known text-MAC pairs of at least two divided blocks each. A simple way to counter this attack is to ensure that each sequence number at the beginning of every message is used only once within the lifetime of the key.

Impersonating attack: Note that the AUTH, SCR and AS maintain a security association with each other. Therefore, neither of them can be used to impersonate the other. Instead, this attack occurs between the STA and AUTH, which causes an authentication failure or misconduct of the principals. Oracle-based impersonating attacks are that the attacker exploits one of principals as an oracle to obtain cryptographic messages in a session since it has no knowledge of K. The attacker applies the obtained messages to the other principal party in another session. For example, it runs a session with an AUTH to obtain a MAC value, impersonating a legitimate STA. It runs another session with an STA and exploits the MAC value on the STA, impersonating the legitimate AUTH. This attack can be countered by exchanging nonce with each other and using a sequence counter.

5. PERFORMANCE EVALUATION

We evaluate the efficiency of MAP via experimentation and simulation, contrasting it with other protocols. We first describe simulation methodology and model and then analyze the MAP's performance benefits via the simulation results and in comparison with other protocols. Finally, we discuss the storage overhead caused by security-context transfer.

5.1. Simulation Methodology

The probe phase, discovering the next AP in WLAN handoffs, takes a large latency (ranging from 50*ms* to 350*ms*), depending on different vendors [22]. Even if the recent effort in [32] to reduce the latency by 84%, the large variance is an obstacle to highlight the effectiveness of our protocol on a real testbed. We therefore use Matlab-based simulation, relying on experimental data. We assume that network traffic is stable with small variations, e.g., the latency of establishing a (re)association with an AP including the probe phase is 30*ms* with 3% jitter, and the round-trip time (RTT) between two communicating servers across a domain is about 20*ms* with 4% jitter. In addition, the RTT between the AP and SCR/AS is less than 3*ms*. We use these values throughout the simulation. In cryptographic computations, we conducted an experiment using three machines: Linux v.2.4.19 iPAQ 206MHz ARM processor with 64 megabyte memory (iPAQ), Linux v.2.4.2 Laptop Mobile Pentium 366MHz processor with 128 megabyte memory (MP2) and Linux v.2.4.23 Desktop Intel Xeon 3Ghz bi-processor with 2GB memory (Xeon). We compiled crypto libraries [12] in gcc v.3.3 with an option of Level-1 optimization.

Table 1. Throughput of hash/symmetric and asymmetric algorithms (in Megabit per second)

Alg.\Pow.	iPAQ	MP2	Xeon
SHA-1	15.8 Mbps	18 Mbps	104.9 Mbps
SHA-256	3.4 Mbps	9 Mbps	64.0 Mbps
SHA-512	0.2 Mbps	4.3 Mbps	24.8 Mbps
MD5	15.8 Mbps	41 Mbps	290.9 Mbps
AES-128	2.7 Mbps	10 Mbps	80 Mbps
RSA enc.	15.1 Kbps	138.9 Kbps	625 Kbps
RSA dec.	0.9 Kbps	4.6 Kbps	21.6 Kbps
RSA sig.	0.9 Kbps	4.4 Kbps	21.2 Kbps
RSA ver.	15.1 Kbps	138.9 Kbps	625 Kbps

Table 1 shows the computation throughput of symmetric-key and public-key algorithms, respectively. With these measurement data, we numerically calculate the time to perform each authentication protocol while ignoring the overhead of running applications for simplicity.

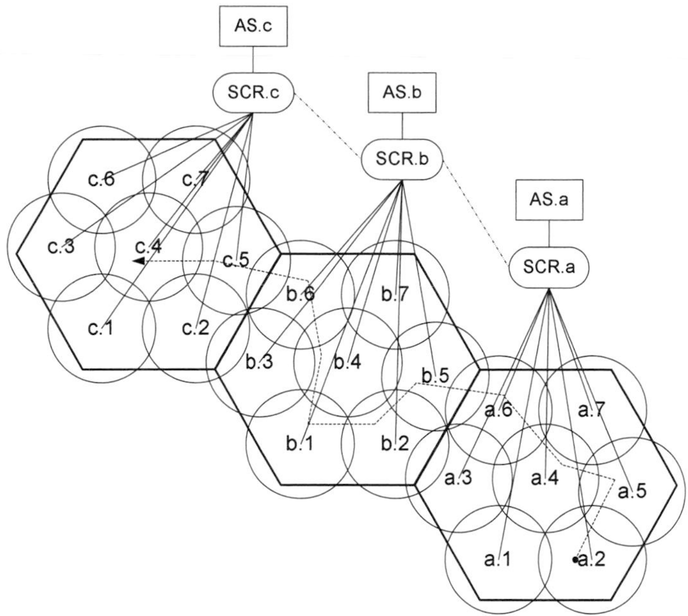

Figure 6. The simulation model for inter-domain handoffs. A circle and hexagon indicate an AP's radio coverage range and a domain, respectively. Each SCR controls its domain and is securely connected with its neighbor. An STA initially associates with a.2 and move around in its local domain (from a.5 to a.6 via a.4). It crosses Domain b and finally associates with c.4 in Domain c.

5.2. The Simulation Model

Figure 6 shows the simulation model we used. Each AS constructs a domain consisting of an SCR and several APs. The SCR and AS may reside on the same machine as mentioned before.

Handoff Pattern

The handoff pattern for STAs is basically random; the STAs cross the boundary after hopping a random number of times. Random pattern is sufficient to evaluate the overall efficiency performance. Nevertheless, to notice the comparative effectiveness of our protocol, we additionally set a regular handoff pattern; after association in the home domain, STAs hop three times and then cross a domain boundary. In a visited domain, every five hops they traverse the domain.

SCR Configuration

Whether or not the "visitor" can use storage resources in a domain affects the performance of its handoffs. There can be three system configurations according to the storage availability in the SCR of the visited domain. First, if only *relaying security context* is allowed, the authentication process takes place in the AS/SCR of the home domain. The SCR in the visited domain serves as a relay agent. Second, if *caching security context* is allowed, the foreign SCR serves as a proxy authentication server. In this case, security context is transferred and stored in the visited domain, which enables avoiding contact with the home server. Third, if *pre-caching security context* is allowed somehow, i.e., security context is transferred to the foreign SCR before the STA arrives, then the latency of fetching security context from the home server/SCR can be eliminated. We will evaluate the caching effect via simulation.

5.3. The Simulation Results

MAP performs an optimized re-authentication procedure based on the security context generated after the initial authentication. It allows one to (1) consolidate the re-authentication procedure (with two-message exchanges, the mutual authentication is completed) and (2) avoiding contact with the home server from the visited domain. Figure 7 clearly shows that from re-authentication, the authentication latency dramatically drops by up to 45% thanks to (1). As a regular handoff pattern, after three hops in the local domain (the first handoff corresponds to the initial authentication in the figure), the STA crosses the domain boundary at every 5 handoffs, which triggers the foreign SCR to request the security context from the home server. As a result, the latency increases in proportion to the RTT between the end-to-end points of two domains. Even if the STA roams in the foreign domain, it shows the same latency performance as in the home domain thanks to (2). In this case, the SCR in the foreign domain supports caching security context. After the 15-th handoff in the figure, the cross-domain authentication encounters the case of relaying security context in the SCR of the visited domain, which triggers the authentication procedure to be performed in contact with the home server for each hop in the visited domain.

Figure 8 shows the results with a random handoff pattern, illustrating the cumulative distributions of the authentication latency for three cases supporting SCR. The figure shows the effect of pre-caching and caching security context to achieve more improvements in time efficiency than just relaying security context which is characteristic of the legacy protocols that are unable to generate security context. For example, more than 70% and 80% of authentication processes in the cases of caching and pre-caching security context, respectively, take less than $36ms$.

We evaluated the increase in storage availability via the number of authentication requests with a random handoff pattern. Figure 9 shows that the higher inter-domain handoff frequency the home SCR has, the higher its storage availability. The x-axis is the ratio of authentication request queries in inter-domain handoffs to the total number of queries, and the y-axis is the ratio of the network traffic in the foreign SCRs. Let AQ_r denote the foreign server's overhead and AQ_l denote the home server's overhead. Then, the ratio of the gain in storage availability with MAP to the overall overhead is expressed as, $1 - AQ_l/(AQ_l + AQ_r)$ which grows as the frequency of the inter-domain handoffs increases.

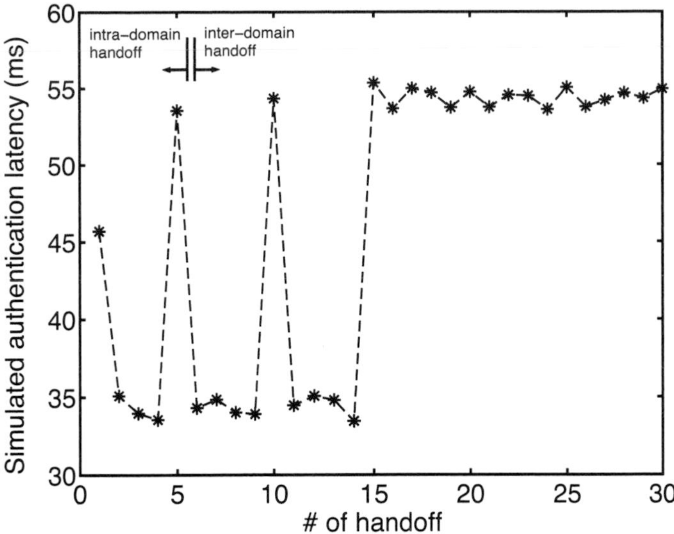

Figure 7. Authentication latency variations in different configurations of foreign servers

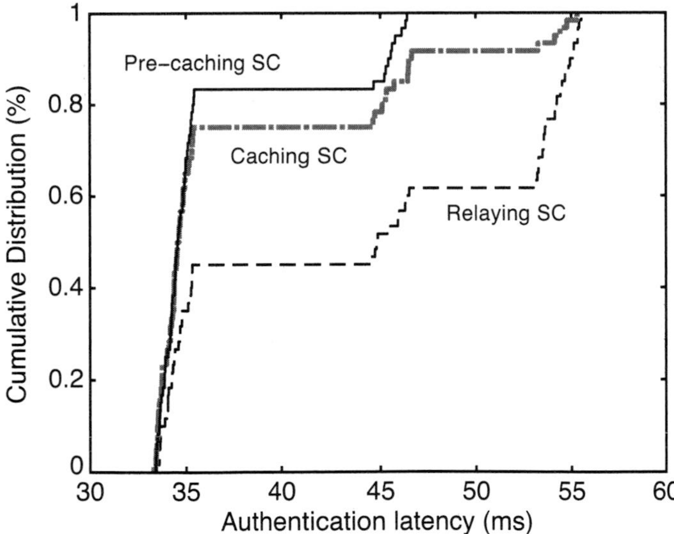

Figure 8. Cumulative distributions of authentication latency under each different configuration. SC stands for security context.

As shown in Figure 10 that plots the results with a random handoff pattern, the performance in authentication efficiency (caching allowed) improves up to 53% over a legacy method (relaying allowed) until the end-to-end domain distance continues

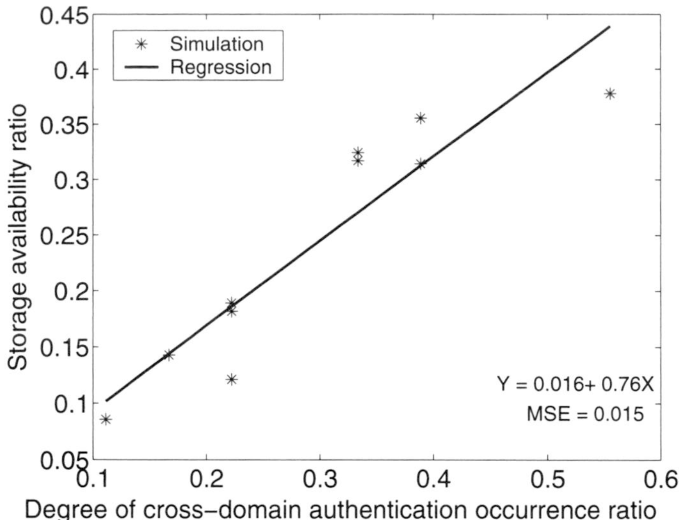

Figure 9. System storage availability affected by the inter-domain handoff authentication occurrence ratio. MSE is Mean Squared Error of the above regression function.

to increase up to RTT=100*ms*. In case of security context pre-cached in the visiting domain, MAP makes a 10% additional improvement with RTT=100*ms*. Therefore, the effectiveness of MAP increases dramatically as the distance gets larger.

5.4. Comparison with Other Protocols

Figure 11 shows the cumulative CPU usage (represented in millisecond) cryptographic primitives of required in ten consecutive times of authentication in symmetry-key-based protocols including MAP, MNS, and Kerberos, and public-key-based protocols including PNS and TLS. We chose one-way hash functions (i.e., MD5 [30], SHA [3]) and block ciphers (i.e., AES [5]) for symmetric-key protocols, and RSA [16] 1024-bit modulus for the public-key protocols. The symmetric-key protocols are shown to be two orders of magnitude faster thanks to the inherent advantage over modulo operations. MAP is faster than the MNS and Kerberos protocols, respectively, by 12.6% and 21.5% CPU usage gains. This is a considerable impact on the performance gain in view of millions of runs for authentication in a single server.

Regarding the number of message exchanges, MAP achieves the cross-domain authentication only with 2-way handshake, the cost of which is minimal, compared to MNS and Kerberos requiring 3-way and 4-way handshakes, respectively. This contributes to the further enhancement of latency performance. Figure 12 shows the comparison of authentication latency of MAP with that of the MNS and Kerberos protocols while mobiles are hopping with a regular pattern. MAP outperforms the others in both inter- and intra-domain roaming. It accounts for 74% of cross-domain authentication

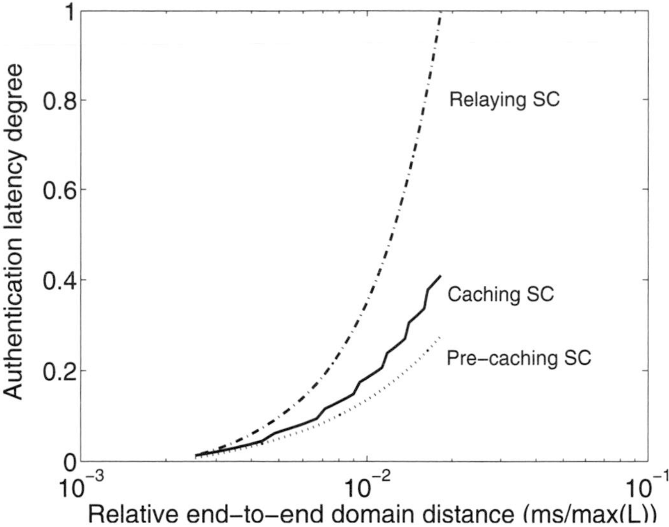

Figure 10. End-to-end domain distance vs. authentication latency. The distance is scaled down at a rate of the maximum authentication latency (max(L)). SC stands for security context.

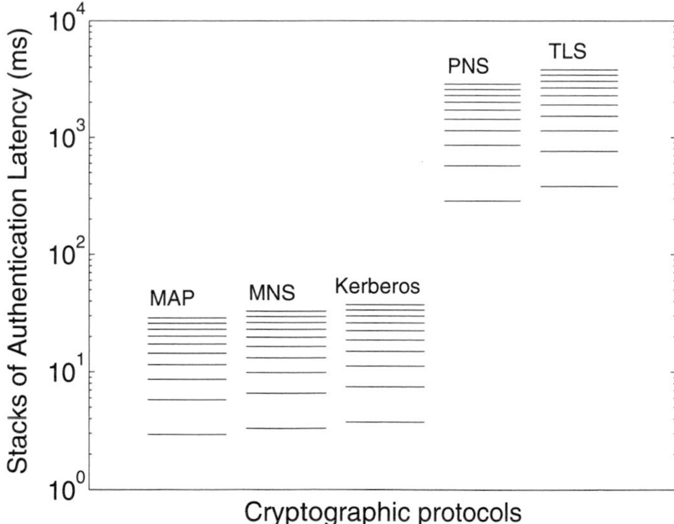

Figure 11. CPU utilization. Ten consecutive times of authentication. MNS and PNS stand for modified symmetry-key-based and public-key-based Needham Schroeder protocols, respectively. TLS is Transport Layer Security protocol.

latency of Kerberos and 85% of that of MNS. It reduces the intra-domain authentication latency by 5% for Kerberos and 7% for MNS.

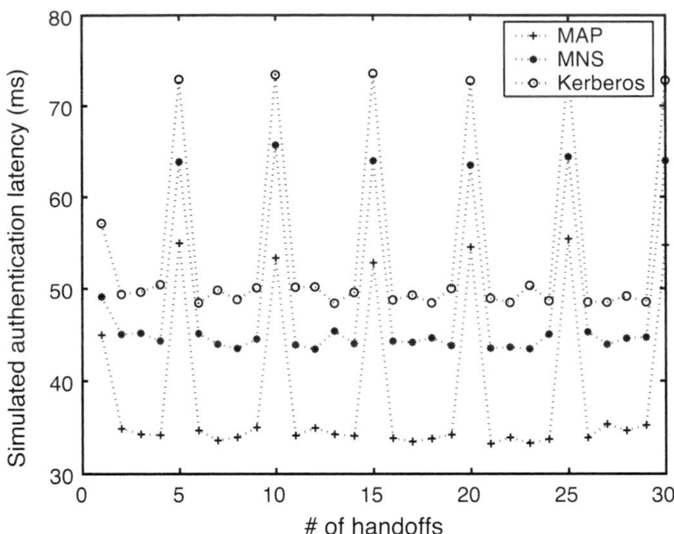

Figure 12. Latency comparison of MAP with MNS and Kerberos. Fetching security context while the mobile crosses the boundary increases the authentication latency.

5.5. Storage Overhead

Security context is transferred and stored in a foreign server (SCR) for cross-domain authentication. It consists mainly of a set of AVPs each of which is composed of nonce (128 bits), MAC (128 bits), DK (128 or 256 bits) and Identity (about 320 bits). In addition, a value (of 40 bits) may be reserved for security context validity and other information. The security context can be of $64 \cdot n+45$ bytes where n is the number of AVPs. Approximately, given a 1 kilobyte security context per STA, manipulating one million STAs requires 1 gigabyte storage capacity, which is usually a small overhead to the server system.

6. RELATED WORK

There have been several studies on how to achieve fast handoffs and enhance the performance of authentication mechanisms, including WLAN protocols.

Michra *et al.* [24, 23] presented a keys distributing method by means of proactive context caching. The idea of proactive caching is for an AP to broadcast its cached context to its neighbor APs in advance by using neighbor graphs and IAPP. However, this method is limited to intra-domain handoffs since APs are required to be functionally identical. Pack *et al.* [29] presented a pre-authentication method that skips the 802.1X authentication phase by distributing the key to a certain number of selected APs and computing the likelihood based on the analysis of past network behavior. Bargh *et al.* [8] presented the applicability of the pre-authentication method for inter-domain handoffs. However, a pre-authentication method creates a higher risk of compromising security.

Wong *et al.* [35] proposed a hybrid protocol based on a certificate containing a symmetric key signed with a public key which is suitable for wireless communications. An asymmetric method for wireless communications presented in [15] uses Diffie-Hellman key exchange combined with Schnorr signatures. In addition, there are several legacy authentication protocols [36, 37, 31, 11, 28, 13] for the general purpose in the literature.

There are several approaches to analyzing the security of authentication protocols. One is the formal methods that model and verify the protocol using specification languages and verification tools [21]. It consists of model checking and theorem-proving methods. Application examples [19, 25, 14, 20] demonstrated the feasibility of formally verifying the authentication protocols with general-purpose verification tools. Also proposed in [9, 34, 7] are modular approaches aiming to establish a sound formalization and a security analysis for the authentication problem.

7. CONCLUSIONS

The cross-domain authentication requires retrieval of security context from the server of the previously-visited or home domain. Contacting a remote server may increase the authentication latency significantly. the longer the end-to-end distance, the larger the latency reduction. The longer the end-to-end distance, the larger the latency reduction. If security context is allowed to be pre-cached/transferred before the mobile arrives, the latency can be reduced significantly. In this chapter we designed and evaluated a mobility-adjusted authentication protocol, MAP, by leveraging symmetric-key cryptography for cross-domain authentication and key generation. MAP can be configured to make tradeoffs between performance and storage usage. MAP introduces three concepts to the cross-domain authentication: (1) a re-authentication mechanism based on a 2-way handshake; (2) the temporary-key generation of the IEEE 802.11i authentication; and (3) security context eliminating the need to contact a remote server. MAP performs best in cases of long end-to-end domain distances and high cross-domain authentication traffic.

8. REFERENCES

1. Authentication Authorization and Accounting IETF WG.
2. Wi-fi alliance. http://www.wi-fi.org/.
3. Secure Hash Standard. In *Federal Information Processing Standards Publication 180-1*. NIST, Apr. 1995.
4. Part 11: Wireless LAN Medium Access Control (MAC) and Physical Layer (PHY) specifications: Specification for Robust Security. In *ANSI/IEEE Std 802.11: 1999(E)*. ISO/IEC 8802-11, 1999.
5. Advanced Encryption Standard (AES). In *Federal Information Processing Standards Publication 197*. NIST, Nov. 2001.

6. Part 11: Wireless LAN Medium Access Control (MAC) and Physical Layer (PHY) specifications: Specification for Robust Security. In *IEEE Std 802.11i/D3.1*. ISO/IEC 8802-11, 2003.

7. William Aiello, Steven M. Bellovin, Matt Blaze, Ran Canettiand John Ioannidis, Angelos D. Keromytis, and Omer Reingold. Efficient, DoS-Resistant, Secure Key Exchange for Internet Protocols. In *Conf. on Computer and Comm. Security*. ACM Press, 2002.

8. M. S. Bargh et al. Fast Authentication Methods for Handovers between IEEE 802.11 Wireless LANs. In *Int. Workshop on Wireless Mobile App. and Services on WLAN Hotspots (WMASH)*, pages 51–60. ACM, Oct. 2004.

9. Mihir Bellare, Ran Canetti, and Hugo Krawczyk. A Modular Approach to the Design and Analysis of Authentication and Key Exchange Protocols. In *30th Symposium on Theory of Computing*, pages 419–428. ACM Press, 1998.

10. Michael Burrows, Martin Abadi, and Roger Needham. A Logic of Authentication. Technical Report 39, Digital Equipment Corporation, Palo Alto Calif., February 1989.

11. Michael Burrows, Martin Abadi, and Roger Needham. A Logic of Authentication. *ACM Trans on Comput. Systems*, 8(1):18–36, 1990.

12. Wei Dai. Crypto++, http://www.eskimo.com/~weidai/ cryptlib.html.

13. Tim Dierks, Alan O. Freier, Martin Abadi, Ran Canetti, Taher Elgamal, Anil R. Gangolli, Kipp E.B. Hickman, and Hugo Krawczyk. Transport Layer Security protocol version 1.0. RFC 2246, Jan. 1999.

14. James Heather and Steve Schneider. Towards automatic verification of authentication protocols on an unbounded network. In *13th Computer Security Foundations Workshop*, page 132. IEEE Computer Society, 2000.

15. Markus Jakobsson and David Pointcheval. Mutual Authentication for Low-Power Mobile Devices. In *Financial Cryptography 2001*, Grand Cayman Island, British West Indies, Feb. 2001.

16. Burt Kaliski. PKCS #1 RSA Encryption Version 1.5. RFC 2313, Mar. 1998.

17. John Kohl and B. Clifford Neuman. The kerberos network authentication service (v5). RFC 1510, Sep. 1993.

18. Hugo Krawczyk, Mihir Bellare, and Ran Canetti. HMAC: Keyed-Hashing for Message Authentication. RFC 2104, Feb. 1997.

19. Gavin Lowe. Breaking and fixing the Needham-Schroeder public-key protocol using FDR. In *Tools and Algorithms for the Construction and Analysis of Systems (TACAS)*, volume 1055, pages 147–166. LNCS, 1996.

20. Paolo Maggi and Riccardo Sisto. Using spin to verify security properties of cryptographic protocols. In *9th SPIN Workshop on Model Checking of Software*, volume 2318, pages 187–204. LNCS, 2001.

21. Catherine A. Meadows. Formal Verification of Cryptographic Protocols: A Survey. In *ASIACRYPT: Int. Conf. on the Theory and Application of Cryptology*, volume 917, pages 135–150. LNCS, Dec. 1994.

22. Arunesh Mishra, Minho Shin, and William Arbaugh. An Empirical Analysis of the IEEE 802.11 MAC Layer Handoff Process. *SIGCOMM: CCR*, 33(2):93–102, 2003.

23. Arunesh Mishra, Minho Shin, and William Arbaugh. Context Caching using Neighbor Graphs for Fast Handoffs in a Wireless Network. In *INFOCOM*, Hong Kong, Mar. 2004. IEEE.

24. Arunesh Mishra, Minho Shin, and William Arbaugh. Pro-active Key Distribution using Neighbor Graphs. *Wireless Comm. Magazine*, 11 Issue 1:26–36, Feb. 2004.

25. J. C. Mitchell, M. Mitchell, and U. Stern. Automated analysis of cryptographic protocols using murϕ. In *Symposium on Security and Privacy*, pages 141–153. IEEE Computer Society, 1997.

26. Roger M. Needham and Michael D. Schroeder. Using encryption for authentication in large networks of computers. *Comm. of the ACM*, 21(12):993–999, 1978.

27. Roger M. Needham and Michael D. Schroeder. Authentication Revisited. *SIGOPS: OSR*, 21(1):7, 1987.

28. Dave Otway and Owen Rees. Efficient and timely mutual authentication. *SIGOPS: OSR*, 21(1):8–10, 1987.

29. Sangheon Pack and Yanghee Choi. Pre-Authenticated Fast Handoff in a Public Wireless LAN based on IEEE 802.1x Model. In *IFIP TC6 Personal Wireless Comm. 2002*, Singapore, Oct. 2002.

30. Ronald L. Rivest. The MD5 Message-Digest Algorithm. RFC 1321, Apr. 1992.

31. M. Satyanarayanan. Integrating security in a large distributed system. *Trans. on Comp. Systems*, 7(3):247–280, 1989.

32. Minho Shin, Arunesh Mishra, and William Arbaugh. Improving the Latency of 802.11 hand-offs using Neighbor Graphs. In *Mobisys*, Boston, Jun. 2004. ACM.

33. R. Shirdokar, J. Kabara, and P. Krishnamurthy. A QoS-based Indoor Wireless Data Network Design for VoIP. In *Vehicular Technology Conf. (VTC'01)*, volume 4. IEEE, Oct. 2001.

34. Duncan S. Wong and Agnes H. Chan. Efficient and Mutually Authenticated Key Exchange for Low Power Computing Devices. In *ASIACRYPT: Int. Conf. on the Theory and Application of Cryptology and Information Security*, volume 2248, pages 272–289. LNCS, 2001.

35. Duncan S. Wong and Agnes H. Chan. Mutual Authentication and Key Exchange for Low Power Wireless Communications. In *MILCOM 2001*, pages 39–43, USA, Oct. 2001. IEEE Press.

36. Thomas Y. C. Woo and Simon S. Lam. A Lesson on Authentication Protocol Design. *SIGOPS: OSR*, 28(3):24–37, 1994.

37. Yuqing Zhang, Chunling Wang, Jianping Wu, and Xing Li. Using SMV for cryptographic protocol analysis: a case study. *SIGOPS: OSR*, 35(2):43–50, Apr. 2001.

11

AAA ARCHITECTURE AND AUTHENTICATION FOR WIRELESS LAN ROAMING

Minghui Shi, Humphrey Rutagemwa, Xuemin (Sherman) Shen and Jon W. Mark
Department of Electrical and Computer Engineering
University of Waterloo
Waterloo, Ontario, N2L 3G1, Canada
E-mail: {mshi, humphrey, xshen, jwmark}@bbcr.uwaterloo.ca

Yixin Jiang, Chuang Lin
Department of Computer Science and Technology
Tsinghua University
Beijing, 100084, P.R. China
E-mail: {yxjiang, clin}@csnet1.cs.tsinghua.edu.cn

A wireless LAN service integration architecture based on current wireless LAN hotspots is proposed to make migrating to new service cost effective. The AAA (Authentication, Authorization and Accounting) based mobile terminal registration signaling process is discussed. An application layer end-to-end authentication and key negotiation protocol is proposed to overcome the open air connection problem existing in wireless LAN deployment. The protocol provides a general solution for Internet applications running on a mobile station under various authentication scenarios and keeps the communications private to other wireless LAN users and foreign networks. A functional demonstration of the protocol is also given. The research results should contribute to rapid deployment of wireless LANs hotspot service.

1. INTRODUCTION

IEEE 802.11b/g wireless LAN products have become a de facto standard component in mobile devices. An increasing number of wireless Internet access services have been appearing in places such as airports, cafes, and bookstores. Annual industry revenues already exceed US$1 billion, and is expected to pass US$4 billion by 2007 [1].

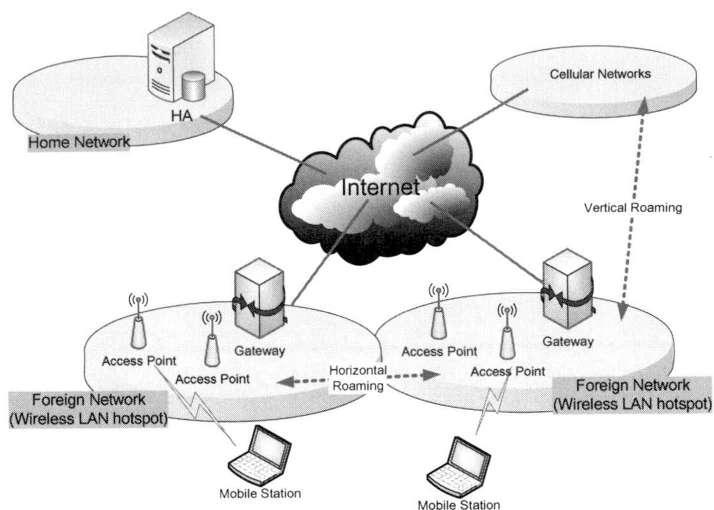

Figure 1. Wireless service integration architecture

In addition, mobile devices with both cellular network and wireless LAN acess are becoming widely available. Demand for integrating multiple mobile computing services into a single entity is preeminent.

Figure 1 shows a global architecture of the public wireless Internet. Mobile IP is used throughout the architecture to support user roaming. The home network has a network prefix matching to that of the home address of the mobile terminal. When the mobile terminal moves from one network to another, such as roaming between foreign networks, it performs registration and updates its registration information with its home agent (HA), either directly or indirectly. The hotspots are mostly based on IEEE 802.11b/g WLAN standard. The layout of the hotspots can be adjacent or distributed in the cellular networks. When the services of WLAN and cellular network are integrated, the mobile users can roam between the heterogeneous networks by seamlessly switching between the associated network access interfaces. Two types of handoff may occur: horizontal handoff and vertical handoff. Horizontal handoff supports user roaming between WLANs or cellular networks, and vertical handoff handles roaming between WLAN and the cellular networks. Since most WLAN hotspots do not overlap, vertical handoff is most common seen when the mobile terminal enters and leaves the WLAN hotspot. This chapter focuses on proposing a high-performance secure cellular network-integratable wireless LAN service framework.

Many wireless Internet service providers (WISPs), such as T-mobile, provide public wireless LAN Internet access at hotspots using a network access server (NAS). The NAS allows only legitimate customers to use the service and provides intra-domain roaming because the hotspots from one WISP share the same customer base. However, it lacks an architecture to provide inter-domain roaming and Mobile IP support. A user cannot access hotspots of service provider B with his/her account from A, even though A and B would like to have a roaming agreement.

In interworking implementation, handoff delay should also be considered. Mobile IP handoff delay can be divided into two parts: movement detection and signaling for registration. Several proposed approaches are actually only effective on registration signaling delays. A micro-mobility approach [2] divides a network in a hierarchical manner, and location management is handled locally when the mobile station moves within a smaller area at the lower hierarchy level. For simultaneous binding in [3], multiple care-of address bindings for the mobile station are maintained and packets destined for the mobile station are transmitted to all care-of addresses to reduce packet loss during handoff. However, it cannot be used in an IEEE802.11 network, because current wireless LAN cards can only access one access point or channel at a time.

On the other hand, Wired Equivalency Protocol (WEP) is a security protocol for WLAN defined in the 802.11b standard and is designed to provide the same level of security as that of a wired LAN. However it has several problems in both transmission privacy and deployment. Various studies show that WEP is vulnerable to several attacks [4, 5], especially in a heavily loaded wireless network. WEP uses a single key shared between the access point and clients. Malicious clients are able to tap into the communication traffic of other clients who are associated with the same access point. Most hotspots do not use data encryption due to this technical limitation. Authentication can be used to negotiate a shared session key to further encrypt data traffic in the session [6, 7, 8, 9]. Although there are many authentication protocols published, they do not generally support Internet applications for wireless mobile devices. For example, the authentication protocols proposed in [19, 20, 21] allow a mobile station to communicate with another one directly, but there is no solution for a mobile station to communicate with a fixed Internet server, which is found in FTP (File Transfer Protocol) applications.

Protected transmission based on Secure Shell (SSH) and/or Secure Sockets Layer (SSL) has been suggested to secure wireless transmission. SSH requires a previously generated public/private key pair, so it may never be applied to authentication between parties who have not contacted with each other before. SSL is not suitable for extension to mobile wireless Internet either, because the operation of SSL relies on certification verification by certificate authority (CA) servers. It is not practical for CA servers to store the certificate of every mobile station because the number of mobile stations is too large (for the same reaso, client authentication in SSL is optional). The home network would not like to register every mobile station to CA servers either. In the case of a wireless LAN hotspot, the service access is controlled by medium access control (MAC) addresses of the mobile stations. Usually there is no key negotiation during the network authentication and Mobile IP registration phases.

The objective of this chapter is to propose a secure wireless LAN service integration architecture and necessary signaling process design. It is divided into three categories:

IEEE802.11 service integration functionality: The architecture should be able to integrate into cellular networks. Since third-generation (3G) and beyond cellular networks use or very likely will continue to use AAA structure and protocol to control network access and manage user accounts, the IEEE802.11 roaming architecture and signaling processes should work with cellular networks.

Wireless network security: The security issues include network access control, user account management, and transmission privacy. The first two items can be taken care of by using the AAA structure. For the third item, a wireless LAN hotspot has no general solution to guarantee data transmission privacy due to the poor design of WEP.

Service quality: It mainly refers to handoff speed and packet loss rate. Naive handoff acceleration solutions do not apply to an IEEE802.11 network interface cards, because they can only talk to one another, so the solutions cannot guarantee no packet loss.

The remainder of the chapter is organized as follows. We first overview AAA mechanism and propose the AAA based infrastructure of wireless LAN roaming. We then present a security mechanism for wireless LAN transmission and related demonstration results, followed by a summary of the work.

2. AAA OVERVIEW

AAA is an architectural framework for configuring a set of three independent security functions in a consistent manner. AAA provides a modular way of performing the following services:

1. **Authentication** is the way a user is identified prior to being allowed access to the network and network services. AAA authentication can be configured by defining a named list of authentication methods, and then applying that list to various interfaces on a per-user or per-service basis.

2. **Authorization** is the process of determining whether the actions, such as accessing a resource is permitted for the corresponding identity. Authorization works by assembling a set of attributes that describe what the user is authorized to perform. These attributes are compared to the information contained in a database for a given user and the result is returned to AAA to determine the user's actual capabilities and restrictions. Remote security servers, such as RADIUS[13, 14, 15] and TACACS+[16, 17] , authorize users for specific rights by associating attribute-value pairs, which define those rights with the appropriate user.

3. **Accounting** tracks the services that users are accessing as well as the amount of network resources they are consuming. When accounting is activated, the network access server reports user activity to the RADIUS or TACACS+ security server in the form of accounting records. This data can then be analyzed for network management, client billing, and/or auditing.

Access control is the way to control who will be allowed to access to the network server and what services they are allowed to use. AAA network security services provide the primary framework to set up access control on the router or access server. The three security functions are used together, for example networks or services, to control which users are allowed access, what functions they are allowed to use and how much resource

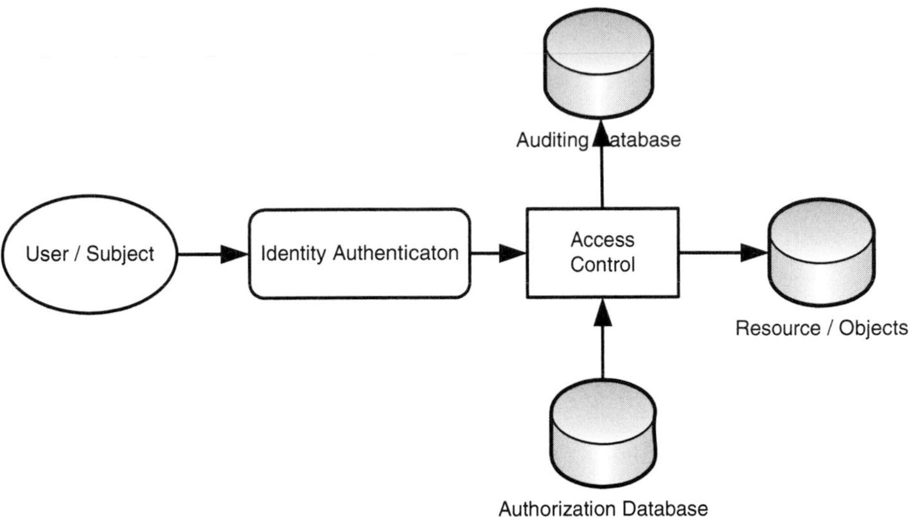

Figure 2. The relations between access control and AAA

they have used. Access Control protects system resources against unauthorized access. The use of system resources is regulated according to a security policy and is permitted by only authorized entities (users, programs, processes, or other systems) according to that policy [10]. In this chapter, AAA is adopted to excise service admission for users in WLAN roaming.

AAA deals with access control of systems and environments based on policies set by the administrators and users of the systems. The access policy may be implied in both the authentication, which can be restricted by the time of day, number of sessions, calling number, etc., and the attribute-values authorized [2]. Access control provides the limited access according to the authorization policies between a subject and objects when a subject access the related resources (objects). Objects mainly include passive entities (file, storage area) while subjects mainly contain active entities (processes, users). Subjects obtain information by accessing objects. Fig. 2 shows the basic relations between access control and AAA services. When a user, or the subject, needs to access the resources, the authentication is first preformed to verify its identity. According to the access policy corresponding to the user's identity, which is stored in authorization database, access control allows the user to access the defined resource. The activity of the subject during the session is logged in the auditing database.

AAA scheme provides the following benefits: (1) increased flexibility and control of access configuration; (2) scalability; (3) standardized authentication methods, such as RADIUS, TACACS+, and Kerberos [18]; (4) multiple backup systems. In many circumstances, AAA uses protocols such as RADIUS, TACACS+, or Kerberos to administer its security functions. If the router or the access server is acting as a network access server, the communication is established between the network access server and

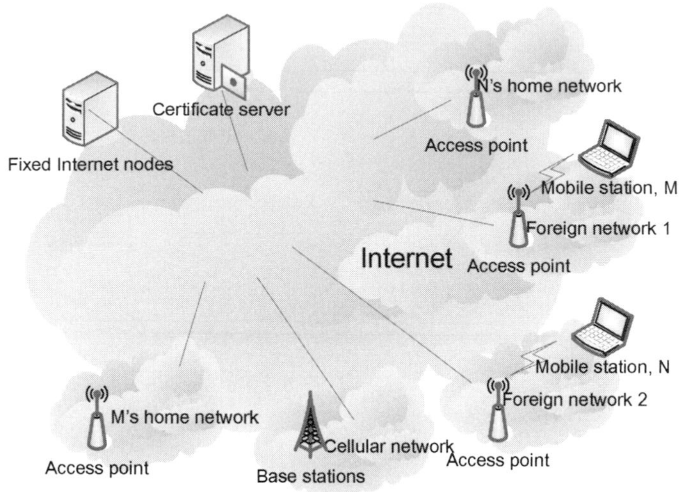

Figure 3. The proposed network structure for IEEE802.11 service integration

the RADIUS, TACACS+, or Kerberos security server through AAA. In wireless LAN environment, there is a strong application requirement for Mobile IP AAA [12].

3. IEEE 802.11 WIRELESS LAN ROAMING

Figure 3 shows the infrastructure of the mobile networks under consideration. The Internet offers much larger bandwidth and lower transmission error rates than wireless links. The home network is considered as a private network, which only allows its users access. The foreign networks are the real WISPs. After completion of a registration process, the mobile station and the corresponding foreign network will share a key for further encrypted communications. Fixed nodes represent common web sites. Authentication is required for accessing some of those sites. The cellular networks and base stations are 3G based. Access points, which form a hotspot, are the attachment points that allow mobile stations to access the network through wireless connection. A mobile station, as a member of its home network, is allowed to access the resources in the home network whenever it is within or outside the home network. CA servers are special servers that issue and verify certificates of fixed nodes or networks upon request so that they have proofs to identify themselves. CA servers are organized in a tree topology and working in a distributed way, so that it is not necessary to connect all Internet servers to one CA server. Mobile stations do not contact CA severs directly because of the large population size. CA shares independent secret key with the servers which it is connected with.

The proposed IEEE802.11 roaming structure is based on an AAA broker with a Remote Authentication Dial-In User Service (RADIUS) server proxy. RADIUS is popular and easier to use for integrating hotspot service into AAA based cellular networks. The broker model is suitable for large-scale and commercial implementation

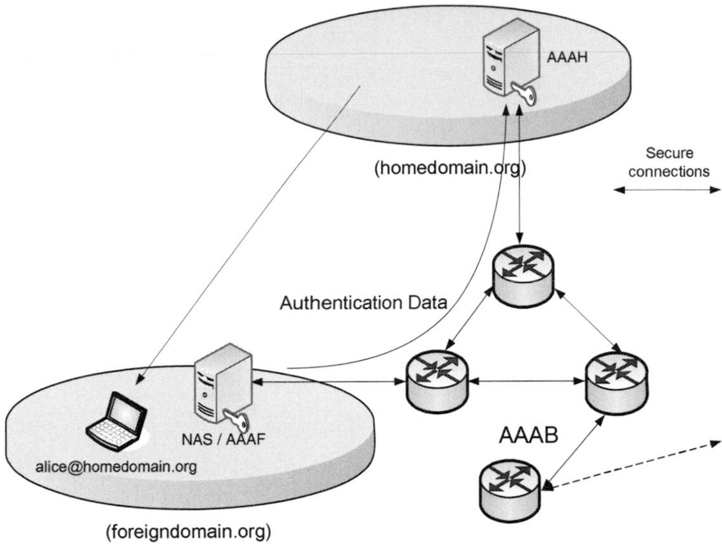

Figure 4. Network structure of AAA brokers

because a RADIUS server can simply have one simple security association or a pre-setup shared secret with the RADIUS proxy. RADIUS proxies forward authentication and accounting requests from different domains to their destination.

3.1. Radius Proxy

RADIUS servers of multiple ISPs can be interconnected via a series of forwarding servers. The RADIUS server retrieves the remote servers domain from the users request that includes the network access identifier (NAI) [22, 23, 24] in the form of identifier@domain_name, which identifies a users name and the domain where the user comes from. Then it forwards the request to the remote server identified by the domain. The remote server also replies through the forwarding server.

A group of forwarding servers with secured communication tunnels between each other are used as AAA brokers (AAABs). Figure 4 illustrates the network structure of AAABs. A mobile user whose NAI is alice@homedomain.org moves from its home domain to another domain (e.g., foreigndomain.org). The NAS located in the foreign domain authenticates the mobile user, and forwards this request via RADIUS protocol to the foreign AAA (AAAF). According to NAI, the AAAF forwards the request to the home AAA (AAAH) through the AAABs.

When the number of domains increases, it is no longer feasible to connect all the AAA servers to one AAAB network. The AAABs will be grouped according to geographical distribution of the network domains. In this way the complexity of each AAA broker can be reduced. The performance of an AAAB cluster is evaluated by the number of hops to forward the AAA request from the originator to the destination.

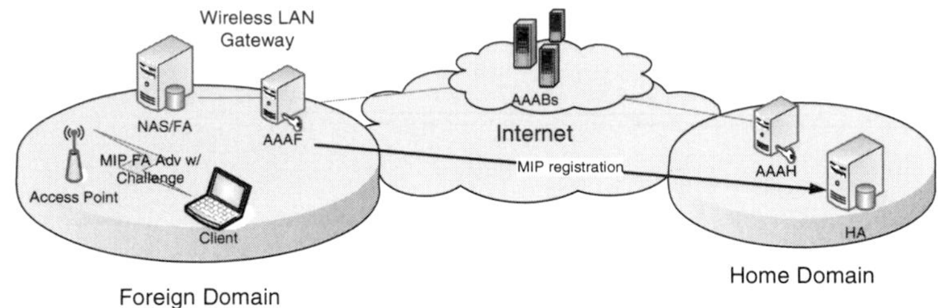

Figure 5. Wireless LAN roaming

3.2. IEEE 802.11 Horizontal Roaming

The IEEE802.11 horizontal roaming architecture is shown in Figure 5. The hotspot is connected to the Internet through a gateway. Each network domain is interconnected by AAABs. In order to provide IP mobility, the functionality of a foreign agent (FA) is placed into the NAS. The FA located in the NAS periodically sends advertisements with challenge packets, and all mobile stations register via the FA. The challenge is a piece of data used to verify if the user device has knowledge of the secret (e.g., a password) without sending it explicitly via a communication link. The architecture is able to process two horizontal roaming scenarios:

The current IEEE802.11 device connects to the network via the NAS: The network can provide IP mobility, however the roaming is not seamless. When a mobile user requests Mobile IP services by sending Mobile IP Registration, the NAS blocks the Registration until the mobile user has been authenticated via the AAA architecture. The NAS prompt the user to enter his or her credential, such as the username and the password for authentication. Once the mobile user is authenticated successfully, the normal Mobile IP registration will retained.

Seamless roaming: Authentication is completely done by the home agent (HA). The mobile station is required to support Mobile IP Challenge/Response extensions with a Mobile-AAA authentication extension so that the user credential can be processed by the program automatically.

In the following, we focus on developing efficient signaling process for the two roaming scenarios. The design shares as many common signaling messages as possible. In order to have further integration with 3G cellular networks, the signaling process should also be able to share with the AAA signaling process for 3G networks. Based on the architecture in Figure 3, Figure 6 illustrates the internal design of an NAS/FA. It has two modes: one for compatibility of current wireless LAN deployment, and the other for seamless wireless LAN roaming.

In the compatible mode, when a mobile station registers, it may use its home address or the mobile station NAI to identify itself in its Registration Request. A mobile station associates itself with an access point and starts sending IP packets, such as Mobile IP requests, to its HA via an FA that relays the Registration Request. After the HA authenticates the request and sends a reply via the same FA, the HA and FA

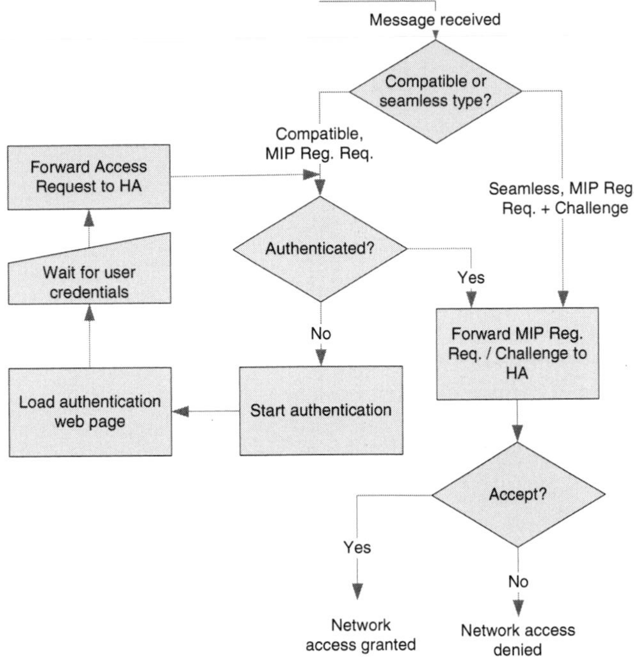

Figure 6. The flow diagram of the NAS/FA

both update their bindings. Sometimes an FA forces all its serving mobile stations to register through it. If a mobile station does not send the user credentials, including NAI and password, along with the Mobile IP request, the user will be redirected to a login page. By extracting the domain portion of NAI, the authentication request will be forwarded to the AAA server of the WISP. After successful authentication, the Mobile IP request is forwarded to the HA of the mobile station and the Mobile IP registration is completed.

In seamless roaming mode, a mobile station associates with the access point and responds to a Mobile IP FA Advertisement packet with a Challenge, and sends the Mobile IP Registration Request with the NAI and Challenge. The user credentials are included in the Mobile IP request. When the FA receives the reply from the mobile station, it realizes that the mobile station can do seamless roaming. It encapsulates the request in the AMR and forwards it to the AAAH and HA. After the HA processes this request, it sends a HAA containing Mobile IP Registration Reply. The AAAH and AAAL forward the encapsulated Mobile IP Registration Request to the FA in the AMA packets. The FA then sends a Mobile IP Registration Reply.

Comparing signaling processes of the two methods, it can be seen that the processes are designed to be quite similar, such as the signaling messages and the signal path.

So some components in the network do not need to differentiate the message type for each mode. Only one signaling processing mechanism needs to be designed. The FAs own local clients still can access the hotspot as they can use AAAF to authenticate themselves.

3.3. Mobile IP Handoff Performance Improvement

In order to roam between a wireless LAN and a cellular network, the mobile station should be equipped with corresponding network access interfaces. The data packets from the corresponding server are routed to the mobile station through its HA. When the mobile station roams to the foreign network, the two network access cards are assigned a temporary care-of address by the FA.

The switching of the two interfaces can be considered a care-of address change in Mobile IP. When the mobile station decides to switch the interface, it informs the HA by updating its current care-of address to the IP address of the other network access interface. The HA redirects the data flow to the new IP address. This method ensures that the process of network access interface switch over is dealt with using the switching process in Mobile IP.

For typical data applications such as web surfing, it is not necessary to use a real-time seamless handoff as for cellular telephony. A gap of a few seconds while a connection is being rerouted should be fine with the applications. However, with the growth of real-time Internet applications, like voice or streaming video, Mobile IP handoff latency and packet loss performance have become more and more critical. In order to provide high-quality applications in a wireless LAN environment, the key issue is to support efficient and seamless network handoff. When a mobile station moves from the coverage of one access point to another, it re-associates with a new AP. This is called a layer 2 handoff. On the other hand, a Mobile IP handoff (layer 3) is the process that takes place when changing FAs. The latency generated by both layer 2 and Mobile IP handoffs should be reduced. In order to reduce the latency of Mobile IP handoff in a wireless LAN, link layer update frames and movement notification packets can be used to assist Mobile IP handoff. A MAC bridge or data tunnel is established between the new FA and old FA servers to improve the latency of Mobile IP handoff in the wireless LAN environment. The pre-registration and authentication data can be sent to the mobile station before it moves, and/or the data packets that arrive at the old FA during movement can be sent to the mobile station via the new FA. Additional flow control should be taken in the handover period, because the connection speed of the old and new access point/base station could be quite different if the mobile station performs handoff between IEEE 802.11 and cellular networks. If the data source is not informed in a timely way, data may block the channel if the device is moving from high speed to low speed connection, or the user cannot get better quality of service otherwise. Therefore, effective congestion control is very important, especially for media streaming service that uses the protocol without an inherent congestion mechanism. Measures should also be taken to ensure that the pre-authentication data transfer between the two FAs is private and unaltered. So the two FAs authenticate each other via a CA server using the

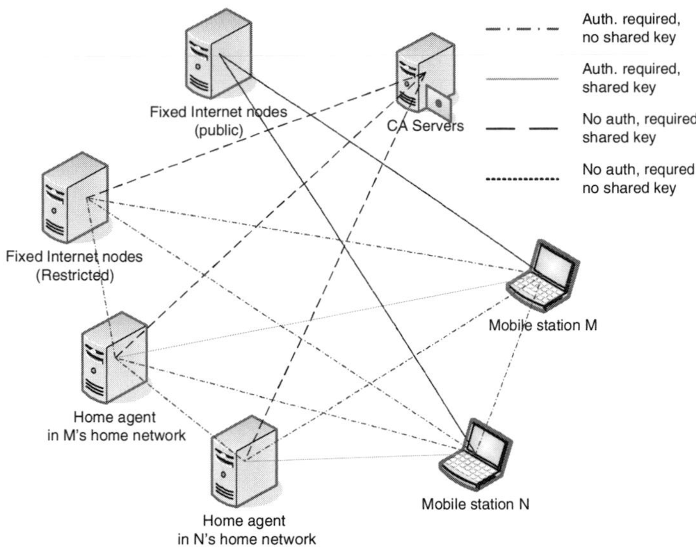

Figure 7. Authentication for Internet applications

protocol proposed in the next section, and a temporary session key can be negotiated to encrypt the pre-authentication data.

4. WIRELESS TRANSMISSION PRIVACY

Although the architecture proposed earlier prevents an unauthorized user from using the service, the wireless transmission is still kept open. Using built-in WEP encryption cannot guarantee data transmission privacy in a public hotspot, since a WEP key is unique for each access point and there is no privacy among the mobile stations associated with an access point. A separate authentication and key negotiation mechanism is required to keep wireless transmission private.

This section presents a protocol that operates at the application layer to avoid any hardware or low-level protocol modification, and the authentication messages are carried in the payload of data packets used in Mobile IP networks. User location updates are transparent to the protocol since user mobility is handled at the network layer. It is an end-to-end solution, so it secures not only the wireless data link in hotspots, but also the entire data path. The FA just forwards the authentication message between the mobile station and its home network, and vice versa.

4.1. Analysis of Authentication for Current Internet Applications

Figure 7 shows various types of authentication with different security requirements, which may occur in applications running on a mobile station. For example, mobile user

M and its home network shares a secret key. Its home network may only be accessed by M. On the other hand, a fixed public Internet site may be visited without authentication. For clarity, these situations are sorted into three categories:

Authenticating parties share a secret key: authentication between a mobile station and its HA. The secret key can be stored in either the mobile station or its Subscriber Identity Module (SIM) card.

Authenticating parties do not share a secret key: authentication between two mobile stations or between a mobile station and a fixed Internet server, and so on. Since the two parties have no common secret key to share, more public key algorithm computations are involved.

Visit the Internet public resource: Since the resource is open to the public, no authentication is needed.

Thus, the parties authenticated with mobile stations are divided into two categories: home network and any authentication parties other than the home network. This design simplifies the protocol and the implementation on the mobile station. In cases other than authentication between a mobile station and its home network, the home network performs the major authentication job and then passes the authentication result to the mobile station.

4.2. Characteristics of Proposed Authentication and Key Negotiation Protocol

In Figure 7, fixed networks are identified by the information issued by the CA server. Identity verification is carried out using the public key encryption and digital signature algorithms. Since CA servers are responsible for large amounts of certificate issuing, the task for CA servers in the proposed protocol is simple, no more than looking up the database and sending the necessary information, such as the public key message, to the corresponding receiver. A mobile station never contacts the CA server in the protocol, since it is not practical for a CA server to record the certificate information of all mobile stations because of their enormous population. The certificate of each mobile station is stored in its home network. Thus, each home network server can be considered as a CA server of its mobile stations.

A CA server works as a bridge connecting the domain servers, such as HAs and fixed servers. A fixed server can be considered as an HA without clients. The proposed protocol puts the corresponding daemon programs in each node. It is designed with the following considerations to compensate for salient features or limitations in both hardware and transmission environments:

- The protocol should be intelligent. The design should enable the protocol to adapt to various application scenarios. The adaptation should be mainly implemented in wired servers.
- The number of different types of message for mobile stations should be limited compared to the home network such that the design simplify the implementation in the mobile station.

- It is desirable to move much of the computation to the corresponding HAs which have more computation power, high speed, and reliable wired network connections. At the mobile station, intensive computations are limited. Only critical data such as secrets and their hash values are encrypted using a public key algorithm. The public key encryption and digital signature algorithms are not used simultaneously in one message.
- The length of messages will be collected for protocol latency evaluation. According to the network structure, the major presence of latency should be in the wireless part, especially when the client is connected to a cellular network. The design goal is that the time taken to transmit all messages in the slowest connection method be less than 3 s.

4.3. A Wireless Transmission Privacy Protocol

The wireless transmission privacy protocol[1] serves as an authentication service provider to other wireless Internet applications. Before an Internet application begins to send data, the mechanism does the authentication first and negotiates a shared key of which the foreign network has no knowledge. At the sender side, all the upcoming data generated by the Internet application with security requirements are encrypted by this shared key. The encrypted or wrapped data are then sent to other data processing blocks. For example, they can be further encrypted by the key acquired by the registration process. At the receiver side, the process is reversed. Thus, a foreign network cannot get plain text even if it holds a key generated during the registration process, and the wireless transmission part is also secured.

There are a few authentication scenarios. We assume that mobile station 1 (MS_1) wants to establish a connection with mobile station 2 (MS_2) via wireless Internet. MS_1 and MS_2 belong to different home networks and have no shared key. This is the most complicated scenario and other scenarios are considered its subsets. The mechanism works this way and is shown in Figure 8. The numbers in the figure represent the sequence of steps.

1. MS_1 finds MS_2's home address and and sets $IP_{auth.desk} = IP_{MS_2}$. MS_1 creates a nonce N with the corresponding hash value Hash(N). The nonce is used to verify the identity of MS_2. N and Hash(N) are encrypted with HA's public key pub_{HA1}. MS_1 sends the authentication request

   $$IP_{MS_1} : E_{key_{HA_1-MS_1}}\{ID_{MS_1}, IP_{auth.dest}, E[pub_{HA_1}, < N, \text{Hash}(N) >]\}$$

   to HA_1. The whole message is encrypted by the shared secret key of MS_1 and HA_1 $key_{HA_1-MS_1}$.

[1] Earlier version of the protocol has been published in Minghui Shi, Xuemin (Sherman) Shen, and Jon W. Mark, "IEEE802.11 roaming and authentication in wireless LAN/cellular mobile networks," *IEEE Wireless Communications Special Issue on Mobility and Resource Management*, vol. 11, issue 4, Aug. 2004, pp. 66-75

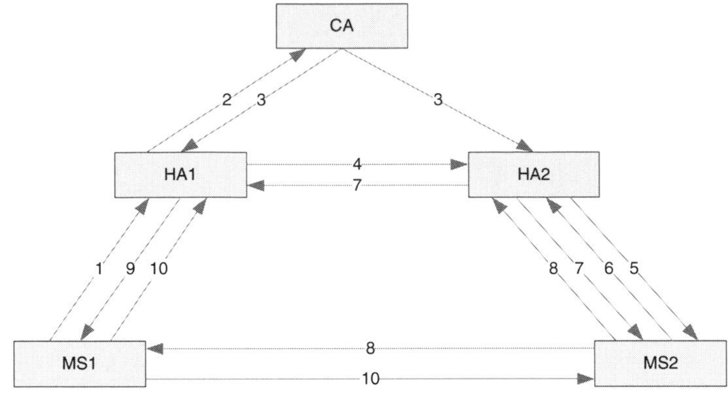

Figure 8. Authentication and key negotiation protocol between two mobile stations belonging to different home networks

2. HA$_1$ decrypts the message from MS$_1$using key_{HA_1-MS1} and $priv_{HA_1}$and gets ID_{MS_1}, N, Hash(N) and IP_{dest}. HA$_1$ realizes that MS$_1$ intends to authenticate with a third party. HA$_1$ is able to find MS$_2$'s HA, HA$_2$, from the IP of MS$_2$. In order to discover if HA$_2$ is legal, HA$_1$ contacts CA for identification information of HA$_2$, such as the public key of HA$_2$. HA$_1$sends

$$IP_{HA_1} : E\{ID_{HA_1}, IP_{HA_2}\}$$

to CA.

3. CA decrypts the message from HA$_1$ by key_{CA-HA_1} and gets ID_{HA_1} and IP_{HA_2}. CA verifies IDHA$_1$. CA searches its database, and finds the public keys of both HA$_1$ and HA$_2$ and the device ID of HA$_2$ ID_{HA_2}. CA does not need to check the information requester strictly since the information CA sends is for public use. What CA needs to do is to ensure the authority and accuracy of the information it sends. CA finds pub_{HA_1}, pub_{HA2} and ID_{HA2}. CA attaches the digital signature of the message and transmits HA$_1$'s public key and device ID

$$IP_{CA} : E_{key_{CA-HA_2}}\{ID_{CA}, ID_{HA_1}, IP_{HA_1}, pub_{HA_1}\} : sig_{CA}$$

to HA$_2$ and HA$_2$'s

$$IP_{CA} : E_{key_{CA-HA_1}}\{ID_{CA}, ID_{HA_2}, IP_{HA_2}, pub_{HA_2}\} : sig_{CA}$$

to HA$_1$.

4. HA$_1$ decrypts the message from CA by key_{CA-HA_1}, and gets ID_{CA}, HA$_2$'s IP IP_{HA_2}, public key $pub_{HA_2}$and device ID ID_{HA_2}. HA$_1$ verifies the validity

of the message sent by CA by its digital signature. Any changes to the message after it was sent can be detected. HA_1 verifies if D_{CA} matches key_{CA-HA_1}. If all validation passes, HA_1 stores the pubHA_2 and IDHA_2 pair. HA_1 generates the temporary session key key_{temp}. HA_1 set $IP_{auth.orig}$ to IP_{MS_1}, $IP_{auth.dest}$ to IP_{MS_2}, and forwards the authentication request and temporary session key

$$IP_{HA_1} : E_{pub_{HA_2}}[key_{temp}, \text{Hash}(key_{temp})] : E_{key_{temp}}\{ IP_{MS_1}, IP_{MS_2}, ID_{MS_1}, pub_{MS_1}\} : sig_{HA_1}$$

to HA_2. The key is encrypted by HA_2's public key. So far, there are two messages in step 3 and 4 sent to HA_2.

5. HA_2 will buffer the latter if the latter comes before the former. By receiving message in step 3, HA_2 can get HA_1's device ID ID_{HA_1}, IP IP_{HA_1}, and public key pub_{HA_1}. HA_2 then verifies the validity of the message in Step 4 by the attached digital signature of HA_1 and decrypt the first part of the message using its private key $priv_{HA_2}$ to get key_{temp}. Then HA_2 can further decrypt the second part of the message to get the $IP_{auth.orig}$, $IP_{auth.dest}$, and information of authentication originator, such as MS_1's public key and device ID in this case. Since HA_2 knows that MS_1 wants to authenticate with MS_2, it initiates the authentication with MS_2, because it is not secure to send the identification information before HA_2 verifies MS_2. HA_2 temporally stores key_{temp} and the information of MS_1. HA_2 send authentication request

$$IP_{HA_2} : E_{key_{Ha_2-MS_2}}\{ID_{HA_2}, IP_{MS_1}\}$$

to MS_2.

6. Similar to step 1, MS_2 starts authentication with HA_2 . MS_2 decrypts the message from HA_2using $key_{HA_2-MS_2}$ and gets ID_{HA_2} and IP_{MS_1}. Because the IP address received is different from its home network's address prefix, MS_2 knows that a third party wants to authenticate with it. MS_2 creates a nonce N and its hash value pair Hash(N). N and Hash(N) are encrypted by HA_2's public key. MS_2 sends the message

$$IP_{MS_2} : E_{key_{HA_2-MS_2}}\{ID_{MS_2}, IP_{auth.dest}, E_{pub_{HA_2}}[N, \text{Hash}(N)]\}$$

to HA_2.

7. HA_2 decrypts the message MS_2 using $key_{HA_2-MS_2}$ and gets ID_{HA_2}. HA_2 decrypts N, Hash(N) and IP_{desk} using its private key $priv_{HA_2}$. HA_2 verifies the identity of MS_2, N and Hash(N). HA_2 sends MS_1's identify information and the session key key_{temp}

$$IP_{HA_2} : E_{Key_{HA_2-MS_2}}\{ID_{HA_2}, \text{Hash}(N), ID_{MS_1}, pub_{MS_1}, E_{pub_{MS_2}}[key_{temp}, \text{Hash}(key_{temp})]\}$$

to MS_2. HA_2 informs HA_1 that MS_2 has accepted the authentication request and tell $\mathrm{HA}_1$the identity information of MS_2. also HA_2 sends

$$IP_{HA_2} : E_{key_{temp}}\{\text{Hash}(key_{temp}), ID_{MS_2}, pub_{MS_2}\} : sig_{HA_2}$$

to HA_1. HA_2's work in the protocol ends here.

8. MS_2 receives the message from HA_2 and decrypts it by $key_{HA_2-MS_2}$. MS_2 verifies Hash(N) and ID_{HA_2}. If they are valid, MS_2 uses its private key $priv_{MS_2}$ to get ke_{temp} and verifies it using Hash(N). MS_2 get the identity information of MS_1. MS_2 sends a confirmation

$$IP_{MS_2} : E_{key_{temp}}\{\text{Hash}(Key_{temp})\} : sig_{MS_2}$$

to HA_2. MS_2 generates a new nonce $newN$, and computes its Hash value Hash$(newN)$. MS_2transmits the acknowledgement to MS_1

$$IP_{MS_2} : E_{key_{temp}}\{newN, \text{Hash}(newN)\} : sig_{MS_2}$$

by using a new nonce $NewN$ and its hash value encrypted by the session key.

9. HA_1 receives the message in step 7 from HA_2. HA_1 verifies the validity of the digital signature signed by HA_2. HA_1 decrypts the first part of the message suing key_{temp} and verify Hash(key_{temp}). HA_1 gets the identity information of MS_2. HA_1sends MS_2's identity information and the session key

$$IP_{HA_1} : E\{ID_{HA_1}, \text{Hash}(N), ID_{MS_2}, pub_{MS_2}, E_{pub_{MS_1}}$$

to MS_1. HA_1's work in the protocol ends here.

10. MS_1 receives the message in step 8 from MS_2 and the message in step 9 from HA_1. Since the decryption of the former message depends on the information contained in the latter message, the former will be buffered until the latter is received. MS_1 get the identity of MS_2 $ID_{MS_2}$and pub_{MS_2}. MS_1then is able to verify and decrypt the message from MS_2. If the signature is valid, and $newN$ and Hash$(newN)$ match, MS_1 sends acknowledgement

$$IP_{MS_1} : E_{key_{temp}}\{\text{Hash}(key_{temp})\} : sig_{MS_1}$$

to HA_1 and replies

$$IP_{MS_1} : E_{key_{temp}}\{\text{Hash}(newN)\} : sig_{MS_1}$$

to MS_2 by sending the hash value of the new nonce.

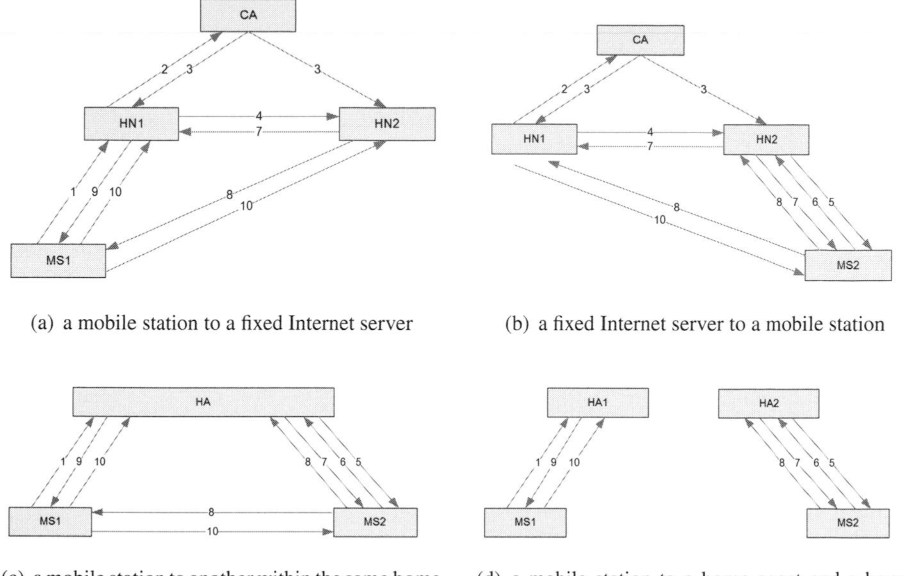

(a) a mobile station to a fixed Internet server

(b) a fixed Internet server to a mobile station

(c) a mobile station to another within the same home network

(d) a mobile station to a home agent and a home agent to a mobile station

Figure 9. Protocol variation in other authentication scenarios

11. MS_2 receives the message from MS_1 and verifies the digital signature of MS_1 and compares Hash$(newN)$ with the original. If they are both correct, the authentication process is now complete.

The protocol is also adaptive to other scenarios. For example, if a mobile station wants to authenticate with a fixed server, we consider HA2 and MS2 as one unit, and steps 5, 6, 7, and 8 are not necessary. The extended scenarios are shown in Figure 9.

4.4. Security Analysis

Security analysis presented here includes data privacy, a built-in feature for dealing with certain security compromises. Device-related information is divided into two categories. Device ID and public key belong to normal sensitive data, which means they will not do harm to the system even if they are leaked. Shared secret key and private key belong to permanent critical information that must not be compromised. The nonce and session key generated in the protocol belong to short-term critical information that can affect the ongoing session in which an attacker can discover communication contents. However, if the permanent critical information is still good, short-term critical information is safe because it is encrypted by the permanent critical information.

In the authentication protocol the exchanged message, except for digital signature in the authentication process, are all encrypted by a shared secret key between the

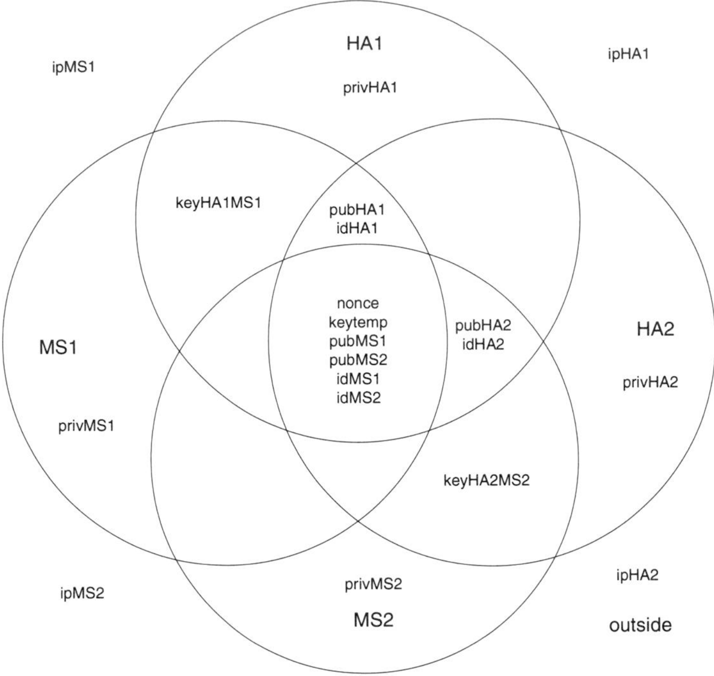

Figure 10. Secret and identification information control

HA and the client, or the session key using a symmetric encryption algorithm. More important information like the session key is further encrypted by a public key algorithm and capsulated by symmetric encryption. The authentication message is not different from the normal payload of a TCP (Transmission Control Protocol) packet, and so the foreign network can route it to the destination. The foreign network and other intruders are not able to discover the information inside because they do not have the shared key; and they must have both corresponding shared key and matched private key to get the session key.

Figure 10 shows the distribution of sensitive data after authentication is completed. The circles in the figure denote the knowledge of the information Normal sensitive information is spread to trusted sites and devices only. The protocol ensures that no sensitive information is released before the information receiver is identified. No permanent critical information is sent in any form during the authentication. Temporary critical data are spread to trusted sites only.

The protocol should be designed to resist certain security compromises. In our authentication protocol, illegal possession of someone's device ID, home IP address, and public key will do virtually no harm to the system. The home network always uses the corresponding shared secret key to process messages according to the carried home

IP address. It is the shared secret key and private key that build up the real or final authentication process.

Comparing the shared secret key and private key, a shared secret key is more likely to be compromised because at least two parties, the home network and mobile station, have a copy of the key. Only one copy exists in the mobile station for the private key case. A private key is never given out because it is not necessary due to the nature of public key algorithms. In our protocol, for example, if the shared secret key is leaked, the intruder can get only the device ID, IP address, and public key, which belong to normal sensitive information and do virtually no harm to the system, because the device ID is used for quick identification, and the public key itself is originally open to the public. But it could be harmful if this key were also used on other occasions such as mobile station registration, since the registration process should be done very quickly to avoid the connection being dropped, so there is no time to execute additional time-consuming public key algorithms.

If a shared secret key is leaked and an intruder tries to use it without a proper private key, the system can detect the compromise of the shared secret, because in our authentication protocol, each entity involved is required to return a hash value that can only be achieved by its private key or attach a digital signature to the message. Once the system detects this flaw, it indicates that the common secret key is leaked and the user should be warned immediately. The system cannot detect the private key flaw though, because without a proper shared secret key, the system cannot look into the message. So only when the shared secret and corresponding private keys are broken simultaneously can the intruder access the network illegally. The system is able to detect a security compromise of the shared secret key, but not of the private key. Fortunately, the private key is unlikely to be leaked due to the nature of public key algorithms.

The protocol is also designed to resist replay attack. Every authentication session between an HA and a mobile station, or two mobile stations or two HAs is completed by using fresh nonces and fresh session key, so replay attack has no effect on it. Since the information exchanged between an HA and a CA server represents facts on the clients identifications and public keys, simply replaying this message does no harm to the system unless the intruder can change the payload and the corresponding digital signature, which is very hard unless the intruder can get the CAs private signature.

In order to totally duplicate component (mobile station, home network server, CA server), the malicious user at least needs a proper home IP address, device ID, shared secret key, and private key to satisfy the authentication protocol completely, or it will be rejected at the corresponding step where the item is checked.

5. AUTHENTICATION AND KEY NEGOTIATION DEMONSTRATION

Figure 11 shows the demonstration program for all the considered scenarios. The demonstration shows the authentication progress, the way in which the protocol self-adapts to each case, and message lengths sent and received by each node. The demonstration uses RSA as the public key algorithm, DES as the symmetric algorithm, and MD5 as one-way hash functions. Since the proposed protocol is

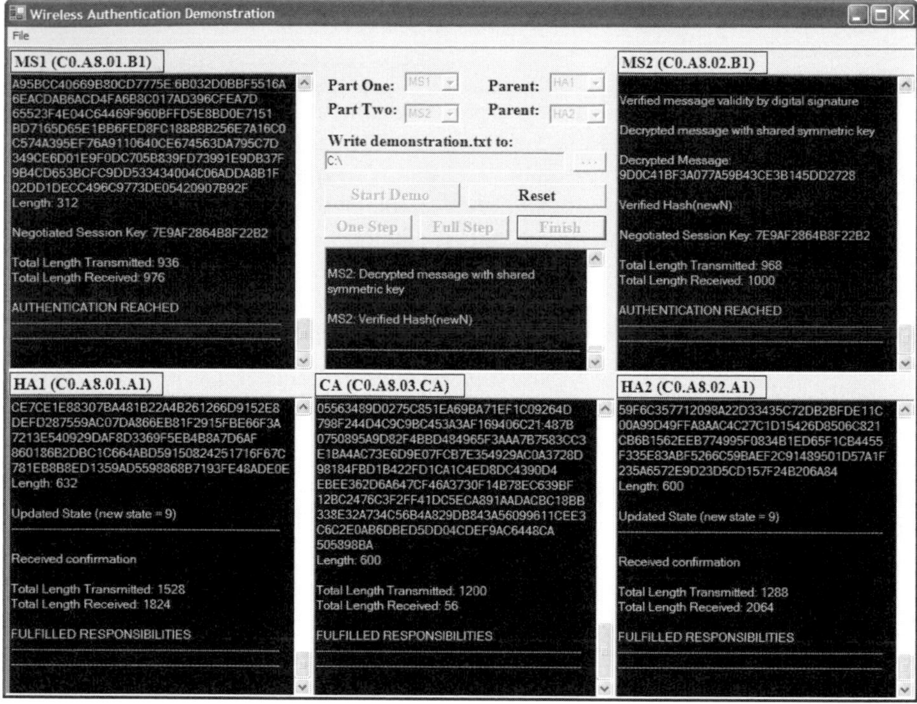

Figure 11. Demonstration of the authentication and key negotiation protocol

cryptographic-algorithm-independent, other stronger or lighter algorithms can be used to accommodate specific application requirement. The demonstration also shows that the total amount of data for the mobile node is less than 2 kbytes. If the slowest network connection speed is 14.4 kb/s in the cellular network with overhead of the transmission considered, the data transmission can be finished in less than 3 s.

6. SUMMARY

In this chapter, a network architecture and a set of signaling mechanisms are proposed to support current available wireless LAN hotspot roaming. The proposed architecture offers a smooth transition of wireless LAN hotspots from non-roaming-supported to seamless-roaming-supported, and therefore previous investment can be protected. Meanwhile, wireless transmission security is carefully considered. An application layer authentication and key negotiation protocol is developed to keep end-to-end transmission secure. The results can enable wireless LAN roaming, enhance wireless communications, and speed up the deployment of public wireless LAN applications.

ACKNOWLEDGEMENT

This work has been supported by a Natural Science and Engineering Council (NSERC) Postgraduate Scholarship and a research grant from Bell University Laboratories (BUL), Canada. We would like thank Matthew Cruickshank for his help in the demonstration programming.

7. REFERENCES

1. C. S. Loredo and S. W. deGrimaldo, "Wireless LANs: Global Trends in the Workplace and Public Domain," *The Strategies Group*, 2002.
2. A. T. Campbell et al., "IP Micro-Mobility Protocols," *ACM SIGMOBILE Mobile Comp. and Commun. Rev.*, vol. 4, Oct. 2001, pp. 45-54.
3. C. Perkins, "IP Mobility Support," *RFC 2002*, Oct. 1996.
4. J. R. Walker, "Unsafe at Any Key Size: An Analysis of the WEP Encapsulation," *IEEE doc. 802.11-00/362*, Oct. 2000.
5. W. A. Arbaugh, "An Inductive Chosen Plaintext Attack Against WEP/WEP2," *IEEE doc. 802.11-01/230*, May 2001.
6. J. Zhang and J. W. Mark, "A Secured Registration Protocol for Mobile IP," *Master's Thesis*, Apr. 1999.
7. Sufatrio and K. Y. Lam, "Mobile IP Registration Protocol: A Security Attack and New Secure Minimal Public-Key Based Authentication," *I-SPAN 99*, Fremantle, Australia, 1999, pp. 364-369.
8. J. S. Stach, E. K. Park, and Z. Su, "An Enhanced Authentication Protocol for Personal Communication Systems," *IEEE Wksp. App.-Specific Software Eng. Tech.*, Dallas, TX, 1999, pp. 128-132.
9. H. Lin and L. Harn, "Authentication Protocols with Nonrepudiation Services in Personal Communication Systems," *IEEE Commun. Lett.*, vol. 3, 1999, pp. 236-238.
10. R. Shirey, "Internet Security Glossary," *IETF RFC 2828*, 2000.
11. A. Westerinen, J. Schnizlein, J. Strassner, M. Scherling, B. Quinn, J. Perry, S. Herzog, A.N. Huynh, M. Carlson, and S. Waldbusser, "Terminology for policy-based management," *IETF RFC 3198*, 2001.
12. IETF AAA Working Group, "Mobile IP AAA Requirements," *IETF RFC2977*, October 2000.
13. C. Rigney, "Remote Authentication Dial In User Service (RADIUS)," *IETF RFC2138*, 1997.
14. C. Rigney, "RADIUS Accounting," *IETF RFC2139*, 1997.
15. Network Working Group, "RADIUS Accounting," *IETF RFC 2866*, June 2000.
16. C. Finseth, "An Access Control Protocol, Sometimes Called TACACS," *IETF RFC1492*, 1993.
17. D. Carrel and L. Grant, "The TACACS+ Protocol: Version 1.78," *IETF Internet Draft, <draft-grant-tacacs-02.txt>*, 1997.
18. J. Kohl and C. Neuman, "The Kerberos Network Authentication Service (V5)," *IETF RFC1510*, 1993.
19. U. Carlsen, "Optimal Privacy and Authentication on a Portable Communication System," *Op. Sys. Rev.*, vol. 28, 1994, pp. 16-23.
20. C. Park et al., "On Key Distribution and Authentication in Mobile Radio Networks," *Proc. Advances in Cryptology-Eurocrypt 93*, Szombathely, Hungary, 1993, pp. 461-465.
21. M. Tatebayashi and D. B. Newman, "Key Distribution Protocol for Digital Mobile Communication Systems," *Proc. Advances in Cryptology-Crypto 89*, Houthalen, Belgium, 1989, pp. 324-333.

22. B. Aboda and M. Beadles, "The Network Access Identifier," *IETF RFC 2486*, Jan. 1999.

23. P. Calhoun and C. Perkins, "Mobile IP Network Access Identifier Extension for IPv4," *IETF RFC 2794*, Mar. 2000.

24. E. Shim and R. D. Gitlin, "Reliable and Scalable Mobile IP Regional Registration," *IETF Internet Draft*, *<draft-shimmobileip-reliable-reg-00.txt>*, Apr. 2001

12

AN EXPERIMENTAL STUDY ON SECURITY PROTOCOLS IN WLANS

Avesh Kumar Agarwal
Department of Computer Science
North Carolina State University
E-mail: akagarwa@unity.ncsu.edu

Wenye Wang
Department of Electrical and Computer Engineering
North Carolina State University
E-mail: wwang@eos.ncsu.edu

Wireless Local Area Networks (WLANs) are vulnerable to malicious attacks due to their open shared medium. Consequently, provisioning enhanced security with strong cryptographic features and low performance overhead becomes exceedingly necessary to actualize real-time services in WLANs. In order to exploit full advantage of existing security protocols at various layers, we study the cross-layer interactions of security protocols in WLANs under different network scenarios. In particular, we present a detailed experimental study on the integration of commonly used security protocols such as WEP, 802.1x and EAP, IPsec and RADIUS. First, we classify individual and hybrid policies, and then, define *security index* and cost functions to analyze security strength and overhead, quantitatively, of each policy. By setting-up an experimental testbed, we measure performance cost of various policies in terms of authentication time, cryptographic cost and throughput using TCP/UDP traffic streams. Our results demonstrate that in general, the stronger the security, the more signaling and delay overhead, whereas, the overhead does not necessarily increase monotonically with the security strength. Therefore, it is suggested to provide substantial security at a reasonable cost of overhead with respect to mobile scenarios and traffic streams. Also, we notice that authentication time will be a more significant factor contributing towards QoS degradation than cryptographic cost, which is critical to real-time service in wireless networks.

1. INTRODUCTION

Wireless Local Area Networks (WLANs) have become prevalent for providing ubiquitous internet access for mobile users. However, security is of utmost concern in WLANs, because interception and eavesdropping of data in transit by malicious users become easier due to shared and broadcast air medium [1], [2]. Therefore, several security protocols such as wired equivalent privacy (WEP), 802.1x port access control with extensible authentication protocol (EAP) support are proposed to address security issues [3], [4], [5], [6], [7]. Moreover, due to strong security provided by IP security (IPsec) in wired networks, it is considered a good option for establishing virtual private networks (VPNs) [8] in wireless network also. However, existing studies demonstrate various types of malicious attacks on these protocols [3], [9], [10]. Consequently, researchers have studied these security protocols individually with respect to cryptographic properties to enhance security in the network [11], [12], [13]. In this study, we explore cross-layer interactions of existing security protocols by integrating the protocols at several layers to enhance security in WLANs.

Moreover, security protocols incur performance overhead due to the configuration and messages at different layers in the network. Meanwhile, many real-time wireless applications have shown an increasing demand for better quality of service (QoS) in real networks [14]. Therefore, it becomes mandatory to determine the performance impact of the security protocols in real-time networks for better QoS. Existing studies in the past have focused mainly on improving cryptographic perspective of security protocols, whereas lacking detailed quantification of performance overhead associated with the protocols [11], [12], [13]. In this study, we provide comprehensive real-time measurements of performance overhead associated with security protocols at various layers in WLANs.

Measurements are very important to determine the realistic view of the performance overhead associated with the security mechanisms. Therefore, to gain fundamental understanding of performance impact due to security protocols, experimental studies are carried out in the past in various network environments [8], [15], [16], [17]. However, these studies have explored security protocols as a stand-alone mode. Moreover, these studies perform experiments in few network scenarios providing less detailed real-time results. In this work, we study the cross-layer integration of security protocols in various non-roaming and roaming network scenarios to gain a deeper understanding of the associated performance overhead. Measurements provided in this study are explained to show how integration of quality of service (QoS) and security service affects system performance.

To achieve above goals, we have setup a real-time experimental testbed. The testbed is a miniature of existing wireless networks, which ensures that our experimental results can be mapped to large-scale wireless networks. The testbed consists of two subnets for configuring various network scenarios. We install open source versions of commonly used security protocols such as 802.1x, EAP, IPsec and RADIUS in the testbed. Security protocols are classified into individual and hybrid security policies to study cross-layer interactions. Moreover, we define *security index* and cost functions to analyze security strength and overhead associated with each security policy, respectively. Authentication time, cryptographic cost and throughput are the performance metrics evaluated under TCP and UDP traffic streams. Moreover, we discuss various attacks on security policies and demonstrate that hybrid security policies (cross-layer integration of security protocols) are less vulnerable than individual security policies.

Our observations demonstrate that there is always a tradeoff between security and performance associated with a security policy, depending upon the network scenario and traffic types. In general, we observe that security policy with higher strength is not always the best option for all scenarios. We find that the cross-layer integration of security protocols may provide the strongest protection, but with more overhead together. Our results demonstrate that in general, the stronger the security, the more signaling and delay overhead, whereas, the overhead does not necessarily increase monotonically with the security strength. Moreover, we observe that IPsec policies provide the best tradeoff between security and performance for authentication time; 802.1x-EAP-TLS policy is the best suitable option for low cryptographic cost and better security strength in many scenarios. In addition, experimental results for throughput reveal that authentication time will be a more significant factor contributing towards QoS degradation in the network than cryptographic cost.

The rest of the chapter is organized is as follows. Section 2 discusses the background of existing security protocols for WLANs. Details of the testbed, network scenarios and classification of security policies are provided in Section 3. Security index and cost functions to analyze the security strength and performance overhead associated with security policies are presented in Section 4. Section 5 explains procedure to carry out experiments. Detailed performance evaluation of experimental results is provided in Section 6. Section 7 discusses vulnerabilities associated with individual security policies with respect to malicious attacks and countermeasures in cross-layer integration of security policies. Finally, Section 8 concludes the chapter.

2. BACKGROUND

To address security issues, many protocols are developed, which operate at different network layers. Wireless Equivalent Privacy (WEP), 802.1x with Extensible Authentication Protocol (EAP), Remote access dial in user service (RADIUS) and IP security (IPsec) are some of the protocols used in wireless networks. We focus on studying these security protocols because they operate at different network layers, which will help us to analyze the overhead introduced by security services across network layers. These protocols are widely adopted in the wireless networks providing a very close analysis, which will be useful for the real-time networks. Brief description of these protocols is as follows:

MAC Layer Protocols: WEP is the very first protocol to be considered for wireless networks. WEP has been identified to be susceptible to many type of attacks [3]. To overcome WEP weaknesses, IEEE 802.1x standard is designed to provide stronger security [4], [5], [6], [7]. 802.1x works at MAC layer and provides port-based access control for wireless nodes. In addition, 802.1x exploits the use of EAP (MD5,TLS), which is used as transport mechanism [4]. Besides considering MAC layer security protocols, we also evaluate network layer and application layer security protocols such as IPsec and RADIUS in the experimental testbed.

Higher Layer Protocols: IPsec is a network layer protocol, originally designed for wired network, which is now being considered for wireless networks due to its strong authentication and cryptographic methods. Further, TSL is a transport layer protocol and successor to Socket Security Layer (SSL), which is the most widely deployed security protocol on the Internet. At application layer, we consider RADIUS protocol, which is based on client-server architecture.

Existing security protocols have some drawbacks and are prone to several attacks. For example, according to previous studies, WEP and 802.1x are susceptible to many types of attacks [3], [9] and [10]. In addition, there are other studies which explain the security aspects of WLANs providing overview of various security protocols such as [2]. To overcome these problems, researchers have come up with many solutions to improve the security aspects of these protocols in recent years. For example, recently a new authentication protocol is proposed for wireless networks in [19]. In addition, other works have proposed solutions to improve security for mobile wireless networks [11], [12] and [13]. Moreover, there are other studies, which focus on performance aspects of security protocols. For example, a performance analysis of different protocols of IPsec is provided in [15]. Similarly, IPsec performance is also analyzed as virtual private networks (VPN) in [8]. In addition, a proposal is provided to implement wireless gateway for WLAN based on IPsec protocol in [16]. But, we observe that most of the research is focused on security aspects with little thoughts given to performance impact of security protocols on system performance. Therefore, we conduct comprehensive experimental analysis to uncover performance issues associated with security protocols in mobile wireless LANs.

Our study focuses on the impact of security protocols on different user's mobility scenarios in combination with IP mobility in WLAN roaming. Moreover, our analysis has considered a wide range of security protocols at different layers such as 802.1x, WEP, SSL other than just IPsec. Unlike previous studies, we focus on the quality of service (QoS) aspects of the network determining impact on QoS when security services are enabled in the wireless networks. To our knowledge, this is the first experimental study on this issue, which analyzes security protocols in various mobility scenarios.

3. IMPLEMENTATION SETUP

In this section, we provide details of the experimental testbed including hardware equipments and software configurations. Figure 1 shows an example of testbed architecture in which two subnets are illustrated. Although, we show only two subnets; with different combinations in hardware and software, virtually we create a heterogeneous environment that captures mobile scenarios of wireless local area networks.

3.1. Hardware Configuration

Mobile IP is used to support mobility and routing in our testbed. A mobile node (MN) is defined as the wireless node, which is able to change its point of attachment [20]. Different mobile nodes used in our testbed consist of iPAQ (Intel StrongARM

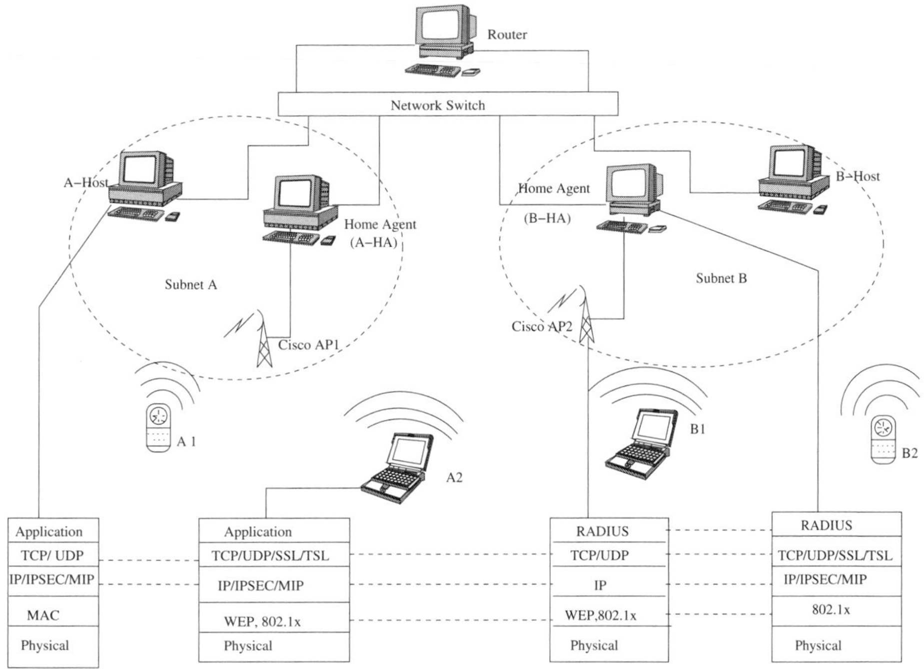

Figure 1. Wireless LAN Testbed.

206 MHZ), Sharp Zaurus (Intel XScale 400 MHz) and Dell Laptop (Celeron Processor, 2.4GHZ). Home agents (HA), A-HA and B-HA, are the gateways in a mobile node's home network (HN) where the mobile node registers its permanent IP address. In our testbed, home agents (HA) are gateways for Subnets A and B and are Dell PC with Pentium IV 2.6 GHZ. Foreign agents (FA) are the gateways in a foreign network (FN) where a mobile node obtains a new IP address to access to the network. Home agents have the functionalities of foreign agents as well in our testbed. They are connected to Cisco Access Points (Cisco Aironet 1200 series) to provide wireless connectivity. In addition, the home agents have functions of IPsec gateways and RADIUS server for IPsec and 802.1x, respectively. An IPsec tunnel is setup between home agents to provide security over the wired segment in our testbed. Hosts A-Host and B-Host act as wired correspondent nodes in Subnets A and B and are Dell PC with Pentium IV 2.6 GHZ. Cisco Catalyst 1900 series is used as a network switch to provide connectivity between two subnets via the router. We have used Netgear MA 311 and Lucent Orinoco Gold wireless cards in all mobile devices.

3.2. Software Configuration

All systems use Redhat Linux 9.0 kernel 2.4.20. We have installed open-source software components for various protocols in the testbed as follows:

- *FreeSwan* open source is installed on home agents and mobile nodes for IPsec functionality [21].
- *Xsupplicant*, which provides 802.1x client functionality, has been installed on mobile nodes [22].
- RADIUS server functionality has been provided by *FreeRadius* and has been installed on home agents [23].
- *OpenSSL* open source software is installed on home agents [24].
- To introduce user mobility in our network, *Mobile IP* implementation from Dynamic is installed on mobile nodes and home agents [25].
- *Ethereal* packet analyzer is used for packet capturing.
- *Iperf* and *ttcp* are used for generating TCP/UDP traffic streams.

3.3. Network Scenarios

Network scenarios are classified into non-roaming ($\mathcal{N}$) and roaming ($\mathcal{R}$) based on user's current location, i.e., whether a user is in its home domain or foreign domain, respectively. To make the description of scenarios clear, we assume that subnet A is the home domain for mobile nodes $A1$ and $A2$; and subnet B is the home domain for mobile nodes $B1$ and $B2$. All scenarios are demonstrated in Figure 2. Non-roaming scenarios, represented as $\mathcal{N}$, are defined as the scenarios when both communicating mobile users are in their home domain. Following are the details of various non-roaming scenario configured in the testbed.

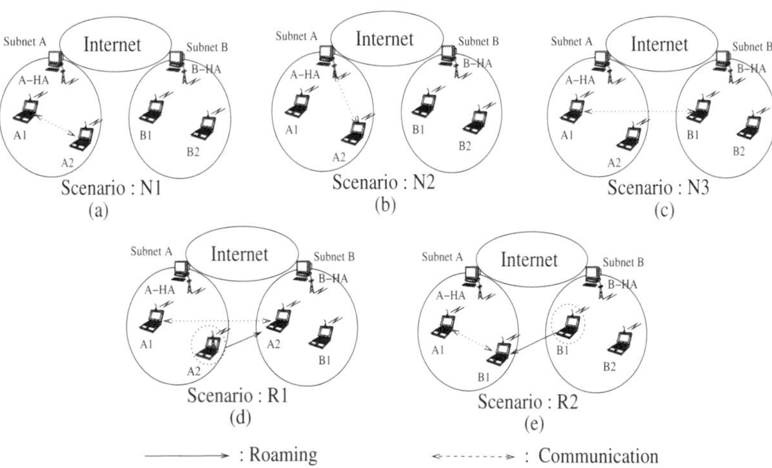

Figure 2. Non-Roaming and Roaming Scenarios.

- **Scenario $N1$:** It deals with the situation when both mobile nodes are in the same subnet, which is their home domain also. For example, when communication occurs between $A1$ and $A2$ as shown in Figure 2(a).
- **Scenario $N2$:** Mobile nodes communicate with their home agent that is acting as an application server providing services to mobile clients in the network. Here, a part of the communication path is wired, which is not the case in scenario $N1$. As shown in Figure 2(b), this scenario occurs when home agent A-HA communicates with $A1$ or $A2$.
- **Scenario $N3$:** It is to capture the impact of security services when participating mobile nodes are in different domains. For example, when $A1$ or $A2$ communicates with $B1$ or $B2$ as shown in Figure 2(c).

When at least one of two communicating mobile users is in a foreign domain, we refer it as roaming scenario, represented as $\mathcal{R}$. The following roaming scenarios are configured in our experimental testbed.

- **Scenario $R1$:** This scenario specifies when one end node, which is in a foreign domain, is communicating with the other node, which is in its home domain, but two nodes are in different domains. It aims to analyze the effect of security services on data streams when one node is roaming. As shown in Figure 2(d), this scenario occurs when node $A2$ roams to subnet B and communicates with $A1$.
- **Scenario $R2$:** This scenario occurs when both nodes are in the same domain but one node is roaming. Therefore, current network is the foreign domain for one node, whereas it is the home domain for other node. It helps us in analyzing performance impact on data streams when roaming node is communicating with a non-roaming node in the same domain. For example, when node $B1$ roams to subnet A and communicates with $A1$ or $A2$ as shown in Figure 2(e).

3.4. Security Policies

Security policies are designed to demonstrate potential security services provided by the integration of security protocols at different layers. Each security protocol uses key management protocols, various authentication, and cryptographic mechanisms. Therefore, a variety of security policies are configured in our experiments by combining various mechanisms of security protocols. Let $\mathcal{P} = \{P_1, P_2, \ldots, P_{12}\}$ represent the set of individual and hybrid security policies configured in the network. Next, we explain these security policies and their significance in detail.

Individual Security Policies

When a policy involves mechanisms in a single security protocol, it is called an *individual security policy*. "No security" means that there are no security services enabled in the network. "No Security" policy helps us in comparing the overhead associated with others in terms of end-to-end response time, throughput and performance overhead. In the following paragraphs we discuss security policies for each security protocol.

- **WEP Policies:** WEP supports 40-bit and 128-bit encryption keys. Although, we carried out experiments on WEP with 40-bit and 128-bit keys, we found that results in both cases showed little variations. Therefore, in this paper, we present WEP only with 128-bit due to longer key size. Although WEP has been shown vulnerable to many attacks [3], we study WEP in this paper for two reasons. First, WEP is still being used in many real-time networks for dynamic session keys along with other security protocols such as EAP-TLS with 802.1x framework [5]. Second, comparing WEP's performance with other security protocols provides a complete study of the performance impact of existing security protocols for WLANs. P_2 is the only individual WEP policy configured in the testbed.

- **IPsec Policies:** IPsec protocol supports a large set of cryptographic and authentication algorithms, and provides strong security. Since we use $Freeswan$ [21] for IPsec functionality, our analysis is restricted to the security services provided by Freeswan open source implementation. Freeswan includes 3DES as an encryption mechanism, and MD5 and SHA as authentication algorithms. Since IPsec tunnel mode is considered better by providing stronger security services than IPsec transport mode, we analyze only IPsec tunnel mode in our setup. P_3 is the only individual IPSEC policy configured in the testbed.

- **802.1x Policies:** In case of 802.1x, we use RADIUS as a backend server maintaining users' secret credentials. EAP is used as a transport mechanism that involves MD5 and TLS modes. Since FreeRadius open source also supports MD5 and TLS, we analyze 802.1x with EAP in both TLS and MD5 modes. Although EAP-MD5 is not considered a very strong authentication mechanism for WLANs [26], it can provide better security when configured with other security protocols. Therefore, we believe that inclusion of EAP-MD5 makes our study complete. Moreover, as a discussing performance aspect of various security protocols is the main objective of this paper, inclusion of EAP-MD5 enables us to provide comprehensive performance measurements of the existing security protocols in WLANs. Policies P_5 and P_6 are two 802.1x individual policies configured in the testbed.

Hybrid Security Policies

When security policies involve mechanisms belonging to multiple security protocols at different network layers, they are called *hybrid security policies*. Such policies are required, if visiting clients have security support at more than one network layer. Therefore, the network can fulfill the needs of a large number of clients. Also, security functionalities required by the network may not be fulfilled by one security protocol, leading to the need for configuration of more than one security protocol in the network.

Our study incorporates security services provided by WEP, IPsec and 802.1x in different ways. Initially we focus on the combination of IPsec and WEP. We first analyze the overhead associated with IPsec (3DES, MD5 and SHA) and WEP (40 or 128 bits), but here we present results only for IPsec (3DES, SHA) and WEP (128 bits) due to page limit. Then we perform experiments with 802.1x and WEP to capture combined effects of all security services at MAC layer and transport layer. Finally, we unite different security services of 802.1x, WEP and IPsec together for analysis. P_4, P_7, P_8, P_9, P_{10}, P_{11} and P_{12} are hybrid security policies configured in our testbed. Integration of different security protocols helps us answer a vital question, i.e., whether it is beneficial to combine security mechanisms at different network layers at the cost of adding extra overhead. A subset of security policies and associated features are shown in TABLE 1.

Table 1. Features of Security Policies.

Policy No.	Security Policies	"A"	"C"	"D"	"N"	"M"
P_1	No Security					
P_2	WEP-128 bit key	Y	Y			
P_3	IPsec-3DES-SHA	Y	Y	Y	Y	Y
P_4	IPsec-3DES-SHA-WEP-128	Y	Y	Y	Y	Y
P_5	8021x-EAP-MD5	Y				
P_6	8021x-EAP-TLS	Y			Y	Y
P_7	8021X-EAP-MD5-WEP-128	Y	Y			
P_8	8021X-EAP-TLS-WEP-128	Y	Y		Y	Y
P_9	P_7 +IPsec-3DES-MD5	Y	Y	Y	Y	Y
P_{10}	P_8 + IPsec-3DES-MD5	Y	Y	Y	Y	Y
P_{11}	P_7 + IPsec-3DES-SHA	Y	Y	Y	Y	Y
P_{12}	P_8 + IPsec-3DES-SHA	Y	Y	Y	Y	Y

In the above table, "A" denotes authentication; "C" denotes confidentiality; "D" denotes data integrity; "N" denotes non-repudiation; and "M" denotes mutual authentication.

4. SECURITY INDEX AND COST ANALYSIS

In this section, we present a simple model to analyze the security strength of various policies. Then, we develop cost functions to evaluate the associated security features of each policy. Our model is based on security features such as authentication, encryption, data integrity, non-repudiation, access control and mutual authentication required by a security policy.

4.1. Security Index (SI)

In this work, we aim to represent the security services that can be configured in security protocols. Our goals regarding quantifications of security of a system are manifolds. Generally, there are requirements to quantify the security even before system is deployed so that an appropriate security policy can be chosen. Therefore, it is not possible to observe the system behavior in advance for security quantification. In addition, the approach for quantifications should be simple and practically feasible with regards to processing time and implementation, so that it can be implemented even in resource constrained environments. Moreover, quantification should have fine granularity to have clear distinction among the strengths of security policies. As existing studies lack one or more goals desired by us, we define security quantification method based on linear sum of weights assigned to various mechanisms in a security policy.

Every security policy provides some security features such as authentication and confidentiality in our experimental study. However, it is difficult to quantify the security strength delivered to a system or a network by a security policy based on its features. This is due to the fact that it is almost impossible to predict that when a system or a network can be compromised in the future during the configuration of a security policy. Generally, it is not easy to be fair in comparing two policies with different features. For example, assume that a security policy P_α consists of 2 features which are very strong, and another security policy P_β has of 4 features which are relatively weak. If we compare two policies with respect to the 2 features of P_α, then we can conclude that P_α provides stronger security than P_β. However, if we compare P_α and P_β with respect to the 2 features not in P_α but in P_β, we find that P_β is better than P_α. The justification of which security policy is better than the other depends upon network requirements, policies installed, and features activated in a network. We define few terms to make discussion clear while defining *security index* and assigning weights to security features.

- Security Feature: Security services, such as authentication, mutual authentication, confidentiality, data integrity and non-repudiations, are defined as security features.

- Security Mechanism: Various protocols, such as EAP-MD5, EAP-TLS, IPsec, WEP and so on, which consists of different algorithms and protocols, are defined as security mechanisms. A security mechanism can provide more than one security features. For instance, EAP-MD5 can provide authentication and data integrity security features.

We define *security index* to quantify and understand the strength of security policies by using weights associated their security feature. By defining *security index*, we aim to achieve the following goals.

- As it is generally intuitive that security policy with more number of features is considered stronger than a security policy with less number of features. Therefore, our definition of *security index* should follow this intuition.

- In addition, if two policies have the same number of security features associated with them, the policy with strong security features is considered stronger than the policy with weak security features. Therefore, our weight assignments to different security mechanisms ensure that policy with strong security features is assigned a higher value of security index than other policies with weak security features.

- Weights assigned to different security mechanisms and the resultant security indices of security policies signify whether a security policy is stronger or not from other policies. Security indices do not imply the absolute security strength of a policy, which is hard to quantify.

- Although, our definition of security index can be applied in different other scenarios not explored here, the weight assignment to various security mechanisms is unique to this study. If security protocols with different security mechanisms are considered, then weights assignments may require the modifications to various weights to accommodate different security mechanisms.

Now, we define *security index* as described below. We consider five security features, $authentication$, $mutual authentication$, $confidentiality$, $data integrity$ and $non\ repudiation$, for assigning weights to different security mechanisms. Let

w_A^i be the weight of a mechanism i on authentication.
w_C^i be the weight of a mechanism i on confidentiality.
w_T^i be the weight of mechanism i on data integrity.
w_R^i be the weight of a mechanism i on non-repudiation.
w_M^i be the weight of a mechanism i on mutual authentication.

Assume a security policy P_α consists of n security mechanisms. Then, *security index* security policy P_α of is a metric which is defined as

$$SI(P_\alpha) = \sum_{i=1}^{n} w_A^i \mathcal{I}_{\mathrm{A}} + w_C^i \mathcal{I}_{\mathrm{C}} + w_T^i \mathcal{I}_{\mathrm{T}} + w_R^i \mathcal{I}_{\mathrm{R}} + w_M^i \mathcal{I}_{\mathrm{M}} \quad (1)$$

In the above expression, $\mathcal{I}_{(\cdot)}$ is an indicator function, which equals to 1 if that particular security feature is provided by the mechanism i, otherwise zero. Next, we explain weight assignments in detail.

The purpose of weight assignment to each security mechanism is to quantify strengths of different security mechanisms with respect to various security features. Weight assignment to these security mechanisms is based on several criteria such as the key length and use of digital certificates used in a particular mechanism, which are explained in detail below.

- Since WEP-128, 802.1x-EAP-MD5, IPsec and 802.1x-EAP-TLS provide authentication feature in the testbed, four different weights are assigned to each of them. Weights assigned to each of these mechanisms are based on their relative strengths. WEP-128 is assigned the lowest weight of 1 due to its weak cryptographic algorithm [3]. 802.1x-EAP-TLS is assigned the highest weight of 4 due to its use of digital certificate for signing private keys. IPsec is assigned weight of 3 which is lower than the weight assigned to 802.1x-EAP-TLS, because IPsec uses public key cryptography without certificates unlike 802.1x-EAP-TLS. Although digital certificate can be used with IPsec as well, but we use IPsec without certificate due to some practical problems in configuring them together in the testbed. On the other hand, 802.1x-EAP-MD5 is assigned weight of 2 which is lower than those of IPsec and 802.1x-EAP-TLS, because it uses weak unencrypted user-password mechanism [27].

- In case of mutual authentication, IPsec and 802.1x-EAP-TLS mechanisms are considered, and are assigned weights of 1 and 2, respectively. The reason for assigning a higher weight to 802.1x-EAP-TLS than IPsec is the same as we described for the authentication feature.

- In addition, WEP-128 and 3DES offer confidentiality feature for various security policies in the testbed. 3DES encryption mechanism is allocated higher weight of 2 than the weight of 1 assigned to WEP-128, because 3DES provides complex and more secure cryptographic algorithm than WEP-128.

- IPsec with SHA/MD5 and 802.1x-EAP-MD5 provide the data integrity security feature. Since SHA uses longer keys than MD5 [28], IPsec with SHA is assigned a higher weight of 2 than those of IPsec with MD5 and 802.1x-EAP-MD5. IPsec with MD5 and 802.1x-EAP-MD5 are assigned the same weight of 1, because both of them use MD5 algorithm.

- We have considered IPsec and 802.1x-EAP-TLS as security mechanisms with respect to non-repudiation. 802.1x-EAP-TLS policy is assigned a higher weight of 2 than the weight 1 of IPsec.

In general, we notice that TLS has been assigned a higher weight than MD5, because it makes use of digital certificates which provide stronger authentication and access control mechanism than MD5 [27]. Note that the weight assigned to each protocol signifies only its relative strength corresponding to other protocols. These weights do not imply any quantification of absolute security strength associated to a security protocol. For instance, if two mechanisms providing authentication feature are assigned weights of 4 and 1, respectively, it does not mean that the mechanism with weight 4 is four times stronger than the mechanism with weight 1 with respect to authentication feature. It only signifies that the mechanism with weight 4 is stronger than the mechanism with weight 1 with respect to authentication feature. The weights assigned to each protocol are shown in Table 2.

Table 2. Weights Associated with Security Protocols.

Security Feature	Security Mechanism	Weight
Authentication (w_A)	WEP-128 (Shared)	1
	802.1x-EAP-MD5	2
	IPsec	3
	802.1x-EAP-TLS	4
Mutual (w_M) Authentication	IPsec	1
	802.1x-EAP-TLS	2
Confidentiality (w_C)	WEP-128	1
	3DES	2
Data Integrity (w_T)	MD5 (IPsec/802.1x-EAP)	1
	SHA (IPsec)	2
Non-repudiation (w_R)	IPsec (ESP)	1
	802.1x-EAP-TLS	2

Next let us look at an example of how SI is obtained. We notice from Tables 1 and 2 that P_{12} (802.1x-EAP-TLS-WEP-128-IPsec-3DES-SHA) consists of 3 mechanisms: IPsec-3DES-WEP, WEP-128 and 802.1x-EAP-TLS. These 3 mechanisms consist of 10 features: 5 by IPsec-3DES-WEP, 2 by WEP-128, and 3 by 802.1x-EAP-TLS. Let i, j and k represent IPsec-3DES-WEP, WEP-128 and 802.1x-EAP-TLS, respectively. By using Table 2, weights of features provided by IPsec-3DES-SHA are $w_A^i = 3$, $w_M^i = 1$, $w_C^i = 2$, $w_T^i = 2$, and $w_R^i = 1$. The corresponding weights of features in WEP-128 are $w_A^j = 1$, $w_M^j = 0$, $w_C^j = 1$, $w_T^j = 0$, and $w_R^j = 0$. With 802.1x-EAP-TLS, we obtain the weight as $w_A^k = 4$, $w_M^k = 2$, $w_C^k = 0$, $w_T^k = 0$, and $w_R^k = 2$. Although 802.1x-EAP-TLS can provide confidentiality and data integrity, but in our testbed, it is used in access control for authentication in wireless network without its confidentiality and data integrity features. Therefore, we do not take into account the confidentiality and data integrity features in 802.1x-EAP-TLS. According to (1), P_{12} has an index of

$$\begin{aligned} SI(P_{12}) &= w_A^i + w_M^i + w_C^i + w_T^i + w_R^i \\ &\quad + w_A^j + w_M^j + w_C^j + w_T^j + w_R^j \\ &\quad + w_A^k + w_M^k + w_C^k + w_T^k + w_R^k. \end{aligned} \tag{2}$$

By substituting the weights of various features, the value SI for policy P_{12} is $3+1+2+2+1+4+2+2+1+1 = 19$. For comparative study, we normalize the SI values of other policies based on the highest value of 19 of security policy P_{12}. Table 3 lists actual and normalized SI (NSI) values of security policies in the increasing order.

Table 3. Security Index.

Policy	P_1	P_2	P_5	P_7	P_6	P_3	P_8	P_4	P_9	P_{11}	P_{10}	P_{12}
SI	0	2	3	5	8	9	10	11	12	13	18	19
NSI	0	10.5	15.8	26.3	42.1	47.4	52.6	57.9	63.2	68.4	94.7	100

4.2. Cost Analysis

Now we analyze performance cost associated with various security policies in terms of authentication time, cryptographic cost and throughput. Metrics, authentication time and cryptographic cost, are associated with authentication phase and encryption/decryption process of a security policy, respectively. On the other hand, throughput helps us in quantifying QoS degradation in a network.

Authentication Time

Authentication time is defined as the total time consumed in an authentication phase of a security policy. We consider authentication time, represented as (C_A), as the cost associated with authentication phase of a security policy. It is due to the fact that time involved in an authentication phase is one of the important factors contributing towards performance impact in a network. Here, we describe steps to calculate the authentication time (C_A) as follows:

- Assume that security policy P_α is configured in the network. Let the total time involved in transmitting, receiving and processing k_{th} packet by P_α during its authentication phase be denoted as $T_k(P_\alpha)$.
- Assume that n packets are exchanged during authentication phase. Then, authentication time can be represented as,

$$C_A(P_\alpha) = \sum_{k=1}^{n} T_k(P_\alpha). \tag{3}$$

Cryptographic Cost

Cryptographic cost represents the performance overhead associated with a security policy. Since we compute the cost of authentication phase of a security policy separately in terms of authentication time, cryptographic cost involves overhead due to other security features, such as encryption/ decryption, data integrity and so on. Below we describe the procedure for calculating the cryptographic cost of a security policy.

Let P_α denote the case that there is no security policy configured in the network and P_α denote security policy when there is some security service configured in the network where $\alpha = \{2, 3, \ldots, 12\}$. Let $t^s(k, P_\alpha)$ denote the time required to process kth packet by a sender s during the configuration of security policy P_α in the testbed. The time duration, $t^s(k, P_\alpha)$, usually involves adding extra header by security policy, encryption of packet and so on. Let $t^r(k, P_\alpha)$ denote the time required to process kth packet by a receiver r during the configuration of security policy P_α in the testbed. The time duration, $t^r(k, P_\alpha)$, usually involves removing extra header of security policy, decryption of packet and so on. Let $t^{sr}(k, P_\alpha)$ denote the time taken by the kth packet in traversing the network between the sender and the receiver during security policy P_α. Therefore, the total time involved in processing the kth packet, denoted by $T(k, P_\alpha)$, between the sender and the receiver during policy P_α is the sum of three time periods defined above, and is given by,

$$T(k, P_\alpha) = t^s(k, P_\alpha) + t^r(k, P_\alpha) + t^{sr}(k, P_\alpha). \tag{4}$$

Assume that n packets are transmitted between the sender and the receiver, then the total time required for processing n packets during security policy P_α is the sum of time involved in processing all n packets, that is,

$$\sum_{k=1}^{n} T(k, P_\alpha) = \sum_{k=1}^{n} [t^s(k, P_\alpha) + t^r(k, P_\alpha) + t^{sr}(k, P_\alpha)]. \tag{5}$$

Consider that the number of bits in each packets may be different, for example, the size of kth packet is b_k bits. Then the total number of bits in n packets, denoted as B_n, is,

$$B_n = \sum_{k=1}^{n} b_k. \tag{6}$$

Now we compute bit rate associated with various security policies to measure the associated cryptographic cost with each policy. Let $R_B(P_\alpha)$ denote the bit rate (bits/sec) that can be experienced during security policy P_α. Using (5) and (6), bit rate for security policy P_α can be obtained as:

$$R_B(P_\alpha) = \frac{B_n}{\sum_{k=1}^{n}(t^s(k,P_\alpha) + t^r(k,P_\alpha) + t^{sr}(k,P_\alpha)}. \quad (7)$$

Assume that $C_C(P_\alpha)$ denotes the cryptographic cost associated with security policy P_α. In this work, we evaluate the cryptographic cost as the difference between the bits rates for security policies (P_α) and (P_1). Then, $C_C(P_\alpha)$ can be calculated as follows:

$$\begin{aligned} C_C(P_\alpha) = & \frac{B_n}{\sum_{k=1}^{n}(t^s(k,P_1) + t^r(k,P_1) + t^{sr}(k,P_1))} \\ & - \frac{B_n}{\sum_{k=1}^{n}(t^s(k,P_\alpha) + t^r(k,P_\alpha) + t^{sr}(k,P_\alpha))}. \end{aligned} \quad (8)$$

Throughput (bits/second) (η)

It is defined as the data transferred per unit time between participating nodes during the configuration of a security policy in the network. According to this definition, we observe that if the data is represented in bits, then the throughput associated with a security policy is same as the bit rate associated with a security policy that we computed previously during the calculation of cryptographic cost. Therefore, throughput (η) during security policy P_α can be represented as follows:

$$\eta(P_\alpha) = \frac{B_n}{\sum_{k=1}^{n}(t^s(k,P_\alpha) + t^r(k,P_\alpha) + t^{sr}(k,P_\alpha)}. \quad (9)$$

5. DATA ACQUISITION

For each security service configured in our testbed, experimental data are collected in two phases. The first phase collects measurements during the initial negotiation of protocols. The second phase focuses on generating streams, and then collecting data such as throughput and response time for different security policies. In addition, the transmission rate for each wireless card has been set to 11Mbps.

In the *First phase*, we concentrate on taking data that is related to initial negotiations, which take place during the handshake stage of any protocol. We use $Ethereal$ network packet analyzer to capture the packets exchanged in handshake phase. Using timestamp option provided in every packet, we record the time difference between the first and last packet of the negotiation phase. Since in our analysis, we interpret initial negotiation phase as the authentication phase, data obtained in this manner is used to compute and compare authentication time for different security services.

The *Second phase* in our study includes generating different traffic streams in the network between two participating nodes. We use $ttcp$ and *Iperf* traffic generators, because they can generate TCP and UDP traffic. Moreover, these utilities provide different types of statistics such as end-to-end delay, throughput, packet loss, and so on. Also, we can verify whether measurements provided by one tool are in consistent with experimental data provided by other tools.

Initially, we generate TCP and UDP streams with different data sizes. But after analyzing experimental data obtained, we observed that, for smaller size data, differences in measurements of security services are not visible, so they are not helpful in the analysis. Then, we focus our measurement on larger data size such as 16MB from which we can notice significant differences in the measurements. The data obtained in this phase is used to investigate and compare network throughput and protocol overhead for different security services configured in the testbed. Moreover, we repeat experiments more than 15 times to obtain accurate measurements. The average values of these measurements are further used in our analysis and comparison.

6. EXPERIMENTAL RESULTS

In this section, we discuss experimental results obtained for afore-mentioned security policies in various mobility scenarios. We provide experimental data for authentication time, cryptographic cost and throughput defined in Section 4. Experimental results presented in this section are particular to the open source software used in the network. As other existing implementations of security protocols may demonstrate varied performances depending upon the software design, coding methods and language used; actual quantitative results may vary slightly as compared to the results presented in this paper. Moreover, to draw useful conclusions, we categorize the security policies in three groups having low, $middle$ and $high$ security strength as described below.

- **Low Security Group:** P_2, P_5, P_6 and P_7 belong to this group and are the security policies with SI values below 45%.
- **Middle Security Group:** Security policies with SI values between 45% and 70%, such as P_3, P_4, P_8, P_9 and P_{11}, are in the middle security group.
- **High Security Group:** Security policies, P_{10} and P_{12}, have SI values approaching to 100% and belong to the high security group.

6.1. Authentication Time

Authentication time is associated with the initial phase of a security policy as defined in Section 4. During this period, a mobile node provides its credentials to the authentication server, such as home agent or foreign agent in the testbed, to access a network. Messages exchanged during the initial phase of a security policy vary with the security mechanisms involved in the policy. Moreover, authentication time for various policies is obtained for non-roaming and roaming mobility scenarios, respectively. Table 5 shows authentication time for individual protocols, whereas Table 4 shows authentication time (C_A in sec) for IPsec and 802.1x policies. Since WEP does not involve exchange of control messages, there is no authentication time involved with it. Since Mobile IP is used for enabling mobility in the testbed, authentication time (C_A) for IPsec and 802.1x involves Mobile IP authentication time as well. In addition, Figures

3 and 4 demonstrate the authentication versus SI. Note that SI values are demonstrated in an increasing order in the figures.

Table 4. Authentication Time.

Policy	Non-Roaming	Roaming
IPsec (sec)	1.405	2.837
802.1x-EAP (MD5) without IPsec (sec)	0.427	2.176
802.1x-EAP (MD5) with IPsec (sec)	1.722	3.471
802.1x-EAP (TLS) without IPsec (sec)	1.822	4.966
802.1x-EAP (TLS) with IPsec (sec)	3.117	6.281

Table 5. Individual Authentication Time.

Protocol	Time (Sec)
Mobile IP(HA)	0.11
Mobile IP(FA)	1.432
IPsec	1.295
802.1x-EAP-MD5	0.317
802.1x-EAP-TLS	1.712

We observe from Figures 3 and 4 that 802.1x-EAP-TLS policies cause the longest authentication time among all policies. This is due to the fact that 802.1x-EAP-TLS uses digital certificate for mutual authentication, which involves exchange of several control packets. We find that a total of 17 control packets are exchanged during the initial phase of 802.1x-EAP-TLS, which is much higher than 8 and 9 control packets exchanged in 802.1x-EAP-MD5 and IPsec authentication phases, respectively. Moreover, IPsec policies generate longer authentication time than 802.1x-EAP-MD5 (without IPsec) policies because of IPsec tunnel establishment. In addition, we can see that the security policies create longer authentication time in roaming scenarios than non-roaming scenarios due to the reauthentication in a foreign network. Besides these

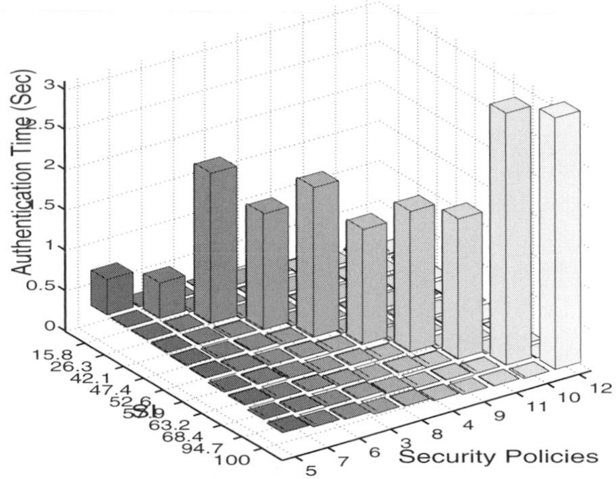

Figure 3. Non-Roaming Scenarios: Authentication Time vs. SI.

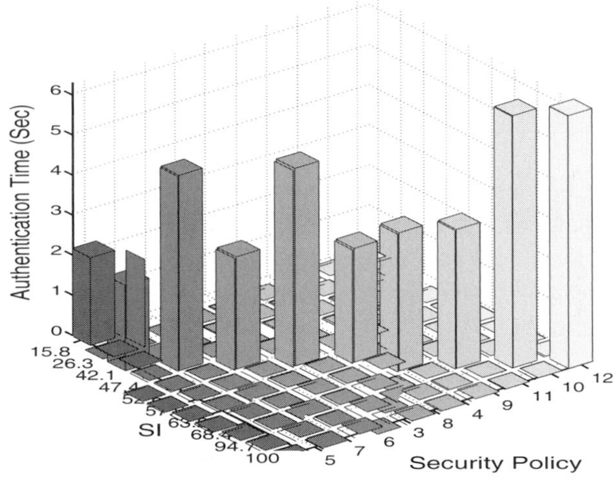

Figure 4. Roaming Scenarios: Authentication Time vs. SI.

general observations, we notice that authentication time does not increase proportionally with respect to the SI of security policies. For example, we recognize that the P_3 (IPsec) induces lower authentication time than the P_6 (802.1x-EAP-TLS) in all scenarios although it has higher SI value than the P_6. Although P_{10} and P_{12} cause longer authentication time than other policies but these policies consist of highest SI values due to more than one levels of security mechanisms involved.

We observe that policy P_{12} (with the highest SI value) consists of longest authentication and incurs around 7 and 3 times longer authentication time than P_5 which has the shortest authentication time in non-roaming and roaming scenarios, respectively. This observation suggests that variations in authentication time values are less in roaming scenarios than in non-roaming scenarios. In other words that even policies with lower SI values induce higher authentication time in roaming scenarios. The reason for this phenomenon is that registration to foreign agent takes a very long time (1.432 sec). Further, authentication time of P_6 is approximately 3 and 1.3 times longer than authentication time of P_7 in non-roaming and roaming scenarios, respectively, which is due to the much higher SI value of P_6 than P_7. Similar behavior can be observed between P_{10} and P_{11}. However, we find that although SI value of P_4 is higher than that of P_8, authentication time of P_4 is less than that of P_8 in both scenarios. Therefore, it can be concluded that authentication time for security policies does not increase monotonically with SI values.

Further, P_{12} with highest SI value incurs around two times longer authentication time than the security policies in the middle group, for instance IPsec policies, in both roaming and non-roaming scenarios. Moreover, we discover that security policies, which are in the middle security group, for example P_3, P_8, P_4 and P_9, do not exhibit much variations in authentication time, and IPsec policies, P_3 and P_4, induce the lowest authentication time (1.4 sec in non-roaming and 2.8 sec in roaming) among them. In addition, authentication time of a security policy in roaming scenarios is about twice of its authentication time in non-roaming scenarios. An exception is 802.1x-EAP-MD5 (without IPsec) policy which exhibits 5 times longer authentication time due to large difference between its authentication time (0.432 sec) and registration time to a foreign agent by mobile node (1.432 sec).

Based on these observations, we conclude that policies in the middle security group provide the best tradeoff between security and performance overhead, and IPsec policies, P_3 and P_4, are the best among them. On the other side, P_{12} (802.1x-EAP-TLS with IPsec) is the best suitable for the network carrying very sensitive data.

6.2. Cryptographic Cost

Now, we discuss cryptographic cost associated with security policies in roaming and non-roaming scenarios. By analyzing cryptographic cost, we capture encryption and decryption time associated with security policies during the data transmission. In addition, we have normalized experimental data for comparing results in various scenarios. Tables 6 and 7 list cryptographic costs in non-roaming and roaming scenarios for low, middle, and high security, respectively. Values presented in italics in Tables 6 and 7, represent the best-recommended security policies in each security group in a particular scenario. On the other hand, values presented in bold face indicate the overall recommended security policy for a particular network scenario.

We notice from Tables 6 and 7 that cryptographic costs associated with policies P_4, P_9, P_{11}, P_{10} and P_{12} in non-roaming scenarios are very close to each other, showing little variations. This is due to the fact that these policies use the same IPsec and WEP

Table 6. TCP Cryptographic Cost (Kbits/sec): Low Security.

Scenarios	No Security	Low	Security		
	P_1	P_2	P_5	P_7	P_6
N1	0	71.10	11.88	*75.88*	11.47
N2	0	70.90	2.09	*101.88*	3.92
N3	0	108.78	7.29	*105.11*	4.33
R1	0	90.43	1.97	*104.81*	6.15
R2	0	208.04	25.60	*232.66*	1.79

Table 7. TCP Cryptographic Cost (Kbits/sec): Middle and High Security.

Scenarios	Middle	Security			High	Security	
	P_3	P_8	P_4	P_9	P_{11}	P_{10}	P_{12}
N1	264.90	**77.21**	302.80	286.89	313.94	291.48	*301.77*
N2	273.45	**57.15**	311.70	347.33	298.78	299.25	*296.13*
N3	304.59	**118.71**	331.68	343.84	378.10	382.87	*343.52*
R1	209.54	*92.19*	216.27	246.49	251.56	259.97	**260.81**
R2	318.32	*230.78*	367.53	393.13	391.12	381.70	**395.29**

mechanisms which are the dominating factors contributing towards their cryptographic costs. Generally, the policies P_4, P_9, P_{11}, P_{10} and P_{12} exhibit 16% higher cryptographic costs than P_3, and around 3.5 times higher than that of P_2, P_7 and P_8. The reason is that policies P_4, P_9, P_{11}, P_{10} and P_{12} have more than one levels of encryption and decryption mechanisms associated with them. Further, we observe that P_5 and P_6 exhibit negligible cryptographic costs, which is due to the fact that these policies do not consist of any encryption/decryption mechanisms associated with them. Although, theoretically, cryptographic costs of policies P_5 and P_6 should be zero, but the small values obtained are due to some external factors in real-time environments. A closer look at the table reveals that cryptographic cost increases corresponding to SI values.

However, we see that P_8 is the policy with a higher SI value but with lower cryptographic cost. Specifically, P_8 exhibits almost half of cryptographic cost of policies P_4, P_9, P_{11}, P_{10} and P_{12}, and almost similar to policies P_2 and P_7.

We also notice the similar behavior for UDP traffic in various non-roaming scenarios in our experiments. However, cryptographic costs of security policies for UDP traffic are less than that of TCP traffic. It is due to the fact that TCP requires acknowledgment for each packet, leading to the transmission of more number of packets through the networks than UDP. So TCP results in higher encryption and decryption processing overhead, leading to increased cryptographic cost. Therefore, the observations suggest that P_8 (802.1x-EAP-TLS with WEP) provides the best tradeoff for both TCP and UDP types of traffic streams in non-roaming scenarios. However, we recognize that in non-roaming scenarios during UDP traffic, difference between cryptographic costs of policies P_9, P_{10}, P_{11}, P_{12} (with high SI values), and P_8 is relatively less. Therefore, P_{12} is a good choice with little extra overhead in these scenarios due to its very strong security features.

Comparing the cryptographic costs in roaming scenarios, we find that cryptographic cost of P_{12} (with the highest SI value) is about two times higher than those of policies P_2, P_7 and P_8, and 25% higher than that of P_3 for TCP traffic in R1 scenario. Whereas, P_{12} exhibits almost twice the cryptographic cost of P_2, P_7 and P_8, and 24% higher than P_3 in R2 scenario during TCP. On the other side, P_{12} demonstrates about 4 times higher overhead than P_2, P_7 and P_8, and almost twice of P_3 for UDP traffic in R1 scenario. In addition, P_{12} shows almost twice the overhead of P_2, P_7 and P_8, and 40% higher than P_3 during UDP traffic in R2 scenario. Moreover, we observe that P_9, P_{10} and P_{11} show cryptographic cost very close to P_{12} with little variations. Cryptographic costs for P_5 and P_6 are negligible in almost all scenarios due to the same reason cited previously. Therefore, we notice that P_8 provides the best tradeoff in all roaming scenarios due to the low overhead associated with it. However, we observe that variations between cryptographic costs of P_{12} and P_8 are small. Therefore, it suggests that P_{12} is also an alternative in roaming scenarios.

6.3. Throughput

To understand the impact of security policies on the network performance, Tables 8 and 9 enumerate throughput exhibited by security policies in different network scenarios for TCP traffic. We follow the similar methodology for representing values in these tables as in the tables for cryptographic cost. The only difference is that Tables 8 and 9 illustrate only the overall recommended policy in each scenario. This is due to the fact that variations in throughput across security policies are not very significant, therefore, we prefer to represent the overall recommended security policy in each scenario rather than illustrating the most suitable security policy in each security group.

We observe that the highest variations in throughput for various security policies during TCP traffic are up to 12%, 13.5%, 14.6% in N1, N2 and N3 scenarios, respectively. Whereas, variations in throughput for UDP traffic are up to 6%, 5.6%, 8% in N1, N2 and N3 scenarios, respectively. On the other side, roaming scenarios, R1 and

Table 8. TCP Throughput (Kbits/sec): Low Security.

Scenarios	No Security	Low	Security		
	P_1	P_2	P_5	P_7	P_6
N1 (1.0e+03)	2.90	2.83	2.89	2.83	2.89
N2 (1.0e+03)	5.64	5.51	5.64	5.45	5.64
N3 (1.0e+03)	2.97	2.86	2.96	2.86	2.97
R1 (1.0e+03)	2.83	2.74	2.83	2.73	2.83
R2 (1.0e+03)	2.86	2.65	2.83	2.62	2.86

Table 9. TCP Cryptographic Cost (Kbits/sec): Middle and High Security.

Scenarios	Middle	Security			High	Security	
	P_3	P_8	P_4	P_9	P_{11}	P_{10}	P_{12}
N1 (1.0e+03)	2.64	2.83	2.60	2.62	2.59	2.61	**2.60**
N2 (1.0e+03)	5.11	5.53	5.04	4.97	5.06	5.06	**5.07**
N3 (1.0e+03)	2.67	2.85	2.64	2.63	2.59	2.59	**2.63**
R1 (1.0e+03)	2.62	2.74	2.62	2.59	2.58	2.57	**2.57**
R2 (1.0e+03)	2.54	2.63	2.49	2.46	2.47	2.48	**2.46**

R2, exhibit variations up to 10% and 16% for TCP traffic, respectively. In addition, from our experiments we find that R1 and R2 demonstrate variations up to 12% and 7% for UDP traffic, respectively. We notice that variation in throughput values are around 10% in most of the scenario and in some scenarios even lower than 10% Therefore, it suggests that the effect of security policies over throughput during data transmission is not very significant. This is based on the fact that we have not taken into account the cost of authentication time for calculating throughput, because throughput for a data stream is calculated using the total time involved in transmission of the entire data after authentication phase is over. Therefore, variations in throughput values presented in this paper are only because of cryptographic costs. Another reason to segregate authentication phase from throughput phase is to measure the authentication cost independently, which would be helpful in comparing authentication cost and cryptographic cost to uncover some useful facts below.

In addition, we believe that, in the future, as hardware becomes faster, cryptographic cost (i.e., time involved in encryption/decryption process) will be reduced further. Moreover, based on our previous observations from Figure 4, authentication time in roaming scenarios is very high, and it may affect mobile applications significantly as user's mobility increases. As we observe that variations in throughput across security policies are almost similar, we speculate that QoS degradation in a network may be more significant due to the authentication cost than the cryptographic cost in the future.

7. SECURITY ANALYSIS

In this section, we discuss security issues associated with security policies against malicious attacks. In addition, we demonstrate the advantages of the cross-layer integration of security protocols in providing enhanced security.

7.1. Authentication Forging

In the low security group, policy P_2 (WEP) offers open or shared key authentication mechanisms. However, P_2 (WEP) is known to be highly vulnerable, and therefore, prone to authentication forging by malicious users. It is because of the small key space and the reuse of the same initialization vector (IV) [3] involved in WEP. Therefore, WEP is not advisable to be used in stand-alone mode in WLANs. Similarly, P_5 (802.1x-EAP-MD5) and P_7 (802.1x-EAP-MD5-WEP-128) in the low security group are vulnerable to authentication forging due to the inclusion of MD5 in which passwords are transmitted in clear-text form [26]. Only policy P_6 (802.1x-EAP-TLS) in the low security group is immune to authentication forging due to digital certificate used in its authentication process. Policies in the middle and high security groups either include IPsec or EAP-TLS mechanisms which use public key cryptography and digital certificate, and therefore, are not vulnerable to authentication forging. Here, we observe that integrated policies in the middle and high security groups are able to overcome the vulnerability exhibited by the policies in the low security group.

7.2. Man In The Middle Attack (MITM)

Policies in the low security group are vulnerable to the man in the middle attack (MITM). For example, policies P_5 (802.1x-EAP-MD5) and P_7 (802.1x-EAP-MD5-WEP-128) transmit response in plain-text form, are vulnerable to MITM attack [26]. Since it is easy to decipher WEP security keys [3], it is possible to perform MITM attack with the policy P_2 as well. On the other hand, security policies with IPsec and TLS protocols are not vulnerable to MITM attack. Although, IPsec employs public key cryptography, which is vulnerable to MITM attack, ISAKEMP key agreement protocol used in IPsec prevents the attack. Whereas, TLS protocol use digital certificate and is not vulnerable to MITM attacks.

7.3. User Access Control (or Unauthorized Participation)

Besides previous attacks, such as authentication forging and MITM, we observe that user access control is not strong enough in the policies in the low security group. Moreover, policies P_3 and P_4 in the middle security group are not appropriate for the access control at the user level as well. It is due to the fact that IPsec mechanism used in policies P_3 and P_4 works at network layer 3, and provides system authentication but not user level authentication [21]. For example, if a system is authenticated using IPsec, any user, authorized or unauthorized, with access to the system can use network resources. Therefore, for user authentication, P_3 and P_4 must be employed with higher layer security protocols. For instance, policies P_9 and P_{11} provide strong security at layer 3 due to IPsec, and enable user access control by using 802.1x-EAP-MD5. However, user access control with policies P_9 and P_{11} is not secure because MD5 transmits passwords in clear-text form [26]. Policies P_8, P_{10} and P_{12} use EAP-TLS mechanism, and provide strong access control due to the use of digital certificates, and dynamic keys which are refreshed periodically during a session as well [27]. Here again, we observe the advantages of integrating security protocols at different layers.

7.4. Fabrication of Messages

Fabrication of message is possible in the network configured with policies in the low security group, such as P_2, P_5 and P_7. It is due to the fact that it is easier to decipher security keys used in these policies. However, policies in the middle and high security groups use IPsec or EAP-TLS with public key cryptography and digital certificate mechanisms, respectively. Therefore, modification of messages in the transit is not possible over these policies.

7.5. Denial of Service (DOS) and Traffic Analysis

All policies are vulnerable to denial of service (DOS) and traffic analysis attacks [29], [30], [31]. The reason is that these attacks are hard to prevent even with very strong cryptographic features.

In addition, WEP is susceptible to dictionary attacks, statistical cryptanalysis, known plaintext, and partial known plaintext attacks. Moreover, due to security issues with 802.1x framework, policies P_5 to P_{12} suffer from packet spoofing attacks [4]. In general, policies P_3 and P_4 provide enhanced security due to the strong cryptographic and key management protocols used in IPsec. Moreover, policies P_{10} and P_{12} are very strong because they use IPsec and EAP-TLS together with WEP [21], [27]. Consequently, P_{10} and P_{12} provide strong authentication, confidentiality, mutual authentication, non-repudiation and data integrity features.

In summary, we observe that when security policies are used in stand-alone mode, they are prone to many attacks. But when we configure protocols at various layers together, the attacks associated with a weak protocol can be prevented more effectively by the strong protocols at other layer. Therefore, the policies in the middle and the high security groups provide enhanced security and prevent vulnerabilities related to a single protocol. Therefore, the cross-layer integration of security protocols seems an advantageous choice in providing better security solutions for many wireless applications.

8. CONCLUSIONS

In this work, we addressed the issue of performance overhead and security strength associated with security protocols in WLANs. Specifically, we studied the cross-layer integration of various security protocols with respect to authentication time, cryptographic cost and throughput in different network scenarios with TCP and UDP data traffic. Moreover, we performed a comprehensive study to obtain the experimental results of performance metrics associated with security policies. We found that IPsec policies, P_3 and P_4, provide the best tradeoff between security and performance regarding authentication time. Moreover, we observed that P_8 (802.1x-EAP-TLS) is the most suitable option for low cryptographic cost and better security strength in many scenarios. However, we also found that P_{12} (802.1x-EAP-TLS with IPsec) is also a suitable option with little extra overhead during some network scenarios.

We noticed that there is always a tradeoff between security and performance associated with a security policy depending upon the network scenario and traffic types. It is seen that security policy with higher strength may not always be the best option for all scenarios with respect to the tradeoff between security strength and performance overhead. We found that cross-layer integration of security protocols at many layers, for example, policies P_9, P_{10}, P_{11}, P_{12} provide very strong security but with higher overhead. Therefore, we suggest that these policies are most suitable for the networks carrying very sensitive data. Moreover, we noticed that the variations in throughput under different scenarios are not very large, and concluded that authentication time may be a more significant factor contributing towards QoS degradation than cryptographic cost as hardware becomes faster in the future.

In summary, our results recommended the appropriate security policy for each scenario. In addition, we provided the quantification of performance overhead. We believe that combination of these results can lay a very strong foundation for designing new security protocols or improving the existing ones. Moreover, the real-time nature of our results can enable network designers make intelligent decision regarding the implementation of security policies in a network. Also, performance measurements for authentication cost, cryptographic cost and throughput are helpful in deciding about which security feature should be enabled in a particular scenario while keeping overhead low. Therefore, our study provide first-hand valuable results, which will be very useful to the design of network protocols for security and flexible quality of service in future mobile wireless networks.

9. REFERENCES

1. T. Karygiannis and L. Owens, "Wireless Network Security 802.11, Bluetooth and Handheld Devices," *National Institute of Technology, Special Publication*, pp. 800–848, November 2002.
2. Y. Zahur and T. A. Yang, "Wireless LAN Security and Laboratory Designs," *Journal of Computing Sciences in Colleges*, vol. 19, pp. 44–60, January 2004.
3. N. Borisov, I. Goldberg, and D. Wagner, "Intercepting Mobile Communications: The Insecurity of 802.11," in *Proceedings of the Seventh Annual International Conference on Mobile Computing And Networking*, July 2001.
4. "IEEE 802.1X," *http://www.ieee802.org/1/pages/802.1X-rev.html*, 2004.
5. "IEEE 802 Standards," *http://standards.ieee.org/getieee802*.
6. A. Hecker and A. H. Laboid, "Pre-authenticated Signaling in Wireless LANs using 802.1X Access Control," in *Proceedings of IEEE Global Telecommunications Conference, GLOBECOM '04*, vol. 4, pp. 2180 – 2184, November-December 2004.
7. A. Hecker and A. H. Laboid, "A New EAP-based Signal Protocol for IEEE 802.11 Wireless LANs," in *Proceedings of IEEE 60th Vehicular Technology Conference, VTC-Fall, 2004* , vol. 5, pp. 3214 – 3218, September 2004.
8. W. Qu and S. Srinivas, "IPSEC-Based Secure Wireless Virtual Private Networks," in *Proceedings of IEEE MILCOM*, pp. 1107–1112, October 2002.
9. D. B. Faria and D. R. Cheriton, "DoS and Authentication in Wireless Public Access Networks," in *Proceedings of ACM Workshop on Wireless Security (WiSe)*, pp. 47–56, September 2002.
10. W. A. Arbaugh, N. Shankar, J. Wang, and K. Zhang, "Your 802.11 Network Has No Clothes," *IEEE Wireless Communications Magazine*, December 2002.
11. S. Kasera, S. Mizikovsky, G. S. Sundaram, and T. Y. Woo, "On Securely Enabling Intermediary-Based Services and Performance Enhancements for Wireless Mobile Users," in *Proceedings of ACM Workshop on Wireless security (WiSe)*, pp. 61–68, September 2003.
12. J. Kong, S. Das, E. Tsai, and M. Gerla, "ESCORT: A Decentralized and Localized Access Control System for Mobile Wireless Access to Secured Domains," in *Proceedings of ACM workshop on Wireless security (WiSe)*, pp. 61–68, September 2003.
13. Y. Matsunaga, A. Merino, T. Suzuki, and R. H. Katz, "Secure Authentication System for Public WLAN Roaming," in *Proceedings of The 1st ACM international workshop on Wireless mobile applications and services on WLAN hotspots*, pp. 113–121, 2003.
14. M. Li, H. Zhu, S. Sathyamurthy, I. Chlamtac, and B. Prabhakaran, "End-to-End Framework for QoS Guarantee in Heterogeneous Wired-cum-Wireless Networks," in *Proceedings of The First International Conference on Quality of Service in Heterogeneous Wired/Wireless Networks, 2004*, pp. 140 – 147, Oct 2004.
15. O. Elkeelany, M. M. Matalgah, K. Sheikh, M. Thaker, G. Chaudhary, D. Medhi, and J. Qaddour, "Perfomance Analysis Of IPSEC Protocol: Encryption and Authentication," in *Proceedings of IEEE Communication Conference (ICC)*, pp. 1164–1168, May 2002.
16. A. Godber and P. Dasgupta, "Secure Wireless Gateway," in *Proceedings of ACM Workshop on Wireless Security (WiSe)*, pp. 41–46, September 2002.
17. G. Hadjichristofi, N. D. IV, and S. Midkiff, "IPSec Overhead in Wireline and Wireless Networks for Web and Email Applications," in *Proceedings of IEEE International Performance, Computing, and Communications Conference, 2003*, pp. 543 – 547, April 2003.
18. K. Wang and S. Tripathi, "Mobile-End Transport Protocol: An Alternative to TCP/IP Over Wireless Links," in *Proceedings of IEEE INFOCOM*, pp. 1046–1053, April 1998.

19. M. D. Corner and B. D. Noble, "Zero-Interaction Authentication," in *Proceedings of IEEE/ACM MOBICOM*, pp. 1–11, September 2002.

20. C. Perkins, "IP Mobility Support," *http://www.ietf.org/rfc/rfc2002.txt*, October 1996.

21. "IPSEC," *http://www.freeswan.org*.

22. "802.1x Supplicant," *http://www.open1x.org*.

23. "RADIUS," *http://www.freeradius.org*.

24. "*OpenSSL*," *http://www.openssl.org*.

25. "Mobile IPv4," *http://dynamics.sourceforge.net*.

26. I.-G. Kim and J.-Y. Choi, "Formal Verification of PAP and EAP-MD5 Protocols in Wireless Networks: FDR Model Checking," in *Proceedings of The 18th International Conference on Advanced Information Networking and Applications*, vol. 2, pp. 264 – 269, March 2004.

27. B. Aboba and D. Simon, "PPP EAP TLS Authentication Protocol," *RFC 2716*, October 1999.

28. A. Satoh and T. Inoue, "ASIC-Hardware-Focused Comparison for Hash Functions MD5, RIPEMD-160, and SHS," in *Proceedings of International Conference on Information Technology: Coding and Computing, ITCC 2005*, vol. 1, pp. 532 – 537, April 2005.

29. C. B. McCubbin, A. A. Selcuk, and D. Sidhu, "Initialization Vector Attacks on the IPsec Protocol Suite," in *Proceedings of EEE 9th International Workshops on Enabling Technologies: Infrastructure for Collaborative Enterprises, (WET ICE 2000)*, pp. 171–175, June 2000.

30. A. Mian and A. Masood, "Arcanum: A Secure and Efficient Key Exchange Protocol for the Internet," in *Proceedings of International Conference on Information Technology: Coding and Computing, ITCC 2004*, vol. 1, pp. 17 – 21, April 2004.

31. J. Rejeb, M. Vohra, and T. Le, "IKE-based Secure Wireless and Mobile Networks," in *Proceedings of The IEEE 6th Circuits and Systems Symposium on Emerging Technologies: Frontiers of Mobile and Wireless Communication, 2004*, vol. 2, pp. 567 – 570, June 2004.

Part IV

SECURITY IN SENSOR NETWORKS

13

SECURITY ISSUES IN WIRELESS SENSOR NETWORKS USED IN CLINICAL INFORMATION SYSTEMS

Jelena Mišić and Vojislav B. Mišić
Department of Computer Science
University of Manitoba
Winnipeg, Manitoba R3T 2N2, Canada
E-mail: {jmisic, vmisic}@cs.umanitoba.ca

High quality healthcare is an important aspect of the modern society. In this chapter we address the security and networking architecture of a healthcare information system comprised of patients' personal sensor networks, department/room networks, hospital network, and medical databases. Areas such as diagnosis, surgery, intensive care and treatment, and patient monitoring would greatly benefit from light untethered devices which can be unobtrusively mounted on patient's body in order to monitor and report health-relevant variables to the interconnection device mounted on the patient's bed. Interconnection device should also have larger range wireless interface which should communicate to the access point in the patient's room, operation room or to the access points within the healthcare institution. The results of measurements will then be stored in central medical database with appropriate provisions for protecting the patient privacy as well as the integrity of personal health records. We review confidentiality and integrity polices for clinical information systems and discuss the feasible enforcement mechanisms over the wireless hop. We also compare candidate technologies IEEE 802.15.1 and IEEE 802.15.4 from the aspect of resilience of MAC and PHY layers to jamming and denial-of-service attacks.

1. INTRODUCTION

Healthcare is an important area for deployment of wireless sensor and personal area networks. The IEEE 1073 Medical Device Communications standards organization is currently in the process of developing the specifications for wireless interface communications. The main objective for this effort is to develop universal and interoperable devices for medical equipment which are transparent to the user and easily re-configurable. The group has recognized that developing new wireless technologies is not an option and is looking instead in deployment of existing wireless technologies belonging to IEEE 802 family in the healthcare applications.

There are many research issues related to sensor and Wireless Personal Area (WPAN) networks in healthcare. First, there are different healthcare applications which monitor vital signs, electrocardiogram signals (ECG), electroencephalogram signals (EEG), as well as signals from other electronic or electro-mechanical devices that may be used in healthcare (dialysis, infusion, . . .). All these applications require some minimum event detection reliability, i.e., the minimum number of data bits per second, as the result of sampling and digitizing analog variables related to patient health or proper functioning of electronic and electro-mechanical devices. Therefore, it is important to pair the medical application with the low rate WPAN technology from the aspect of sufficient bandwidth as well as from the aspect of supported security mechanisms. As the bandwidth requirements of different variables and devices vary, it is probably unrealistic to assume that any given WPAN technology can cope with requirements of different medical applications. Then, among the candidates for one application it is necessary to address several issues:

1. We need to define the security policies to be utilized for management and use of patient medical records within the clinical information system. The policies should aim to protect the confidentiality and integrity of data from its very entry into the system at the patient WPAN.

2. To this end, it is necessary to develop appropriate security and network architecture for sensor networks, WLANs, and WPANs that might be deployed within the clinical information system. That architecture (or architectures) will serve as the foundation upon which individual health applications can monitor the health of individual mobile patients without harming their health or life habits.

3. We need to consider secure location management whenever the patient changes location within the hospital, either because the patient walks around, or his/her bed is moved from one room to another.

4. We need to look at the security issue of denial of service at the physical and MAC layers (jamming) which can cut the flow of patient's data to the monitoring station. This problem is related to the interference issues since every mobile patient or patient's bed presents independent WPAN(s). They might interfere among themselves, and with WLAN running in the room or WPAN carried by the doctor/nurse.

5. We need to provide secure interconnections among different WPANs among themselves and with WLAN. The efficiency of interconnecting devices will determine the scalability of our secure healthcare network design.

6. The packet delay issue, which is related to the Medium Access Control (MAC) protocol used in particular technology. It is also necessary to look at the packet size for given technology since all measured health variables, which are analog

and hence have to undergo analog-to-digital conversion as well as encryption, produce a stream of bits with a constant rate. In such cases, packetization delay becomes an issue.

We begin the chapter by reviewing clinical information security policies. Then, we propose networking and security architecture of clinical information system which includes patient sensor networks, wireless local area networks which belong to the departments, and the central medical database where results of patient examinations are held. Enforcement of policy rules using cryptographic mechanisms over networking infrastructure is discussed, followed by a discussion of the classification of medical applications and pairing with WPAN technologies. We also compare some candidate technologies for wireless sensor networks from the aspects of MAC and physical layer security and sensing reliability. A brief summary concludes the chapter.

2. SECURITY POLICY FOR HEALTHCARE SENSOR NETWORKS AS PART OF CLINICAL INFORMATION SYSTEMS

Sensor networks in medical applications are the edge component of the clinical information system. The wireless data flows with health variables measurements are part of 'personal health information' and must be protected from the aspect of integrity and patient privacy before they can be stored in the patient medical record. Actually, health sensing information forms the most important part of the medical record. The security policies for medical records have been extensively studied; they have to be carefully designed in order to (a) limit the number of actors, clinical physicians and others, that can access the patient record, and (b) control the operations over the record [2, 4]. The policies are typically expressed as a number of security rules, including the following:

1. Each medical record has an associated access control list which names the individuals and groups that may read, update, and append the information to the record. The system must restrict the access to those identified on the access control list.

2. One of the clinicians on the access control list (called the *responsible clinician*) must have right to add other clinicians to the access control list.

3. The responsible clinician must notify the patient of the names on the access control list whenever the patient medical record is accessed.

4. Each time the record is accessed, the name of the clinician, the date and time, and the manner of access have to be recorded.

The purpose of previous four access rules is to control the confidentiality of the medical record. Patient must consent to the treatment and he/she must have the access to his/her record at any time. Moreover, the patient must be informed whenever any clinician accesses the record. In all previous cases, if the patient is incapacitated to make the decisions for him- or herself, the authority rests with the legal guardian or another person with the appropriate power of attorney.

Integrity of the patient's medical record is protected by the following set of rules:

Creation When a new medical record is created, the clinician creating the record should have access to it, as should the patient. If the medical record is created due to the referral from another referring clinician, he/she should also be authorized to access the record.

Deletion Clinical information cannot be deleted from the medical record until the predefined time period has passed.

Confinement Information from one medical record may be appended to a different medical record if and only if the access control list of the second record is a subset of the access control list of the first.

Aggregation Aggregation of patient data must be prevented.

Enforcement Any computer system that handles medical records must have a subsystem that enforces previous rules.

The need for wireless sensor networks which are integrated in the medical information system presents a big challenge to the implementation of aforementioned rules. Unfortunately, previous access principles can't be implemented in the network environment through simple access lists. Instead, we will need to use some cryptographic techniques which we will discuss in the next section.

3. SECURITY ARCHITECTURE OF THE WIRELESS LAYER OF THE MEDICAL INFORMATION SYSTEM

Let us consider the medical information system infrastructure including the wireless sensor networks, as shown in Fig. 1. Important parts of the architecture are the patient security processor (PSP) which is attached to bed and the wireless access point in the patient room. The PSP is module that implements networking as well as security-related functions.

From the networking aspect, PSP is the coordinator of sensing nodes which belong to the patient's Personal Area Network (PAN) and participates in the Medium Access Control function of the nodes. For example, for IEEE 802.15.1 technology (Bluetooth) PSP will be executed on the piconet's master and for IEEE 802.15.4 PSP will be executed on the cluster's coordinator.

From the security aspect, it generates the symmetric encryption key by which all data packets with sensed health information are encrypted. It distributes the symmetric key to the sensing nodes by encrypting it with public key which is common for all sensing nodes. Sensing nodes are pre-configured with the private key by which they

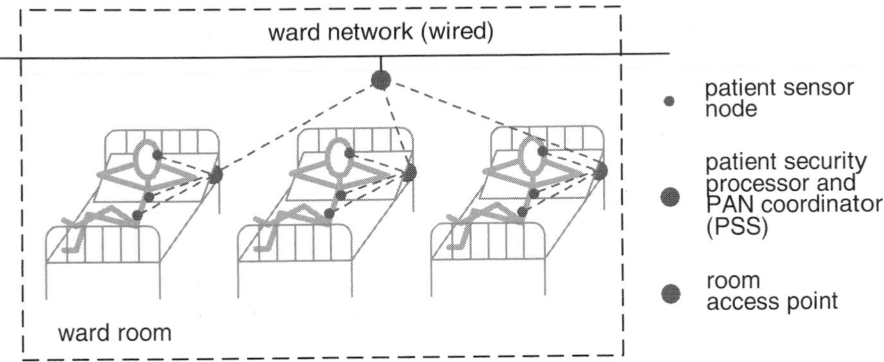

Figure 1. Security architecture of wireless part of medical information systems.

can decrypt the symmetric key. Sensing nodes will send packets with encrypted payload and completely authenticated to the patient security processor which forwards them, possibly aggregated, to the room access point.

The patient room access point is further connected to the central medical record database through a suitable wired network. The access point forwards encrypted and authenticated packets to the central database. Data packets which carry measurements of personal health variables must be authenticated and encrypted in the way which we discuss below. From the networking point of view access point is interconnection device which interconnects Personal Area Network technology (IEEE 802.15.1 or IEEE 802.15.4) with the hospital network which might be implemented using wireless LAN and mesh technologies.

Medical personnel might carry their own PAN nodes and communicate directly to medical health database through the patient's room access point.

Security of medical applications over sensor networks has to be protected at every networking layer. At the physical and MAC layer there exists possibility of denial of service attack by generating to much interference or by generating unnecessary traffic. Therefore, MACs should be evaluated from this perspective also. Payload of packets with sensed data should be encrypted when needed. Also, in some situations, patient's location should be hidden as well. Given the hierarchical application architecture, there should exist layered security architecture with different keys and possibly different encryption algorithms at WPAN, room and hospital level. The encryption standards used at particular level should match importance and vulnerability of the data. For example the traffic at the WPAN level has to be encrypted at the MAC level but the traffic between access points should be protected by IPSec.

However, security measures will affect the delay and throughput of sensed health data and this impact has to be carefully evaluated. Initial work on performance evaluation of IPSec is presented in [9] but much more needs to be done for multi-tier communication architecture built over WPANs. We plan to develop multi-layer security architecture which will match confidentiality and integrity of the sensed data and evaluate the performance of overall application architecture.

4. ENFORCEMENT OF PRIVACY AND INTEGRITY RULES

In order to protect from the attacks from the outside world, all hospital equipment and personnel must possess the secret 'hospital/department/room' key K_H. This key is used to sign and authenticate network packets generated by the equipment and personnel belonging to specific medical department. Authentication is achieved by calculating the hash function over the packet with measurement data, hospital/department/room key and timestamp T_s with time of packet generation. For hash function, we adopt Secure Hash Algorithm (SHA) [27]. Let us denote i-th packet containing measurements of some health variable as P_i, its Medium Access Control header as H_i and its payload as D_i. Packet authentication code for packet $i(\text{PAC}_i)$ can then be calculated as

$$PAC_i = H(K_H, T_{s,i}, H_i||D_i).$$

4.1. Patient privacy

Aforementioned access policy rules require that only patient and clinicians have access to patient's medical record and that patient must be informed of any access to his/her record. Therefore, this small group must have dedicated secret session key K_p (p comes from patient who is the principal of the group), but no one from this group must have the capability to derive the key without the participation of other members. Particularly, participation of the patient is necessary in all accesses. This key will be used as encryption key of an symmetric encryption system such as 3-DES (Data Encryption Standard) or AES (Advanced Encryption Standard). Operations of encryption and decryption with patient's key will be denoted as $E_{K_p}()$ and $D_{K_p}()$ respectively. Encryption using public key cryptography takes long time and generates high packet payload which is a problem for existing candidate technologies for wireless sensor networking.

Process of generating patient's key requires attention. If the patient is unable to participate in the decisions regarding his/her healthcare, then his part of the key generation must be done either by proxy person or by central hospital authority. Clinicians who are supposed to participate in the key generation are responsible (principal) clinician and referring clinician. Therefore we assume that minimum three entities must participate in the generation of patient's key.

One approach which we adopted for patient's key generation is the concept of secret sharing with threshold. Secret is divided into n parts called shadows and in order to recover it, m shadows are needed. This idea was first independently proposed in [28] and [5]. It was further elaborated in [3, 18] and nice overview of the work in this area is given in [29].

Priority among the users can be modeled by giving important user more shadows. For example, for emergency cases central hospital authority together with responsible (principal) clinician should be able to reconstruct the patient's key. The basic mathematical idea behind the key generation among m entities is to create the system of m equations with m variables by using the polynomial with random coefficients. For example for $m = 3$ we start from the polynomial:

$$F(x) = (ax^2 + bx + K_p) \bmod p$$

where p is public random prime number, $a, b < p$ are secret random numbers and K_p is patient's symmetric key. Assume that each participant j in key generation has some numerical representation of his/her identity ID_j. Then the shadows become

$$\begin{aligned}
F(ID_{pt}) &= (aID_{pt}^2 + bID_{pt} + K_p) \bmod p && \text{patient's shadow} \\
F(ID_{pc}) &= (aID_{pc}^2 + bID_{pc} + K_p) \bmod p && \text{principal clinician's shadow} \\
F(ID_{rc}) &= (aID_{rc}^2 + bID_{rc} + K_p) \bmod p && \text{referring clinician's shadow} \\
F(ID_{ca}) &= (aID_{ca}^2 + bID_{ca} + K_p) \bmod p && \text{central authority's shadow}
\end{aligned}$$

To generate the patient's key and start the measurement of health variables, three shadows are needed and must be presented to PSP. For the start of measurement, patient's shadow, principal clinician's shadow and central authority's shadow are sufficient. Three shadows are also needed in order to decrypt the medical record from the medical database which is also encrypted with K_p. In this case, central authority should be excluded and key should be recovered from patient's shadow, principal clinician's shadow and referring clinician's shadow. In this case, patient will be always notified when his/her record is accessed and he/she will be sure that record is not changed. Shadows should be changed frequently.

4.2. Timestamping the sensed records as results of patient's examination

The fourth access rule calls for recording of all accesses for the purpose of auditing. Auditing requires that accesses are recorded together with the date, time and name of each person who accessed the record. This problem can be solved by linking current record of access (timestamp, list of persons involved) with previous records as proposed in [19, 20, 21, 29, 4]. It is also facilitated by the fact the central medical database can be associated with the trusted timestamping server. Server builds a tree of hashes of timestamping requests received for given time period (second, minute). Server further sends to the medical database signed hashes from the leaf generated by the opening of patient's record till the root of the tree. Assume that information about patient's i record is n-th leaf in the tree counting from the root and it has format:

$$R_{ID_{pt},n} = T_n, L_n, K_p, ID_{pc}, ID_{rc}.$$

where L_n denotes the record lifetime. Let us denote $H_n = H(R_{ID_{pt},n})$. Let us also assume that timestamping server has public/private key pair K_t and that encryption with public and private key is denoted as V_{K_t} and S_{K_t} respectively. Then timestamping server will associate information about access to the patient's record with:

$$S_{K_t}(H(H_0, H_1, H_2 ... H_n)).$$

where H_0 represents the hash of the information at the root of the tree and H_i are hashes of the access information along the path to the root of the tree.

Timestamping is also related to the deletion principle which states how long patient's medical record must be kept before deletion. The lifetime of the patient's examination record L_n which is entered into medical database must be also protected using the timestamping service. Patient's record lifetime can be determined staring from the moment when record is generated. If the particular record is missing but its hash exists in the timestamping tree, the integrity of the patient's record is corrupted.

4.3. Enforcement of the confinement principle

Patient must be informed when clinician non-familiar to his/her medical record accesses the record. On the other hand responsible clinician must be able to add other clinicians to the access list. In that case the number of secret shadows has to change (increase) and central clinical authority has to increase the number random parameters in the equation which determines secret shadows. For example, if second clinician has to be added to the access list, the system of secret shadow equations becomes:

$$\begin{aligned} F(ID_{pt}) &= (aID_{pt}^3 + bID_{pt}^2 + cID_{pt} + K_p) \bmod p \text{ patient's shadow} \\ F(ID_{pc}) &= (aID_{pc}^3 + bID_{pc}^2 + cID_{pc} + K_p) \bmod p \text{ principal clinician's shadow} \\ F(ID_{sc}) &= (aID_{sc}^3 + bID_{sc}^2 + cID_{sc} + K_p) \bmod p \text{ second clinician's shadow} \\ F(ID_{rc}) &= (aID_{rc}^3 + bID_{rc}^2 + bID_{rc} + K_p) \bmod p \text{ referring clinician's shadow} \\ F(ID_{ca}) &= (aID_{ca}^3 + bID_{ca}^2 + bID_{ca} + K_p) \bmod p \text{ central authority's shadow} \end{aligned}$$

In this case four out of five shadows are needed to generate or access the patient's examination record so this presents (4,5)-threshold scheme.

4.4. Enforcement of the aggregation principle

Aggregation of patients' records must be prevented in the case the principal/second clinician becomes corrupted. This is mostly prevented by sharing the secret encryption key through the shadows. Another helpful thing would be to encrypt the database records [11, 10]. The index filed can be the hash of last name of the patient concatenated with his/her ID number. Data fields must be encrypted by the secret key assembled from m secret shadows. In this way, the list of the patients is hidden as well as their medical records.

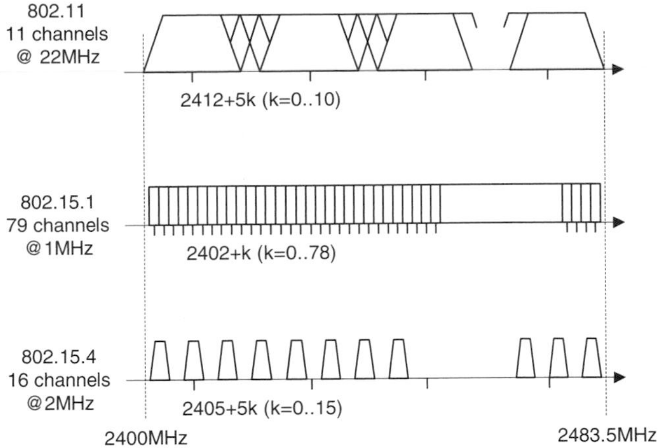

Figure 2. Spectrum usage for various WPAN technologies running in ISM

5. IMPACT OF THE WIRELESS PAN TECHNOLOGIES

We plan to evaluate current WPAN standards namely, IEEE 802.15.1, and 802.15.4 and their interworking among themselves and with IEEE 802.11b WLANs as major candidates for implementations of healthcare sensor networks. We agree with [7] that the success of wireless sensor networks as a technology rests on the success of the standardization efforts to unify the market and avoiding the proliferation of proprietary, incompatible protocols that, although, perhaps optimal in their individual market niches, will limit the size of overall wireless sensor market.

5.1. Classification of healthcare applications and pairing with WPAN technologies

We will analyze a number of healthcare applications from the aspects of bandwidth and delay. For example electrical signals from the heart are sampled at the rate of 500 samples per second and each sample is digitized to 8 bits giving data flow of 4000bps. Furthermore, samples must be taken from several points on the body. Each flow can not be delayed more than few hundreds of milliseconds and flows must be synchronized. We will look at the following issues which are of foremost importance for sensor networks and which follow from the requirement for controlled event detection reliability at the network sink and use them as criteria to match the technology with the application.

1. How much is physical layer immune to the interference errors? We note that all candidate technologies run in Industrial Scientific and Medical (ISM) band between 2400 and 2483.5MHz. They use different modulations at the physical layer, for example, the networks compliant with the 802.15.4 and 802.11 use Direct Sequence Spread Spectrum (DSSS), while those compliant with the

802.15.1 (Bluetooth) use Frequency Hopping Spread Spectrum. Therefore dynamic channel allocation algorithms and interference mitigation techniques will be needed to avoid excessive interference at the physical layer. Some work on interference mitigation between 802.151 and 802.11b is reported in [14] but much more is needed for the interworking with 802.15.4. The channel layout for all the technologies under consideration is given in Fig. 2.

2. The MACs for candidate technologies can be classified as TDMA with polling and CSMA-CA. Node access delay has to be evaluated for both MAC classes under varying number of nodes and packet rate from node. Is acknowledged transfer necessary for achieving desired event reliability and which packet spacing it induces? How much of the buffering is reasonable to have at the source nodes? For specific MAC, maximum effective bandwidth left to the application has to be evaluated and paired with the delay.

5.2. Design and evaluation of interconnection devices

There will be a need to interconnect different WPAN and to interconnect WPAN/WLAN networks in order to regulate the scale of power, distance, and bandwidth-related issues. The example of location of interconnecting devices (bridges) is given in Fig. 1.

These devices have to be designed in the scope of MAC, channel and buffering issues and their performance has to be evaluated. The operation of interconnection device is very important for the overall network design since it affects the end-to-end delay and the scalability of the overall design. Some work in this area exists for interconnection of Bluetooth piconets. The work in [32, 31] requires computation of non-overlapping rendezvous points for bridges while the work in [25] allows bridge to visit the piconet at will trading delay for scalability. However, little is known about interconnections between IEEE 802.15.4 with other WPANs and WLANs.

5.3. Reliable event detection

The rate at which data is propagated from source nodes at patient's body to monitoring devices at the patient's bed and monitoring room (sink) must be high enough to obtain the desired event detection reliability R, which is commonly defined as the number of data packets required per second for reliable event detection at the sink [1]. At the same time, sensor nodes operate on battery power which means that energy efficiency must be maintained.

Reliable event detection using minimal energy resources requires simultaneous achievement of several sub-goals. First, packet loss along the path from source to the sink has to be minimized; at the Physical (PHY) layer, packets can be lost due to noise and interference, while at the Medium Access Control layer (MAC) layer, losses may be incurred by collisions. Second, packet waiting has to be minimized, including queueing delays experienced in various devices along the data path towards the sink, but also delays due to potential congestion in the network. Congestion control has to be addressed as a cross-layer problem and solved at the MAC level since excessive active nodes have to be turned off [23]. Finally, packet propagation should take place along the shortest paths, while avoiding congested nodes and paths; this is the responsibility of the network layer.

Given that the protocol stack on sensor nodes—which operate on battery power and have limited computational capabilities—has to be as simple as possible, we conclude that simultaneous minimization of packet losses and improvement in efficiency (with the goal of maximizing the lifetime of the network) necessitate that some of the aforementioned functions of different layers are performed together. In other words, cross-layer optimization of network protocol operation is needed; the feasibility of this optimization is determined by the communication technology used to implement the network. This problem has classically been treated only as a graph-theoretical problem where only connectivity has been addressed [30, 15] or addresses the collision based MAC [8] although it does not allow the active node to get into the sleep state and achieve load balancing among the nodes. Second group of proposals [17, 12] tries to regulate the event sensing reliability but without looking at MAC and PHY properties at all. The congestion problem is particularly important in networks which use a collision-based MAC protocol such as CSMA-CA, e.g., in 802.15.4 [16]. The decrease in throughput due to congestion may lead the coordinator to the erroneous conclusion that the number of active nodes is too low. Therefore it is important to look at the cross-layer implementation of power/congestion control in wireless sensor networks.

5.4. Handling patient's mobility

The patient wearing wireless sensors will either walk within the hospital or he will lie in his bed while the bed is moved to another room. Therefore, sensed data will have to be sent to new access points and experience new level of interference and congestion. The handover procedure can be handled at the MAC layer or at the networks layer. The handover between 802.11 MACs is analyzed in [22]. However there is an open issue about how the handover between 802.15.4 and 802.11 or between 802.15.1 and 802.11 has to be executed, and how much of data flow interruption will occur. We plan to design and analyze secure MAC layer handover procedures between involved WPAN and WLAN technologies.

Besides MAC layer handover, it can happen that network layer handover is also needed if the IP subnets covering access points have changed. Once when MAC layer handover is finished, the mobile node (bridge on the patient's bed) has to discover the network layer information on the link, i.e. the new care-of-address router and network prefix. Foreign routers periodically advertise this in Router Advertisement using mobile IPv6. When mobile node learns the new care-of-address it registers this address with its home agent. We have to model and evaluate the acceptability of latencies and packet losses during secure network layer handover.

6. COMPARISON BETWEEN TWO TECHNOLOGIES REGARDING THE DEPLOYMENT IN SENSOR NETWORKS

After individual descriptions of IEEE 802.15.1 and IEEE 802.15.4 we will give direct comparison of their properties against the criterion of feasibility of their deployment in sensor networks.

6.1. How much is physical layer immune to the noise errors

Both standards, 802.15.1 and 802.15.4 with 250kbps rate, operate in 2450MHz band known as Industrial, Scientific and Medical – ISM. This band is already hosting wireless LAN/PAN standards such as 802.11b and 802.15.1 (Bluetooth) and a lot of interference is expected. It is also worth mentioning that Bluetooth packets can be 1, 3 or 5 slots long which results in payload sizes of 17, 121, 224 bytes for DM 1, 3 and 5 packet types respectively with Forward Error Correction (FEC) or in payload sizes of 27, 183 and 339 bytes for DH-type , 1, 3 and 5 packet types without FEC. On the other hand, 802.15.4 does not have FEC and allows maximum packet size of 127 bytes. This packet size includes all headers from physical and MAC layer which minimum size is 15 bytes giving the actual maximum payload size of 112 bytes. Therefore, it makes sense to compare these two technologies only in the case of payload size of 27 bytes (DH1).

As mentioned, Bluetooth uses FHSS and is very resilient to interference. According to the exhaustive simulation results reported in [33] when 10 fully loaded piconets each with 7 slaves are placed in the room with dimensions 10m x 20m, (and interfere with each other) packet error rate for DH1 packets was 0.03. When the same experiment was repeated with 100 co-located piconets, packet error rate was 0.3.

IEEE 802.15.4 standard in the 2450 MHz range (ISM band) uses 16-ary quasi-orthogonal modulation technique. Four data bits represent one modulation symbol and that symbol is further encoded into 32 bit chip sequence. There are 16 nearly-orthogonal Pseudo-Noise chip sequences. Each chip sequence is modulated onto the carrier using offset quadrature phase shift keying (O-QPSK). Since the chip rate is 2Mcps and raw data rate is 250kbps the maximum supported ratio of bit energy to the noise power spectral density of $\frac{E_b}{N_0} = 8$. According to the properties of QPSK, the Bit Error Rate is determined using known expression given for example in [13]. Therefore, without the interference, we should expect BER slightly less than 10^{-4}. This is confirmed in the section 6.1.6 of the standard where Packet Error Rate (PER) of 1% is expected on packets which have 20 bytes including MAC and physical level headers. However, in the presence of interference in the ISM band, it is more realistic to expect BER around 10^{-3} and Packet Error Rate more than 28% for packets with 27 bytes of payload and 15 bytes of headers. (Packet Error Rate can be calculated as $PER = 1-(1-BER)^X$ where X is packet length including MAC and physical layer header expressed in bits).

Although, Zürbes' experiment can not be directly translated into BER, 10 co-located piconets present interference probably much larger than what the physical layer of 802.15.4 can handle.

6.2. The access delay

Bluetooth has a polling based MAC protocol and its access delay depends on the order in which master polls the slaves and on the amount of packets which are exchanged between master and slave in one visit. Mathematically speaking, packet service time directly depends on the piconet cycle time i.e. the time needed for the master to visit each slave. It has been shown [26] that under low traffic exhaustive scheduling (where master exchanges packets with slave as long as one of them has packets in the queue) offers the lowest access delay compared with other limited round-robin polices where master can exchange at most M packets per one polling cycle. However, under high loads exhaustive scheduling is not the best one compared to limited polices and fairness issue raises since one station can keep the master busy for a long period of time. Under limited policies every station has equal amount of bandwidth and piconet cycle time is limited. Therefore, if one or more slaves have excessive traffic their packets will suffer from the large delay, but the other slaves with lower traffic will not.

6.3. Can wireless sensor network reach the regime when delays are unacceptable?

Bluetooth piconet can reach such regime only if duration of piconet cycle becomes extremely long and this can happen only under exhaustive scheduling of slaves. This can represent also a security problem, since one malicious node can bring the whole piconet down.

IEEE 802.15.4 network can reach this saturation regime if the number of nodes and packet arrival rates exceeds certain limits. For example, for packet size of 30 bytes (including PHY and MAC headers) saturation is reached with 30 nodes each having packet arrival rate of 3 packets per second (total of 45 bytes per second). Under packet size of 90 bytes, saturation is reached with 15 nodes with packet arrival rate of 3 packets per second. Saturation also can represent a security problem since a couple of malicious nodes can quickly bring the network down as shown in [24].

6.4. How much of the buffering is reasonable to have at the source nodes?

Assuming that the entire measurement can fit in a single packet, transmitting several packets from the node buffer means that some slightly older information is sent. Also, the inter-packet time will be less than in the case where each node sends one packet only. Therefore, exhaustive scheduling of active sensor's periods with large buffers increases spatial and temporal correlation of sensed data. This fact is important in the applications where controlled reliability means controlled inter-packet spacing or in applications with security concerns where mal-functioning node with exhaustive scheduling can inject large amount of bogus data into the network. Therefore, buffer sizes at the nodes should not exceed several packet sizes.

6.5. What is the effective bandwidth left to the application, i.e. what is the maximum possible event detection reliability for particular MAC?

The main concern in sensing applications of Bluetooth is that the major part of the traffic is targeted toward the network coordinator, i.e., in the uplink direction. Because of the Bluetooth polling mechanism, the downlink packet slots—which are still necessary—will be empty. This wastes the bandwidth and limits the maximum throughput of the network to something of the order of 723kbps out of 1Mbps in Bluetooth version 1.2 with DH5 packets. The recent Enhanced Data Rate option in Bluetooth version 2.0 allows for maximum data rates over 2Mbps [6], however the same conceptual problem remains.

On the other hand, the 802.15.4 standard allows for maximum raw data rate of 250kbps, i.e. about one-quarter of that obtainable under Bluetooth version 1.2. Due to the random backoff countdown procedure and the need to listen to the medium before attempting transmission, the traffic intensity of one node affects the activities of the others. Under large traffic volume (which may be expected when the network has many nodes), there will be many collisions and many deferred transmissions. This usually results in severe congestion and all nodes experience large delays. In severe cases when saturation occurs, the network throughput drops to few percent of the raw data rate. Since the backoff window can not exceed value of 31 and the packet size is limited to 127 bytes, an 802.15.4 network can easily reach saturation regime. Our results show that the highest throughput of around 25% of the theoretical maximum occurs for packet size of around 90 bytes (including PHY and MAC headers), with five active stations in the network (we did not check for networks with smaller number of stations), and under the superframe size of 48 backoff periods. This puts a limit of effective data rate of 62.5 kbps per cluster, or around 12.5 kbps per node. However, with fifteen active nodes the total throughput drops to only about 18%, and this drop continues in proportion with the increase of the cluster size.

7. SUMMARY

In this chapter we have addressed security and networking architecture of the clinical information systems with emphasis on the wireless hop. Wireless hop includes sensor networks and possibly wireless local area or mesh networks. We have reviewed confidentiality and integrity polices for clinical information systems and proposed the policy enforcement mechanisms which cover the wireless hop. We have also compared two candidate technologies: IEEE 802.15.1 (also known as Bluetooth) and IEEE 802.15.4, from the aspect of resilience of MAC and physical layers to the jamming and denial-of-service attacks.

8. REFERENCES

1. O. B. Akan and I. F. Akyildiz. ESRT: Event-to-Sink Reliable Transport in Wireless Sensor Networks. In *IEEE/ACM Transaction on Networking (to appear)*, October 2005.
2. R. Anderson. A security policy model for clinical information systems. In *Proc. of the 1996 IEEE*

Symposium on Security and Privacy, pages 34–48, 1996.

3. C. Asmuth and J. Bloom. A modular approach to key safeguarding. *it*, 29(2):208–210, 1979.
4. M. Bishop. *Computer Security – Art and Science*. Pearson Education, Inc., Boston, MA 02116, 1st edition, 2003.
5. G. R. Blakely. Safeguarding cryptographic keys. In *Proceedings of the National Computer Conference, American Federation of Information Processing Societies*, volume 48, pages 313–317, 1979.
6. Bluetooth SIG. *Draft Specification of the Bluetooth System*. Version 2.0, Nov. 2004.
7. E. H. Callaway, Jr. *Wireless Sensor Networks, Architecture and Protocols*. Auerbach Publications, Boca Raton, FL, 2004.
8. A. Cerpa and D. Estrin. Adaptive self-configuring sensor network topologies. In *Proceedings Twenty-First Annual Joint Conference of the IEEE Computer and Communications Societies IEEE INFOCOM 2002*, volume 3, pages 1278–1287, New York, NY, June 2002.
9. O. Elkeelany, M. M. Matalgah, K. P. Sheikh, M. Thaker, G. Choudry, D. Medhi, and J. Qaddour. Performance analysis of IPSec protocol: Encryption and authentication. In *Proceedings of IEEE International Conference on Communications ICC 2002*, pages 1164–1168, 2002.
10. J. Feigenbaum, M. Liberman, and E. Grosse. Cryptographic protection of membership lists. *Newsletter of the International Association of Cryptologic Research*, 9:16–20, 1992.
11. J. Feigenbaum, M. Liberman, and R. N. Wright. Cryptographic protection of databases and software. *Distributed Computing and Cryptography, J. Feigenbaum and M. Merritt eds.*, pages 161–172, 1991.
12. J. Frolik. QoS control for random access wireless sensor networks. In *Proc. WCNC 2004*, Atlanta, GA, Mar. 2004.
13. V. K. Garg, K. Smolik, and J. E. Wilkes. *Applications of CDMA in Wireless/Personal Communications*. Prentice Hall, Upper Saddle River, NJ, 1998.
14. N. Golmie. Bluetooth dynamic scheduling and interference mitigation. *ACM/Kluwer Journal on Special Topics in Mobile Networking and Applications (MONET)*, 9(1):21–31, 2004.
15. H. Gupta, S. Das, and Q. Gu. Connected sensor cover: self organization of sensor networks for efficient query execution. In *Proceedings 2003 ACM International Symposium on Mobile ad hoc networking & computing*, volume 1, pages 189–200, Annapolis, MD, June 2003.
16. Standard for part 15.4: Wireless MAC and PHY specifications for low rate WPAN. IEEE Std 802.15.4, IEEE, New York, NY, Oct. 2003.
17. R. Iyer and L. Kleinrock. QoS control for sensor networks. In *Proc. ICC'03*, volume 1, pages 517–521, Anchorage, AK, May 2003.
18. E. D. Karnin, J. W. Greene, and M. E. Hellman. On sharing secret systems. *it*, 29(2):35–41, 1983.
19. R. Merkle. Method of providing digital signatures, Jan. 1982.
20. R. Merkle. A digital signature based on a conventional encryption function. In *Proceedings of the Advances in Cryptology - CRYPTO'87*, pages 369–378, 1988.
21. R. Merkle. A certified digital signature. In *Proceedings of the Advances in Cryptology - CRYPTO '88*, pages 218–238, 1990.
22. A. Mishra, M. Shin, and W. Arbaugh. An empirical analysis of the IEEE 802.11 MAC layer handoff process. *SIGCOMM Comput. Commun.*, 33(3):93–102, 2003.
23. J. Mišić, G. R. Reddy, and V. B. Mišić. Activity Scheduling based on cross layer information in Bluetooth sensor networks. *Computer Communications*, to appear, 2006.
24. V. B. Mišić, J. Fung, and J. Mišić. Mac layer security of 802.15.4-compliant networks. In *Proc. WSNS'05, held in conjunction with IEEE MASS05 2005*, Washington, DC, Dec. 2005.

25. V. B. Mišić, J. Mišić, and K. L. Chan. Walk-in scheduling in Bluetooth scatternets. *Cluster Computing*, 8(2/3):197–210, 2005.

26. Mišić, J. and Mišić, V. B. *Performance Modeling and Analysis of Bluetooth Networks: Network Formation, Polling, Scheduling, and Traffic Control.* Boca Raton, FL: CRC Press, July 2005.

27. National Institute of Standards. *Digital Signature Standard.* US Department of Commerce, 1994.

28. A. Samir. How to share a secret. *IEEE Computer*, 22(11):612–613, 1979.

29. B. Schneier. *Applied Cryptography.* John Wiley & Sons, Inc., New York, N.Y., 2nd edition, 1996.

30. C. Schurgers, V. Tsiatis, S. Ganeriwal, and M. Srivastava. Topology management for sensor networks: exploiting latency and density. In *Proceedings 2002 ACM International Symposium on Mobile ad hoc networking & computing*, volume 1, pages 135–145, Lausanne, Switzerland, June 2002.

31. G. Tan and J. Guttag. A locally coordinated scatternet scheduling algorithm. In *Proceedings of the 26th Annual Conference on Local Computer Networks LCN 2002*, pages 293–303, Tampa, FL, Nov. 2002.

32. W. Zhang and G. Cao. A flexible scatternet-wide scheduling algorithm for Bluetooth networks. In *Proc. 21st IEEE International Performance, Computing, and Communications Conference IPCCC 2002*, Phoenix, AZ, Apr. 2002.

33. S. Zürbes. Considerations on link and system throughput of Bluetooth networks. In *Proceedings of the 11th IEEE International Symposium on Personal, Indoor and Mobile Radio Communications PIMRC 2000*, volume 2, pages 1315–1319, London, UK, Sept. 2000.

14

KEY MANAGEMENT SCHEMES IN SENSOR NETWORKS

Venkata Krishna Rayi
Department of Computer Science
The University of Memphis
Memphis, TN 38152 USA

Yang Xiao
Computer Science Department
University of Alabama
101 Houser Hall
Box 870290
Tuscaloosa, AL 35487-0290 USA
E-mail: yangxiao@ieee.org

Bo Sun
Department of Computer Science
Lamar University
Beaumont, TX 77710 USA
E-mail: bsun@cs.lamar.edu

Xiaojiang (James) Du
Department of Computer Science
North Dakota State University
Fargo, ND 58105 USA
E-mail: Xiaojiang.Du@ndsu.edu

Fei Hu
Computer Engineering Department
Rochester Institute of Technology
Rochester, NY 14623 USA
E-mail: fxheec@rit.edu

In the near future, sensor networks are going to be a part of everyday life. Traffic monitoring, military tracking, building safety, pollution monitoring, wildlife monitoring, patient security are some of the applications in sensor networks. Sensor networks vary in size and can consist of 10 to 1,000,000 sensor nodes. They can be deployed in a wide variety of areas, including hostile environments, demanding secure measures for data transfer. Sensor nodes used to form these networks are resource-constrained, which makes these types of security applications a challenging problem. A basic technique to protect data is encryption; but, due to resource constraints, achieving necessary key agreement for encryption is not easy. Many key establishment techniques have been designed to address this challenge, but which scheme is the most effective is still debatable. Our work aimed to generate a brief knowledge about different key management schemes and their effectiveness. We noticed that no key distribution technique is ideal to all the scenarios where sensor networks are used; therefore the techniques employed must depend upon the requirements and resources of each individual sensor network.

1. INTRODUCTION

Distributed Sensor Networks (DSNs) are going to be widely used in the near future due to their breadth of applications by military, exploration teams, researchers, and so on. It is not possible to use general wireless techniques for DSNs since they are resource-constrained and security measures are required. Distribution techniques that are applicable employ assorted key management methods, such as public key cryptography, and require numerous communication and computation capabilities. Therefore, it is important to examine the different requirements, constraints and evaluation metrics of sensor networks as well as single network-wide key scheme, which is the simplest of key management techniques, before discussing the various germane key management techniques.

1.1. Requirements

Sensor networks must arrange several types of data packets, including packets of routing protocols and packets of key management protocols. The key establishment technique employed in a given sensor network should meet several requirements to be efficient. These requirements may include supporting in-network processing and facilitating self-organization of data, among others. However, the key establishment technique for an secure application must minimally incorporate authenticity, confidentiality, integrity, scalability, and flexibility.

- *Authenticity*: The key establishment technique should guarantee that the communication nodes in the network have a way for verifying the authenticity of the other nodes involved in a communication, i.e., the receiver node should recognize the assigned ID of the sender node.
- *Confidentiality*: The key establishment technique should protect the disclosure of data from unauthorized parties. An adversary may try to attack a sensor network by acquiring the secret keys to obtain data. A better key technique controls the compromised nodes to keep data from being further revealed.

- *Integrity*: Integrity means no data falsification during transmissions. Here in terms of key establishment techniques, the meanings are explained as follows. Only the nodes in the network should have access to the keys and only an assigned base station should privilege to change the keys. This would effectively prevent unauthorized nodes from obtaining knowledge about the keys used and preclude updates from external sources.
- *Scalability*: Efficiency demands that sensor networks utilize a scalable key establishment technique to allow for the variations in size typical of such a network. Key establishment techniques employed should provide high-security features for small networks, but also maintain these characteristics when applied to larger ones.
- *Flexibility*: Key establishment techniques should be able to function well in any kind of environments and support dynamic deployment of nodes, i.e., a key establishment technique should be useful in multiple applications and allow for adding nodes at any time.

1.2. Constraints

One of the challenges in developing sensor networks is to provide high-security features with limited resources. Sensor networks cannot be costly made as there is always a great chance that they will be deployed in hostile environments and captured for key information or simply destroyed by an adversary, which, in turn, can cause huge losses. Part of these cost limitation constraints includes an inability to make sensor networks totally tamper-proof. Other sensor node constraints that must be kept in mind while developing a key establishment technique include battery life, transmission range, bandwidth, memory, and prior deployment knowledge.

- *Battery Life*: Sensor nodes have a limited battery life, which can make using asymmetric key techniques, like public key cryptography, impractical as they use much more energy for their integral complex mathematical calculations. This constraint is mitigated by making use of more efficient symmetric techniques that involve fewer computational procedures and require less energy to function.
- *Transmission Range*: Limited energy supply also restricts transmission range. Sensor nodes can only transmit messages up to specified short distances since increasing the range may lead to power drain. Techniques like in-network processing can help to achieve better performance by aggregating and transmitting only processed information by only a few nodes, thereby saving the dissipated energy.
- *Bandwidth*: It is not efficient to transfer large blocks of data with the limited bandwidth capacity of typical sensor nodes, such as the transmitter of the UC Berkeley Mica platform that only has a bandwidth of 10Kbps. To compensate,

key establishment techniques should only allow small chunks of data to be transferred at a time.

- *Memory*: Memory availability of sensor nodes is usually 6-8Kbps, half of which is occupied by a typical sensor network operating system, like TinyOS. Key establishment techniques must use the remaining limited storage space efficiently by storing keys in memory, buffering stored messages, etc.

- *Prior Deployment Knowledge*: As the nodes in sensor networks are deployed randomly and dynamically, it is not possible to maintain knowledge of every placement. A key establishment technique should not, therefore, be aware of where nodes are deployed when initializing keys in the network.

1.3. Evaluation Metrics

A key establishment technique is not judged solely based upon its ability to provide secrecy of transferred messages, but must also meet certain other criteria for efficiency in light of vulnerability to adversaries, including the three Rs of sensor networks: resistance, revocation, and resilience. Though scalability may be considered an evaluation metric, it is not discussed here since we have included it in DSN requirements.

- *Resistance*: An adversary might attack the network by compromising a few nodes in the network and then replicating those nodes back into the network. Using this attack the adversary can populate the whole network with his replicated nodes and thereby gain control of the entire network. A key establishment technique must resist node replication to guard against such attacks.

- *Revocation*: If a sensor network become invaded by an adversary, the key establishment technique should provide an efficient way to revoke compromised nodes, a lightweight method that does not use much of the network's already limited capacity for communication.

- *Resilience*: If a node within a sensor network is captured, the key establishment technique should ensure that secret information about other nodes is not revealed. A scheme's resilience is calculated using the total number of nodes compromised and the total fraction of communications compromised in the network. Resilience also means conveniently making new inserted sensors to join secure communications.

1.4. Single Network-Wide Key

Using a single network-wide key is by far the simplest key establishment technique. In the initialization phase of this technique, a single key is preloaded into all the nodes of the network. After deployment, every node in the network can use this key to encrypt and decrypt messages. Some of the advantages offered by this technique include minimal storage requirements and avoidance of complex protocols. Only a single key is to be stored in the nodes' memory and once deployed in the network, there is no need for a node to perform key discovery or key exchange since all the nodes in communication range can transfer messages using the key which they already share.

Though a single network-wide key may seem advantageous, the main drawback is that compromise of a single node causes the compromise of the entire network through the shared key. This scheme counters several constraints with less computation and reduced memory use, but it fails in providing the basic requirements of a sensor network by making it easy for an adversary trying to attack.

1.5. Organization of of this chapter

The key establishment technique employed in a given sensor network should take into consideration all the requirements, constraints, and evaluation metrics discussed. In our work we have assessed different types of key establishment techniques, each ranging in efficiency by providing various necessary characteristics. In Section 3 of this chapter we explain the Basic Scheme for key establishment in sensor networks [1]; Section 4 discusses three more-efficient schemes, two of which are extensions of the Basic Scheme, the Q-Composite Scheme [2] and the Multipath Key Reinforcement Scheme, and one of which is the Random Pairwise Scheme; Section 5 describes the Polynomial-Based Key Predistribution and two efficient instances of that scheme [3]; Section 6 details SPINS and its two building blocks (SNEP and μTESLA) [5]; Section 7 elaborates on LEAP and its implementation [6]; Section 8 clarifies a key management scheme using the deployment knowledge [7]; Section 9 concludes this essay with a brief review of all the schemes evaluated.

2. BASIC SCHEME

In this section we specify the Random Key Predistribution Scheme proposed by Eschenauer and Gligor [1], which we refer to as the Basic Scheme. First we will describe the structure and features of the Basic Scheme and then how it may be evaluated using two of the three Rs of efficient sensor networks, revocation and resilience (through rekeying). We will then analyze the pros and cons of the Basic Scheme for key establishment in a sensor network.

2.1. Key Distribution

In the Basic Scheme, key distribution is divided into three stages: key predistribution, shared-key discovery, and path-key establishment. In the *key predistribution stage*, a large key pool of $|S|$ keys and their identifiers are generated. From this key pool, K keys are randomly drawn and pre-distributed into each node's key ring, including the identifiers of all those keys. At the point that each node has K keys and the identifiers of those keys, trusted nodes in the network are selected as controller nodes, and all the key identifiers of a key ring and the associated sensor identifier on controller nodes are saved. Following this, the i-th contoller node is loaded for each node with the key that is shared with that node. This key predistribution process ensures that, though the size of the network is large, only a few keys need to be stored in each node's memory, thereby saving storage space. These few keys are enough to ensure that two nodes share a common key, based on a selected probability.

After the *key predistribution stage*, we move to the *shared-key discovery stage*. Once the nodes are initialized with keys, they are deployed in the respective places where they are needed, such as hospitals, war fields, etc. After deployment, each node tries to discover its neighbors with which it shares common keys. There are many ways for finding out whether two nodes share common keys or not. The simplest way is to make the nodes broadcast their identifier's list to other nodes. If a node finds out that it shares a common key with a particular node, it can use this key as the communication link. This approach does not give the adversary any new attack opportunities and only leaves room for launching a traffic analysis attack in the absence of key identifiers. More secure alternate methods exist for finding out the common keys shared between two nodes though. For example, for every key on a key ring, each node could broadcast a list α, $E_{K_i}(\alpha)$, $i = 1, ..., k$ where α is a challenge. The decryption of $E_{K_i}(\alpha)$ with the proper key by a recipient would reveal the challenge α and establish a shared key with the broadcasting node [1].

A link exists between two nodes only if they share a key, but the *path key establishment stage* facilitates provision of the link between two nodes when they do not share a common key. Let us suppose that node u wants to communicate with node v, but they do not share a common key between them. Node u can send a message to node y saying that it wants to communicate with v; this message is then encrypted using the common key shared between u and y and, if node y has a key in common with v, it can generate a pairwise key K_{uv} for nodes u and v, thereby acting like a key distribution center or a mediator between the communication of nodes u and v. As all the communications are encrypted using their respective shared keys, there will not be a security breach in this process. After the *shared-key discovery stage* is finished there will be a number of keys left in each sensor's key ring that are unused and can be put to work by each sensor node for path key establishment.

A compromised sensor node can cause a lot of damage to a network and therefore, revocation of a compromised node is very important in any key distribution scheme. In the Basic Scheme, node revocation is conducted by the controller node. When a node is revoked, all the keys in that particular node key ring have to be deleted from the network. Let us assume that the controller node has knowledge about a compromised node in the network and broadcasts a message to all the nodes in the network, the message will include a list of the key identifiers of the compromised node's key ring. To sign the list of key identifiers, the controller node uses a signature K_e and then encrypts its message with K_{c_i}, which is the key that the controller node shares with the nodes during the *key predistribution stage*. Once each node receives the message, it decrypts the message using the key they already share with the controller node. When the signature is verified, the nodes search their key rings for the list of identifiers provided in the message and, if there is any match, they are deleted from the key ring. After the matching keys are completely deleted from all the nodes, there may be links missing between different ones and they then have to reconfigure themselves starting from the shared key discovery stage so that new links can be formed between them. As only few keys are removed from the network, the revocation process only affects a part of it and does not include much communication overhead.

The keys used in a sensor network must be rekeyed to lessen the chance that an adversary may access all of the network keys when a few nodes and their keys are captured. Rekeying effectively increases a network's resilience without incurring much communication and computation overhead.

2.2. Analysis of the Basic Scheme

Let us assume that the probability of a common key existing between two nodes in the network is p, and the size of the network is n. The degree of a node d is derivable using both p and n since the degree of any node is simply the average number of edges connecting that node with other nodes in its neighborhood; therefore, $d = p \times (n-1)$. First we have to find the value of d such that a DSN of n nodes is connected with a given probability P_c. We then must calculate the key ring size k and the size of the key pool $\mid S \mid$.

According to Random Graph Theory, a random graph $G(n,p)$ is a graph consisting of n nodes and p representing the probability of establishing a link between two nodes. Erdos and Renyi [12] showed that there exists a probability state p, which moves from state zero to state one for large random graphs. The function that defines p is called the threshold function of a property. If we are given a desired probability (P_c) for graph connectivity, then p is given as

$$P_c = \lim_{n\to\infty} \Pr[G(n,p) \text{is connected}] = e^{e^{-c}} \tag{1}$$

$$p = \frac{\ln(n)}{n} + \frac{c}{n} \tag{2}$$

where c is a real constant.

Then, to calculate the key ring size k and the size of the key pool $\mid S \mid$, we need to first note that wireless constraints limit the number of nodes in a range to be smaller than n, represented by the value n'. Now the probability of sharing a key between two neighbor nodes varies to $p' = d/(n'-1)$, for a given d value. Also, p' can be denoted as the difference between the total probability and the probability that two nodes do not share a common key; i.e., $p' = 1 - Pr$ [two nodes do not share any key] and, thus,

$$p' = 1 - \frac{(1-\frac{k}{p})^{2(p-k+\frac{1}{2})}}{(1-\frac{2k}{p})^{(p-2k+\frac{1}{2})}} \tag{3}$$

where $|S|$ is the size of the key pool and k is the key ring size.

Eschenauer and Gligor have shown that for a pool size $S = 10,000$ keys, only 75 keys need to be stored in a node's memory to have the probability that they share a key in their key rings be $p = 0.5$. If the pool size is ten times larger, i.e., $S = 100,000$, then the number of keys required is still only 250. Thus, the Basic Scheme is a key management technique that is scalable, flexible and can also be used for large DSNs. Trade-offs in the Basic Scheme can be made between sensor memory and connectivity but, it does not provide the node-to-node authentication property that ascertains the identity of a node with which another node is communicating. This property is very useful when revoking misbehaving nodes from the network and also helps in resisting the node replication attack.

Many key management schemes are proposed as extensions of the Basic Scheme to make it even more secure and reliable.

3. EFFICIENT KEY ESTABLISHMENT TECHNIQUES

Chan, Perrig, and Song [2] have introduced three scheme variations that more efficiently perform key establishment and meet more requirements, constraints, and evaluation metrics of a DSN. These schemes include two expansions of the Basic Scheme (Q-Composite Random Key Predistribution and Multipath Key Reinforcement) and a variation of the commonly known Pairwise Scheme (Random Pairwise Key) that are efficient in a particular DSN environment. Each comes with a different kind of trade-off and is not, therefore, widely applicable. The Q-Composite Scheme achieves security under small scale attacks while being vulnerable under large scale attacks and is useful in DSNs where large attacks are easily detected. The Multipath Reinforcement Scheme offers good security with additional DSN communication overhead for use where security is more of a concern than bandwidth or power drain. Compared to the Q-Composite and the Multipath, the Random Pairwise Scheme offers the best security features in its perfect resilience to node capture with the only drawback being limited scalability.

3.1. Q-Composite Random Key Predistribution Scheme

In the Basic Scheme, two nodes share a unique key for establishing a communication link. A given network's resilience to node capture can be improved by increasing the number of common keys that are needed for link establishment. The Q-Composite Random Key Predistribution Scheme does just this by requiring that two nodes have at least q common keys to set up a link [2]. As the amount of key overlap between two nodes is increased, it becomes harder for an adversary to break their communication link. At the same time, to maintain the probability that two nodes establish a link with q common keys, it is necessary to reduce the size of the key pool $| S |$, which poses a possible security breach in the network as the adversary now has to compromise only a few nodes to gain a large part of S. So the challenge of the Q-Composite Scheme is to choose an optimal value for q while ensuring that security is not sacrificed.

In the *key predistribution stage* of both the Basic and Q-Composite Schemes, k random keys are picked from S and initialized in each node's key ring. In the *shared-key discovery phase*, each node has to find the common keys which it shares with other nodes by either making all the nodes broadcast their key identifiers or by selecting a slower and more secure method of posing puzzles such as the Merkle Puzzle [18]. For this puzzle method, each node issues m client puzzles to each neighboring node and any node that comes up with the correct solution to the puzzle is identified as sharing the associated key. After this, the two schemes differ in that the Q-Composite Scheme requires each node identify neighboring nodes with which they share at least q common keys while the Basic Scheme only requires one shared key. This restriction in the Q-Composite Scheme allows the number of keys shared to be more than q but not less, represented by the value q. At this stage in the process, nodes will fail to establish a link if the number of keys shared is less than q; otherwise, they will form a new communication link using the hash of all the q keys, i.e., $K = hash(k_1||k_2||...||k_q)$.

S, the size of the key pool, is the critical parameter that must be calculated for the Q-Composite Scheme to be efficient. If S is large, then the probability is decreased that two nodes share a common key and therefore can communicate. However, if S is decreased, an adversary's job may be easier as he can gather most of the keys in the key pool by capturing only a few nodes. Thus, S must be chosen such that the probability of any two nodes sharing at least q keys is $\geq p$.

Chan, Perrig, and Song's [2] method to calculate S is

$$p(i) = \frac{\begin{pmatrix} |S| \\ i \end{pmatrix} \begin{pmatrix} |S|-i \\ 2(m-i) \end{pmatrix} \begin{pmatrix} 2(m-i) \\ m-i \end{pmatrix}}{\begin{pmatrix} |S| \\ m \end{pmatrix}^2} \tag{4}$$

where $p(i)$ is the probability that any two nodes have exactly i, which is the number of keys in common; and m is the key ring capacity for a given node. There are $\begin{pmatrix} |S| \\ i \end{pmatrix}$ ways to pick i and $\mid S \mid -i$ is the remaining keys in the key pool after i is picked. There are $\begin{pmatrix} |S| \\ m \end{pmatrix}$ different ways to pick m and $\begin{pmatrix} |S| \\ m \end{pmatrix}^2$ total number of ways for both nodes to pick m.

Also, to assign the remaining keys $2(m-i)$ distinct keys are picked from the key pool for each node and the number of ways to do this is $\begin{pmatrix} |s|-i \\ 2(m-i) \end{pmatrix}$. There are $2(m-i)$ ways to partition the keys equally between the two nodes.

Let P_c be the probability of any two nodes sharing sufficient keys to form a secure connection. Therefore, $Pc = 1-$(the probability that the two nodes share insufficient keys to form a connection) or

$$P_c = 1 - (p(0) + p(1) + ... + p(q-1)) \tag{5}$$

Now the largest $\mid S \mid$ such that $P_c \geq p$ is chosen.

The evaluation of the Q-Composite Scheme can be done by verifying its resilience against node capture. Even though this scheme does not provide resistance against node replication or a means for node-to-node authentication since the keys from the key pool are used more than once, it does improve resilience against node capture when an adversary has successfully captured some other nodes in the DSN. As the same keys are used repeatedly in a network, a situation may arise in which two nodes effectively have their communications exposed due to the compromise of another two nodes that share the same key(s). Chan, Perrig, and Song [2] have calculated the probability that a secure link that is made between two uncompromised nodes will be compromised as

$$\sum_{i-q}^{m} \left(1 - \left(1 - \frac{m}{|S|}\right)^{x}\right)^{i} \frac{p(i)}{p} \tag{6}$$

where x is the number of nodes captured, i is the number of keys in common, m is the number of keys in the key ring of a node, and p is the probability of setting up a secure link.

The Q-Composite Scheme offers greater resilience compared to the Basic Scheme when a small number of nodes have been captured in the network. The amount of communications that are compromised in a given DSN with the Q-Composite Scheme applied is 4.74 percent when there are 50 compromised nodes, while the same DSN with the Basic Scheme applied will have 9.52 percent of communications compromised. Though the Q-Composite Scheme performs badly when more nodes are captured in a DSN, this may prove a reasonable concession as adversaries are more likely to commit a less expensive smaller attack and preventing smaller attacks can push an adversary to launch a larger attack, which is far easier to detect with vast node failure.

Still, random key pre-distribution schemes like Q-Composite and Basic cannot be securely used for large networks because they use keys more than once, which results in the compromise of a larger fraction of communications when just a few nodes are compromised. Since random key pre-distribution schemes are not scalable, the maximum network that can be supported should be measured using the *Limited Global Payoff Requirement*, which states that given a DSN is secure, an adversary should not learn anything about the communications of nodes in the network other than those of captured nodes. Let f_m be the maximum compromise threshold past where the adversary gains an unacceptable high confidence of guessing the sensor readings of the entire network. If x_m is the number of nodes compromised, then the total fraction of secure links compromised after the key setup phase due to this x_m nodes being compromised is $f(x_m)$, if the total fraction of secure links compromised reaches the threshold value with the x_m nodes being compromised ,i.e., $f_m = f(x_m)$, then Chan, Perrig, and Song [2] have calculated the maximum allowable size of the network to be

$$n \leq 2x_m \left(1 + \frac{1}{f_m}\right) \tag{7}$$

For example, when $p = 0.33$, $f_m = 0.1$, and $m = 200$, the maximum supportable network size for a Q-Composite Scheme ($q = 2$) is $1,415$ nodes. Compared to the $1,159$ node maximum of the Basic Scheme, the advantage is obvious. Thus, the Q-Composite Scheme is more efficient than the Basic Scheme in providing more resilience to node capture and significantly increases the maximum allowable size of a DSN. Since both schemes fail to provide node-to-node authentication or resistance against node replication, it is important to review other schemes that work more efficiently in DSNs requiring such security measures.

3.2. Multipath Key Reinforcement Scheme

The idea of using a multipath to reinforce links in a random key establishment scheme was first explored by Anderson and Perrig [10]. Chan, Perrig, and Song [2] further developed the Multipath Key Reinforcement Scheme for establishing a link between two nodes of a given DSN that is stronger than that in the Basic Scheme. The links formed between nodes after the key discovery phase in the Basic Scheme are not totally secure due to the random selection of keys from the key pool allowing nodes in a DSN to share some of the same keys and, thereby, possibly threaten multiple nodes when only one is compromised. To solve this problem, the communication key between nodes must be updated when one is compromised once a secure link is formed. This should not be done via the already established link, as an adversary might decrypt the communication to obtain the new key, but should be coordinated using multiple independent paths for greater security.

If node A needs an updated communication key with node B, all possible disjointed paths to B must be used. Assume that there are h such disjointed paths from node A to node B. Then A generates h random values $(g_1, g_2, ..., g_h)$, each equal to the size of an encryption key, and sends one down each available disjointed path to B. When node B has received all h random values, it computes the new encryption key at the same time as node A does forming a new, secure communication link, or

$$k' = k \oplus g_1 \oplus g_2 \oplus .... \oplus g_h \tag{8}$$

where k is the original key.

With the new link in place, the only way an adversary can decrypt the communications is to compromise all the nodes involved in the formation of the key. The higher h is, the more paths and nodes involved and the greater the security of the new link. This increase in DSN communications causes excessive overhead in finding multiple disjointed paths between two nodes. Also, as the size of a path increases, it may grow so long as it leaves a chance for an adversary to eavesdrop, which makes the whole path insecure. A *2-hop approach* to the Multipath Key Reinforcement Scheme considers only 2-link paths to minimize the overhead of path length by using disjointed paths that are only one intermediate node away from the two original nodes (A and B).

To take such an approach, first the number of common neighbors between the original nodes must be calculated in a planar deployment of sensors. In [2], Chan, Perrig, and Song calculated the overlap of communication radius between two nodes in a network to be $0.5865\pi r^2$ in a planar deployment where r is the communication range of sensors; therefore, the expected number of common neighbors with whom both nodes share a secure link is $0.5865p^2n'$, where p is the probability of sharing sufficient keys to communicate and n' is the number of neighbors in each node. Expressed as $0.5865d^2/n = k$, both nodes share a secure link with an expected k neighbors, e.g., if $d = 20$ and $n' = 60$, then $k = 3.91$. To assess the efficiency of the *2-hop approach*, the new probability for compromising the link between two nodes needs to be derived. If an adversary's basic probability of compromising the link is b, then the probability of compromising at least one hop on any given 2-hop path is the probability of compromising hop 1 in the path plus the probability of compromising hop 2 in the path minus the probability of compromising both hops in the path or $2b - b^2$ [2]; hence, the final probability of breaking the link will be $b' = b(2b - b^2)^k$.

If $b = 0.1$ and the number of neighbors (k) is 3, then the chance of eavesdropping after reinforcement improves to 6.86×10^{-4}, that is about 1 in $1,458$. Chan, Perrig, and Song [2] calculated the total additional communication overhead incurred to be at least $2 \times 0.5865p^2n'$ times more in the 2-hop approach compared to the normal setup. If, for instance, $p = 0.33$ and $n' = 60$, additional overhead can be at least 7.66 times. Given these results, the minimal network overhead of finding the neighbors that share a common key becomes a reasonable trade-off when using a *2-hop* Multipath Key Reinforcement Scheme to increase the security of DSNs, though this scheme remains constrained by certain vital factors, including the deployment density characteristics of the DSN.

3.3. Random Pairwise Key Scheme

Node-to-node authentication not only helps to reduce overhead since sensor nodes instead of a base station take actions when a node is compromised in the DSN, but also entails that each node use a unique identity, which helps nodes to identify exactly which ones are compromised. We have seen that though the Basic Scheme is somewhat efficient, it does not provide node-to-node authentication. The Q-Composite extension and 2-hop Multipath approach do not provide node-to-node authentication too. The Pairwise Key Establishment Scheme, however, is one of the most efficient key establishment schemes in DSNs because it does offer many additional features compared to other schemes, including node-to-node authentication and resilience against node replication.

For a DSN of n nodes in the Pairwise Scheme, the key predistribution is done by assigning each node a unique pairwise key with all the other nodes in the network, or $n - 1$ pairwise keys, which are retained in each node's memory so that each node can communicate with all the nodes in its communication range. With each node sharing a unique key with every other node in the network, this scheme offers node-to-node authentication. Each node can verify the identity of the node it is communicating with. This scheme also offers increased resilience to network capture as a compromised node does not reveal information about other nodes that are not directly communicating with the captured node. Through increased resilience, the scheme minimizes the chance for node replication. The drawback with the Pairwise Scheme is the additional overhead needed for each node to establish $n - 1$ unique keys with all the other nodes in the DSN and maintain those keys in its memory. Utilizing such a scheme makes a DSN size-prohibitive since, as the number of nodes in the network increases, so do the number keys that must be stored in each node's memory. If there is a network of $10,000$ nodes, then each node must store $9,999$ keys in their memory. Since sensor nodes are resource-constrained, this significant overhead limits the scheme's applicability, but it can be effectively used for smaller networks.

Chan, Perrig, and Song [2] developed the Random Pairwise Scheme as an extension of the Pairwise Scheme to help overcome this drawback. They stated that not all $n-1$ keys are required to be stored in a node's key ring. As we have already seen with the Basic Scheme, not all nodes must be connected as long as node connections meet some desired probability P_c, which dictates that only np keys are needed to be stored in a given node's key ring (n being the number of nodes in the network and p being the probability that two nodes can communicate securely). Given this, if k is the number of keys in a node's key ring, the maximum allowable network size can be determined with $n = k/p$ for the Random Pairwise Scheme.

In the initialization stage of this scheme, n unique node identifiers are created, each paired with m other randomly selected node identities. A pairwise key is then generated for each such pair. Both the generated pairwise key and the identity of the other node that shares the key are stored in each node's memory. Additional identifiers can be generated to allow for scalability of the DSN, i.e., the number of nodes may originally be fewer than n created to allow for adding nodes. In the key discovery phase, each node broadcasts its identity to the other nodes in the network. For example, if node A wants to communicate with other nodes in the network, it broadcasts its identity to other nodes in the network; if the neighboring nodes share a pairwise key with node A, they perform a cryptographic handshake with A, thereby forming a secure communication link. This process of broadcasting can also be extended beyond the communication range of a node by making the intermediate nodes rebroadcast the node identity to a certain number of hops, which in turn helps in increasing the maximum allowable size of the DSN. This process of range extension must be done cautiously as it leaves a vulnerable opening for an adversary to perform a 'denial of service attack'. The attack involves the adversary's introducing foreign nodes into the DSN to generate random node identities that flood the DSN with rebroadcasted identities, making the whole scheme slow and inefficient. This type of attack can be avoided by restricting the number of hops for range extension.

Revoking compromised nodes from a DSN helps avert various attacks such as denial of service, implanting clones, dropping legitimate reports, etc. Revocation of sensor nodes through the base station can be a slow process due to the high latency in communications with the sensor nodes. To overcome this difficulty, Chan, Perrig, and Song [2] also developed a distributed node revocation method for the Random Pairwise Scheme. Assume that the scheme can detect compromised nodes. If node A finds a certain node B to be compromised then it casts a public vote against B. If a threshold of t such votes have been cast against node B by other nodes in the DSN, node A will disconnect all its communication with B. This process continues until all the nodes in the DSN break their links with B, thereby deleting B from the network. All the nodes that vote against B are called the 'voting members' of B and, as B shares exactly k pairwise keys with other nodes, there will k voting members of B.

This voting method of node revocation must have certain important properties to function properly: make the broadcasted public votes without replay value, disallow a voting member from forging another vote, provide a means for each voting member to verify the validity of the votes that are being broadcasted, etc. In this voting method, each of the k voting members of B is initialized with a random key K_i, and should know the hash values of the remaining $k-1$ voting members. To revoke node B node from the network, node A broadcasts its K_i key. All other voting nodes verify the key by calculating the hash value of the key. Once verified, the key is replaced with a flag signifying the vote has already been used.

Given this process, the nodes in the DSN must store an additional $k-1$ hash values, a voting key and the pairwise keys, which drastically increases the overhead in a sensor node's memory. Chan, Perrig, and Song [2] proposed using a Merkle Tree [11] to authenticate the k hash values to reduce overhead by requiring verification and storing of only one hash value, which reduces the memory size required on the node, but also increases the size of the voting information to $O(log(k))$ as each node must still recall which ones have already been received from public vote to remove possible replay.

Other precautionary measures with the Random Pairwise Scheme taking the public voting approach include the critical issue of choosing the threshold or t value. If t is high, there may not be enough neighboring nodes to revoke a node that has been compromised; however, if t is low, then a group of compromised nodes may cause the revocation of many legitimate nodes. For instance, a DSN of $1,000$ to $10,000$ nodes should have a t value from 1 to 5. Also, since this approach requires a node have at least t neighbors in its communication range to be revoked, an adversary can attack the DSN by selectively disrupting a given node such that only $t-1$ legitimate nodes are able to communicate with it so that it cannot be revoked.

Beyond problems with t value, in the public voting approach each and every vote that is cast is being transmitted to all the nodes in the network, which may lead to a 'denial of service attack'. To solve this, only the voting members should be required to rebroadcast votes between each other while the remaining nodes are forced to ignore the communication thereby decreasing the degree of vulnerability to falsely rebroadcasted identities. Additionally, the node that first receives the correctly verified vote rebroadcasts it only a fixed number times to increase the probability of successful transmission to neighboring voting members.

In similar action, an adversary that tries to compromise a fixed number of nodes can compromise a significant portion of a DSN when public voting is used to perform distributed node revocation since each node can potentially cast a vote against k others. To prevent this problem, only nodes that establish direct communication are given the ability to revoke a compromised node by distributing masked revocation keys to voting members in a non-working form. Each node would then complete the key discovery phase by sharing the secret key only with other nodes with whom they already share a pairwise key connection.

4. POLYNOMIAL POOL-BASED KEY PREDISTRIBUTION

Every key distribution scheme previously discussed has one or more trade-offs or compromises to be considered and what they fundamentally lack is greater probability of key establishment despite part of the DSN being compromised. To this end, Liu and Ning [3] proposed the Polynomial Pool-Based Key Predistribution Scheme that offers several efficient features the other schemes lack, including:

- any two sensors can definitely establish a pairwise key when there are no compromised sensors;
- even with some nodes compromised, the others in the DSN can still establish pairwise keys;
- a node can find the common keys to determine whether or not it can establish a pairwise key and thereby help reduce communication overhead.

In the initialization stage of the Polynomial Pool-Based Scheme, the setup server randomly generates a bivariate t-degree polynomial $f(x, y)$ over a finite field F_q, where

$$f(x, y) = \sum_{i,j=0}^{t} a_{ij} x^i y^j \tag{9}$$

The value of q is a prime number which can accommodate a cryptographic key. The equation $f(x, y)$ has the property $f(x, y) = f(y, x)$. The setup server then generates a polynomial share of the equation for every node in the sensor network; e.g., node i in the network receives an $f(i, y)$ share and node j receives an $f(j, y)$ share. If both nodes i and j want to establish a common key $f(i, j)$ between them, then node i can compute the common key by computing $f(i, y)$ at node j and then node j can compute $f(j, y)$ at node i for the common key $f(i, j)$. This methodology is secure and reveals nothing about the communication between other nodes until t nodes have been compromised, making it t collusion resistant where the t value depends upon the memory available in the sensors.

Each node in this scheme must store a t-degree polynomial which occupies $(t + 1)log(q)$ storage space. Increasing the size of the DSN increases the chance of compromising more than t nodes, but modifications based on the Basic Scheme can earn good results. For this, instead of using a single t-degree polynomial, a pool of polynomials is used. During the initialization phase, randomly selected polynomials are deployed into each node's memory. When there is only one polynomial remaining in the pool, the scheme falls back to the Polynomial Pool-Based Key Distribution; but, if all of the polynomials are 0-degree, then distribution resembles the Basic Scheme.

In the *key predistribution stage* of the Polynomial Scheme, the setup server generates a set of bivariate t-degree polynomials over a field F_q. Each polynomial is then assigned with a particular ID for the server. A subset of these polynomial shares are then picked up by the server and placed in each of the DSN's nodes. While polynomial placement is the main issue of this stage, in the *key discovery stage* each sensor node finds a node with which it shares the same bivariate polynomial and both nodes establish a common key. The complex issue is to find whether two nodes share the same polynomial or not, for which there are two techniques: *predistribution* and *real-time discovery*.

In the predistribution approach, the knowledge of the nodes with which each node will share a polynomial is pre-loaded. This is a basic method in which each node carries the node IDs of those with which they share a polynomial. The concessions of this method are that it does not offer the flexibility of adding new nodes into a DSN and it leaves the network vulnerable to attack. Since information is predistributed in this approach, an adversary may attack a node and gain access to the stored data, which would help in targeting certain nodes in the DSN. Conversely, nodes must uncover with which others they share a polynomial after deployment when applying the real-time discovery method. This discovery can be done by broadcasting the IDs of the polynomials that nodes share or by challenging the nodes with puzzles that are only solvable if the nodes share part of the bivariate t-degree polynomial. Though this handles problems faced with the predistribution approach, real-time discovery increases the communication overhead of the DSN, which makes weighing these factors critical in choosing a method.

After key discovery, if two nodes do not find a common polynomial share, they must communicate through a path key. If node P wants to communicate with node Q and the two nodes do not have a common polynomial share, node P must find a path through which it can communicate with node Q and either node can then send a request to establish a pairwise key for communication. The problem with this stage is that intermediate nodes should be able to communicate with both nodes and, similar to the previous stage, there are two customary techniques for finding intermediate nodes: *predistribution* and *real-time discovery*.

In the predistribution approach, the setup server preloads each node with information such that, if a node is given an ID, each node can find a path to it. In this stage, the predistribution method suffers from the same problems as faced in key predistribution-no scalability and vulnerability to attack. With real-time discovery, nodes try to find a communication path on-the-fly. A source node sends a message to adjoin intermediate nodes that it wants to establish a pairwise key with the destination node and, as the source node already shares a common key with the intermediate nodes, there is no security threat in this communication. If an intermediate node of the source node shares a common key with the destination node, then a communication path has been discovered between the source and destination nodes through which they may discover a common key. Again, the concession of real-time discovery is additional overhead of communication.

4.1. Random Subset Key Predistribution [3]

A Polynomial Pool-Based Scheme using a random subset key assignment is an extension of the Basic Scheme [1], in which random keys are selected from a large key pool and then assigned to each node in a DSN. In the Random Subset Scheme, random polynomials are selected from a polynomial pool and assigned to each node in a DSN to avoid the Basic Scheme's vulnerability in possibly using a key in more than one node. With Random Subset, the pairwise keys generated by each node are unique and based upon the each node's ID. If no more than t shares of the same polynomial have been disclosed, it is very difficult to attack the communication between two nodes.

The Random Subset works similarly to the Polynomial Pool-Based in the three stages of key establishment. In the key predistribution stage, the setup server generates a set F of s-bivariate t-degree polynomials and then initializes each node with a subset of s' polynomials from F. In the key discovery stage, each node attempts to determine the nodes with which they share a common key by employing the real-time discovery technique as information in not preloaded in the nodes prior deployment. In the path key establishment phase, a source node sends a message to its intermediate nodes seeking to establish a connection with a destination node and, if an intermediate node shares a common key with both the source and destination nodes, then a communication path is formed between the two. Generally, the communication range of the source node is limited to lessen vulnerability.

The probability of two sensors sharing the same bivariate polynomial is the same as the probability of the two sharing a common key as described in the Basic Scheme discussion,

$$p = 1 - \prod_{i=0}^{s'-1} \frac{s - s' - i}{s - i} \tag{10}$$

This can also be applied to calculate the probability that any two sensors can establish a common pairwise key using both the key discovery and path discovery stages. If there are d neighbors to a node and any one of them can act as the intermediate node, the probability that one of them share a common key with the source and destination will be p^2; therefore, the probability that the sensor nodes establish a pairwise key in either the key discovery or the path key establishment stage will be

$$P_s = 1 - (1-p)(1-p^2)^d \tag{11}$$

If $p = 0.3$ and $d = 30$, then $P_s = 0.959$. Assuming that an attacker has compromised N_c sensors in the network, where $N_c > t$, the scheme is known to be secure until the adversary compromises fewer than t sensors. With a pool of F polynomials, the probability that a polynomial is used i times and its probability of being compromised when more than t nodes are compromised must be calculated to determine security efficiency. Given this, the probability that a polynomial is chosen for a sensor node will be s'/s and the probability that this polynomial is chosen exactly i times among the N_c compromised nodes is

$$p(i) = \frac{N_c!}{(N_c - i)|i|} \left(\frac{s'}{s}\right)^i \left(1 - \frac{s^i}{s}\right)^{N_c - i} \tag{12}$$

Thus, the probability of a polynomial being compromised is

$$P_c = 1 - \sum_{i=0}^{t} p(i) \tag{13}$$

Though this makes the DSN more secure, it is still vulnerable to attack because if an adversary somehow knows the distribution of polynomial shares, specific nodes can be targeted for attack to compromise communications. It is enough for an adversary to compromise $t+1$ particular nodes to compromise a polynomial and, in effect, the DSN. This problem is addressed by restricting the use of polynomial shares to a maximum of $t+1$ times in the network so that an attacker must now compromise all the $t+1$ nodes to compromise a polynomial. Though efficient, use of random subsets like this decreases the potential size of a DSN. The maximum number of nodes in a network when random subsets are implemented is $(t+1)s/s'$; however, using this scheme is unnecessary as it is relatively difficult for an adversary to compromise $t+1$ selected nodes.

Polynomial Pool-Based Key Predistribution using random subsets offers greater security and flexibility when compared to other schemes until a certain number (60 percent) of compromised nodes has been reached at which point any scheme would prove ineffective. Compared to the Random Pairwise Scheme, which offers perfect resilience to node capture as no key in the network is used twice, the Polynomial Pool-Based offers the same resilience if a polynomial share is used no more than t times. Also, the Polynomial Pool-Based Scheme offers certain advantages over Random Pairwise in that sensors can be added dynamically without consulting the already deployed sensors while dynamically deploying nodes in Random Pairwise demands that the server has predesignated unassigned space for additional nodes, which may never be deployed. Because of this, the Random Pairwise Scheme can only offer limited scalability, while the more attractive Polynomial Pool-Based Scheme allows for undetermined network growth.

4.2. Grid-Based Key Predistribution

A Polynomial Pool-Based Scheme [3] using a grid-based key assignment offers all the attractive properties of the Polynomial Pool-Based key predistribution and guarantees that two sensors can establish a pairwise key when there are no compromised nodes and the nodes can communicate with each other. Even if some nodes are captured, there will still be a great chance for key establishment between uncompromised nodes using this approach, which also reduces DSN communication overhead. With grid-based key predistribution, a sensor node can determine whether it can establish a pairwise key with another node or not, and can say which polynomial should be used for key establishment.

If a DSN consists of N sensor nodes, the approach involves constructing an $m \times m$ grid with a set of $2m$ polynomials, calculated as $\{f_i^c(x, y), f_i^r(x, y)\}, i = 0, ..., m-1$, where the value of m is the square root of N; each row i in the grid is associated with a polynomial $f_i^r(x, y)$ and each column of the grid is associated with a polynomial share $f_i^c(x, y)$. The setup server distributes an intersection in the grid to each node, and then distributes the polynomial shares of that particular column and row to the node to provide each node with the information required for key discovery and path key establishment. Although the Grid-Based Scheme can be extended to n-dimension, Liu and Ning [3] considered only a 2-dimension with many polynomials, so that is the example discussed here.

In the first stage of key establishment, the setup server generates $2m$ t-degree bivariate polynomials over a finite field F_q and assigns each node to an unoccupied intersection in the grid for deployment in the DSN. If the intersection is $< i, j >$, then the node ID is $< i, j >$. The server provides each node with its ID and the row and column polynomial shares of that grid intersection. To facilitate path discovery, all nodes are densely placed in a rectangular area in the grid.

In the second stage, polynomial share discovery, if node i wants to establish a pairwise key with node j, it checks for common rows or columns with j, i.e., $c_i = c_j$ or $r_i = r_j$. The pairwise key can be established using the polynomial shares of a row or column that matches. Should none match, then nodes i and j must find an alternate path to each other in the path key establishment stage. To do so, node i finds an intermediate node through which it can establish a pairwise key with node j. Even if some intermediate nodes are compromised, node i can still find a path to node j since there are many connecting paths in the grid between the two nodes; but, as the number of compromised nodes increases, so does the length of the path.

In this case the nodes remember the graph composing the grid; however, with large networks, it is not feasible for a node to remember the entire graph or run an algorithm for finding the path between the nodes. Discovering key paths using two intermediate nodes limits the demands on the nodes for this scheme to function in large DSNs also. If, for instance, there are two sensor nodes attempting to establish a path key between them, the source node S determines the set of N nodes with which it can communicate, and then selects some nodes randomly from that set. S also generates a random number r and a counter c. Node S sends each node U of the subset N a message containing the IDs of S and D, the counter value c and K_c in an encrypted form.

Here the value of $K_c = F(r, c)$ where F is a pseudo random function. Encryption of the message is done using the pairwise key that S shares with the intermediate node U. After receiving the message from S, node U checks for the authentication of the message and if the message is authenticated, U attempts to find a non-compromised node V. It then sends to V the message sent by S in the encrypted form using the pairwise key which it shares with V. If V receives the message and discovers that it can establish a pairwise key with D, it sends the message to D in encrypted form using the shared key. Once the destination node D receives the message, it knows that node S wants to establish a pairwise key with it and then sends S the counter value c and the new communication key $K_{s,d} = K_c$.

With grid-based key predistribution, an adversary may try to attack the connection between two nodes by either compromising the pairwise key or by preventing the two nodes from establishing a shared key. If an adversary wishes to attack the entire DSN, the foe may attempt to lower the probability of establishing a pairwise key between nodes. This may be done through attacking a pair of nodes and finding their common key without actually compromising the nodes by compromising the polynomial which the two nodes share. To discover the polynomial share, the adversary must compromise at least $t+1$ nodes as stated previously. The DSN may avert such an attack even when the adversary successfully finds out the polynomial which two nodes share by the nodes' establishing a common key through path key establishment.

From the scheme we can see that there are still $m-1$ nodes that can help nodes U and V establish a common key. An adversary must compromise at least one node in each pair to arrest path key establishment; thus, the adversary must compromise $t+1$ nodes to learn the pairwise key and $t+m$ sensor nodes to prevent two from establishing a pairwise key via intermediates. An adversary can also compromise the polynomial shares in a pool by knowing the subset assignment mechanism. Supposing that the adversary has compromised some l polynomials from the pool, there are about ml sensors with at least one polynomial share disclosed. The attacker has compromised about $(t+1)l$ sensor nodes, but only affects the common keys in ml sensors, including those of the compromised nodes. The adversary may also attack the sensors randomly to disrupt path establishment and thereby make key establishment an expensive process. If P_c nodes have been compromised, then the probability that exactly k polynomial shares on a particular polynomial are disclosed is

$$P(k) = \frac{m!}{k!(m-k)!} p_c^k (1-p_c)^{m-k} \tag{14}$$

The probability that one particular polynomial is compromised would be calculated as

$$P_c = 1 - \sum\nolimits_{i=0}^{t} p(i) \tag{15}$$

This grid-based approach to the Polynomial Pool-Based Scheme has reasonable overhead when compared to other schemes. Each node must store 2 bivariate t-degree polynomials and IDs of the compromised nodes with which it can establish a pairwise key; therefore, the total overhead for each node is, at most, $2(t+1)log(q) + 2(t+1)l$ bits. The DSN overhead is almost null when there is direct key establishment between nodes. There is slight communication overhead when two nodes must find a common key through path key establishment and this overhead increases with each additional node compromised. The approach offers many attractive properties other schemes do not, including nice resilience to node capture until a certain percentage of nodes are compromised (60 percent). The Basic Scheme and Q-Composite Scheme offer this same resilience, but the grid-based approach offers less overhead on both DSN communication and computations. Compared to Random Pairwise Scheme, the grid-based method offers the same degree of security when the same number of sensors and storage overhead are considered. More than any other scheme, the grid-based approach offers greater probability of key establishment when there are no compromised nodes as well as greater probability of key establishment with some nodes compromised.

Finally, there will be a greater chance for nodes to establish a pairwise key with others without communication overhead as the sensors are deployed in a grid-like structure.

Thus, the Polynomial-Based Key Predistribution and the two instantiations of the scheme, polynomial pool-based and grid-based, provide some attractive and efficient properties for key establishment in DSNs. In the near future, these schemes might be extended as part of the research performed in sensor networks. For example, the grid-based approach may be extended into n-dimension or hypercube-based. Also, research must be done for DSNs with Polynomial-Based Key Predistribution scheme and mobile properties.

5. SPINS: SECURITY PROTOCOLS FOR SENSOR NETWORKS

Perrig, Szewczyk, Tygar, Wen and Culler[5] at UC Berkeley presented a suite of security protocols optimized for sensor networks that they called 'SPINS'. The suite is built upon two secure building blocks, each performing individual required work: SNEP and μTESLA. SNEP offers data confidentiality, authentication, integrity, and freshness, while μTESLA offers broadcast data authentication. The μTESLA protocol, used on regular networks, is modified as a SPINS for use in resource-constrained DSNs. SPINS incorporates TinyOS (operating system) in each node, all of which communicate with a base station. Most DSN communications pass through the base station and involve three communication types: node-to-base station, base station-to-node, and base station-to-all nodes.

The main goal of SPINS protocol is to design a key establishment technique based on SNEP and μTESLA to prevent an adversary from spreading to other nodes in the network through a compromised node. Each node in this scheme shares a secret key with the base station that is initialized before deployment. The following are some of the representations in this scheme used to illustrate how this works:

- A and B are two communicating nodes in the network;
- N_a is generated by node A;
- X_{ab} is the master key shared between nodes A and B;
- K_{ab} and K_{ba} are the encryption keys shared between A and B, which are derived from the master key X_{ab};
- K'_{ab} and K'_{ba} are the secret *MAC* keys shared between A and B, which are derived from the master key X_{ab};
- $\{M\}K_{ab}$ denotes the encryption of message *M* with key K_{ab};
- $MAC(K'_{ab}, M)$ denotes the computation of *MAC* for message *M* with *MAC* key K'_{ab}.

5.1. SNEP: Data confidentiality/authentication/freshness

A combination of two schemes forms SNEP including a counter for semantic security and a bootstrapping scheme. Using this combination, SNEP is able to offer a number of advantages and only adds 8 bytes per message by reducing the communication overhead of the network. It uses a counter, like many other protocols, to offer authentication and freshness, but does so using means that also provide *semantic security*. Two counters are shared between nodes attempting to communicate with each other for which some of the source node's cryptographic techniques send the shared counters with a message to the destination node. General encryption can be used as a simple form of confidentiality, but is not sufficient to protect messages; whereas, *semantic security* offers far greater security by making it harder for an adversary to derive the original data even after obtaining one or more encrypted messages. In DSNs, sending messages with a counter can cause overhead; but, the energy can be saved by sharing the counter between both nodes and incrementing it each time the destination node receives a message. As with other schemes, for better security the same keys should not be used again and again. In SNEP, independent keys are used for encryption and *MAC* operations. The secret key shared between source node A and destination node B is used for deriving the encryption and *MAC* keys for each direction. The encrypted data has the form $E = D(K, C)$ where D is the data, K is the encryption key and C is the counter. The MAC is $M = (K', C||E)$.

In SNEP then, the total message that node A sends to B is: $A \rightarrow B : D(K_{ab}, C_a)$, $MAC(K'_{ab}, C_a||D(K_{ab}, C_a))$.

The *semantic security* property is satisfied as each time the message is encrypted, the counter value is incremented to a different value; thus, though the same message is encrypted, an adversary would not be able to decode the message.

With SNEP, an adversary does have a chance of performing a DOS attack by constantly sending the requests for counter synchronization, but this can be prevented either by sending the counter value with each encrypted message or by attaching a short MAC to the message that does not depend on the counter. Data authentication is done using the *MAC*. The counter value in the message prevents an adversary from replaying old messages, which would cause confusion and overhead in a DSN. As the counter value is kept at both ends of communication and the ID is not transferred with every message, communication overhead is negligible. The counter scheme also allows achieving weak freshness. If the counter value is verified correctly, it reveals the sequence of the messages, but only guarantees the sequence of messages, not that the reply from node B is caused by the message from A. To achieve strong freshness that includes delay estimation, a nonce must be included with messages. To achieve strong freshness, node A sends a nonce N_a along with a reply message to B, which resends the nonce with a reply message. This process can be optimized by implicitly using the nonce in the MAC computation; therefore, the entire SNEP protocol with strong freshness is: $A \rightarrow B : N_a, R_a$ and $B \rightarrow A : \{R_b\}_{(K_{ba}, C_b)}, MAC(K'_{ba}, N_a||C_b||\{R_b\}_{(K_{ba}, C_b)})$

If the *MAC* correctly verifies, node A will know that the reply from B is a reply to its message. In this method, it is assumed that both communicating parties know the counter value so that it need not be sent with every message; though, in reality, messages might get lost or tampered and cause inconsistencies in the counter value. Protocols needed to synchronize the counter value include bootstrapping the counter value in the following manner: $A \rightarrow B : C_a$ with $B \rightarrow A : C_b, MAC(K'_{ba}, C_a||C_b)$ and $A \rightarrow B : MAC(K'_{ab}, C_a||C_b)$

The counter value need not be encrypted since the protocol needs strong freshness for which both communicating parties use the counter as nonce. Also, *MAC* need not include the names A and B as the keys they use, K_{ab}, state which nodes are participating in the communication. If node A realizes that the counter C_b of node B is not synchronized, it may request the counter of B with a message including N_a for strong freshness, or $A \rightarrow B : N_a$ and $B \rightarrow A : C_b, MAC(K'_{ba}, N_a||C_b)$.

5.2. μTESLA: Authenticated broadcasts

Authenticating broadcasted data is a critical issue in DSNs, but previous solutions to this problem suffer from too much communication and computation overhead, and therefore, are not so useful in resource-constrained DSNs. TESLA, one of these solutions, provides an inefficient scheme for broadcasting data with authentication by using the digital signatures technique, which adds 24 bytes of overhead to each message that are typically only allotted a pocket size of 30 bytes. Thus, using TESLA can cause almost all of the packet size to be occupied for the code only. Also, TESLA discloses the key with every message packet it sends and receives, which can use a great deal of a DSN's energy. Finally, TESLA authenticates keys using a one-way key chain, which is not possible to be stored in each sensor node. Perrig *et al.* modified TESLA for authenticating broadcasted data in a way that involves no significant overhead. Called μTESLA, this innovative method reduces energy needed by authenticating data using asymmetric mechanisms. Also unlike TESLA, which discloses the key every time a packet is sent or received, μTESLA does so only once in an epoch. The only limit with μTESLA is that it restricts the number of authenticated senders as it is expensive to store the one-way key chain in a sensor node.

μTESLA is able to provide the asymmetric cryptographic type of authenticated broadcast through delayed disclosure of symmetric keys. For broadcasting authenticated information between the base station and nodes of a DSN, μTESLA requires that the base station and nodes are loosely time synchronized and that each node knows an upper bound on the maximum synchronization error. When the base station wants to sends a packet to all the nodes in a given network, it computes a *MAC* on the packet beforehand. Since all the nodes in the network are sure that only a base station can compute the *MAC*, the *MAC* key is not disclosed at this point in time so they will not be vulnerable to attacks from an adversary. The packets sent to the nodes are stored in their buffers until the base station discloses the corresponding keys. Once disclosed, the keys can be authenticated by the nodes' using the one-way function *F*. If a key is correct, a node can use it to authenticate the packet stored in its buffer.

Each *MAC* key is a sequence of keys generated by the function F. The sender chooses the last key K_n of the chain randomly and then generates the one-way key chain by repeatedly applying F. Supposing the base station has sent packets P_1 and P_2 in the time interval t_1, P_3 and P_4 in t_2, P_5 in t_3, and P_6 in t_4, the nodes receiving the packets cannot verify their authentication immediately, so the nodes store them in buffers. Packets sent in a particular time interval are authenticated using the key that corresponds to that time interval. Let the difference of the time interval be two in this case, the receiver node is loosely time synchronized with the base station and knows key K_0. Assuming that all the messages sending the key information about packets $P_1 - P_5$ are lost and only the message that carries the key information about packet P_6 arrives, the receiver node can still authenticate the keys of the other packets by deriving the key information supplied for P_6. Thus, though some of the packets may have been lost, the nodes can still authenticate them using the keys received. To do this, μTESLA has multiple phases that perform a particular job each, including Sender Setup, Broadcasting Authenticated Packets, Bootstrapping New Receivers, and Authenticating Broadcast Packets.

- *Sender Setup*: In this phase, the sender wanting to broadcast messages in the DSN generates a one-way key chain, randomly selects the last key K_n, and generates the other values by applying the one-way function F on the chain for generating a length n. As F is a one-way function that any node can compute, the keys are generated forward but not backward; i.e., given $K_{j+1}, K_0, ..., K_j$ can be computed, not K_{j+2}.

- *Broadcasting Authenticated Packets*: The sender of the packet uses the particular key for the corresponding time interval. For example, in the time interval I the sender uses the key K_1 and in the time interval $I + 1$, the node uses the key K_2. The packets sent in the particular time interval are authorized using the corresponding key. In the time interval $(I + X)$, the sender reveals the key K_1.

- *Bootstrapping New Receivers*: The keys in a one-way key chain are self-authenticating. If the receiver has one key in the key chain it can efficiently authenticate the other keys in the chain. If the receiver has value K_j in the key chain, it can easily authenticate K_{j+1}. Also, the sender and the receiver are required to be loosely time synchronized with the receiver having the knowledge of the sender's time disclosure schedule. Authenticating the key chain and having loose time synchronization establishes strong freshness and point-to-point authentication. A receiver R sends a nonce N_R in the request message to the sender S, which replies to the message containing the following components: $T_S \rightarrow$ the current time of the sender, $K_i \rightarrow$ a key in the one way key chain, $T_I \rightarrow$ starting time, $T_{int} \rightarrow$ duration of time interval, and $\delta \rightarrow$ disclosure delay. The secret key shared between the node and the base station is used as the key for the MAC.

- *Authenticating Broadcast Packets*: An adversary sometimes knows the key used in the time interval I and may also have knowledge about the one-way key chain, so the receiver should ensure that the packet received is from an authenticated sender and not from an adversary before the key is released by that sender. This is achieved through loose synchronization of the sender and receiver. If the packet is legal, the receiver stores it; if it is spoofed, it is dropped. Once the receiver verifies the key, it authenticates the packets with the key and replaces that new key with the key it already has.

5.3. Considerations for SPINS

SPINS is one of the more efficient schemes for DSNs with advantages including smaller code size, efficient performance, universal design, and low overhead. The scheme uses less of a sensor node's memory; i.e., while crypto routines occupy 20 percent of the space, μTESLA occupies 574 bytes and 2 Kbtes is the acceptable total used memory. The scheme's performance is also efficient, as the bandwidth of the DSNs is adequate for the cryptographic primitives which SPINS uses. Additionally, most of the SPINS design is universal and can be used in other networks of low-end devices. Finally, the communication costs for SPINS are small, with security properties like data freshness, authentication and confidentiality only adding an overhead of 6 bytes in a 30-byte packet, which allows for inclusion in each and every packet. SPINS can offer even greater advantages when restrictions on bandwidth and memory are slightly relieved.

Broadcasting and authenticating data are not that easy for individual nodes, as storing a one-way key in a node's memory is not possible, computation of the keys using a function generates much network overhead, and each node does not share a common key with every other node in the network. However, there are two solutions for this problem. Firstly, the base station is used by a node to transmit all data that has to be broadcasted to other nodes. Secondly, the node broadcasts the data to the base station while the base station generates the authenticating keys using the one-way key chain. It is efficient to implement the cryptographic primitives in a single block cipher as DSNs are resource-constrained and, therefore, can not afford additional overhead for security. Yet, a strong cryptographic base is necessary for SPINS.

- *Block Cipher*: Using RC5 can be very efficient in DSNs because of its small size and high efficiency. Moreover, as an algorithm it has been subject to scrutiny under many attacks. Using TEA could also work for block ciphers, but it is not subject to cryptanalysis scrutiny. DES and other algorithms are not usable for block ciphers due to their large size and high computation requirements that cannot be met in DSNs.

- *Encryption function*: The counter (CTR) mode of block ciphers can use the same function for both encryption and decryption, and the size of the cipher text is the same as the data in this mode. These two properties make this mode very useful while working in the encryption function of SPINS. Also, CTR mode offers semantic security, which is a strong cryptographic property already discussed. To use the CTR mode, both the sender and receiver nodes must maintain counters in their memory and possess an efficient way to synchronize the counters if needed. One advantage of maintaining a counter at both ends is that the messages now will not have an overhead of carrying the counter with them.

- *Freshness*: Using a counter and incrementing it every time a message is sent automatically provides weak freshness. For strong freshness, the sender must create a nonce and should include it in the request message to the receiver. SPINS uses a MAC function for generating random numbers and a counter is created to keep track of those created.

- *Message authentication*: Not only is a good encryption function necessary for data, but also a secure *MAC* is needed. As the block cipher is used more than once, *CBC-MAC* is used for *MAC*. An efficient way of message construction must be used to achieve authentication and message integrity. The construction $\{M\}_k, MAC(k', \{M\}_k)$, in which M is the data, K is the encryption key, and K' is the *MAC* key, is secure and protects the nodes from decrypting erroneous ciphered text.

6. LEAP: EFFICIENT SECURITY FOR LARGE-SCALE DSNS

One of the important mechanisms in sensor networks, in-network processing, is not considered in the previous schemes. This critical issue must be handled while dealing with the resource-constrained property of DSNs. Most of the data has to be collected by an aggregator node and then passed on to other nodes in a DSN; however, data fusion through in-network processing can be used to save net- work energy and reduce communication overhead. The key establishment techniques that have been discussed so far do not support an In-Network Processing approach because the nodes in this method are unable to communicate with each other before transmitting data. Passive Participation is a form of In-Network Processing in which a sensor node takes certain actions based on messages from other nodes. Zhu, Setia, and Jajodia [6] devised a scheme called LEAP that would allow for data fusion, In-Network Processing and Passive Participation.

Besides offering basic requirements like confidentiality and authentication, LEAP supports various communication patterns, including unicast (addressing a single node), local broadcast (addressing a group of nodes in a neighborhood), and global broadcast (addressing all the nodes in a DSN). Sometimes DSNs are deployed in an adversary's arena and, where most of the time compromised nodes are undetected, LEAP provides survivability such that compromising of some nodes does not cede the entire network. LEAP is energy efficient since it supports techniques like In-network Processing and Passive Participation that greatly reduce network communication overhead and, in turn, increase node battery life. Furthermore, LEAP ensures that messages transferred are not fragmented, which would increase packet losses in transmission as well as make protocol implementation more complex and difficult.

LEAP is based on the theory that different types of messages exchanged between nodes need to satisfy different security requirements. All the packets transferred in a sensor network need to always be authenticated where a sensor node knows the sender of the data since an adversary may attack a DSN with false data at any time. On the other hand, confidentiality, like encryption of packets carrying routing information, is not always needed. Different keying mechanisms are necessary to handle the different types of packets. For this, Zhu, Setia, and Jajodia [6] establish LEAP with four types of keys that must be stored in each sensor: individual, pairwise, cluster, and group. Each key has its own significance while transferring messages from one node to another in a DSN and by using these keys, LEAP offers efficiency and security with resistance to copious attacks such as the worm hole and the sybil.

- *Individual Key*: This is a unique key that is shared between the base station and each sensor node. Sensor nodes use this key to calculate the MACs on their messages to the base station like alert signals (reports on abnormal nodes). In the same way, a base station can use an individual key to send messages to each and every node in the network.

- *Pairwise Shared Key*: This is a unique key that is shared between each node and its neighboring node in the network. A node can use it to transfer individual messages like sharing a cluster key or sending data to an aggregator node.

- *Cluster Key*: This is a key that is shared between a node and its neighboring nodes, and is very important since it supports In-network Processing and Passive Participation. A node may elect not to send a message to the base station if its neighboring node is sending the same message with a better signal, a discovery that is only possible to implement if a node shares a common key with its neighboring nodes. With such a cluster key, a node can select which messages to transfer, thereby reducing the system communication overhead.

- *Group key*: The base station shares this key with all the nodes in the network to send queries to them. Group key used requires an efficient rekeying mechanism for updating it as there is a chance for an adversary to know the key whenever a node is compromised.

6.1. Efficiently Establishing LEAP[6]

Establishing Individual Keys

Every node in a DSN shares a unique key with the base station that is preloaded into each node's memory before being deployed. The individual key K_u^m for node U is calculated as $K_u^m = f_{K^m}(u)$. For this, f is a pseudo-random function and K^m is the master key known only to the controller. There is no need for the base station to store all the individual keys, because the base station generates them on the fly whenever it attempts to communicate with a node

Establishing Pairwise Shared Keys

The most common key used in a DSN is the pairwise that is shared between each node and its neighboring node. Sensor nodes are randomly scattered in an area; therefore, the key establishment technique used should guarantee that nodes discover neighboring nodes when deployed. Because sensor nodes are static, the key establishment technique does not have to consider deployment knowledge of others before node deployment. When an adversary obtains a sensor node, it is assumed that the node cannot be compromised before time t_{min}. Whenever a node is deployed in a DSN, it requires some minimum time to identify neighbors and establish keys with them, which will be t_{est}. It is expected that $t_{min} > t_{est}$; otherwise, the adversary could easily capture all the nodes in the DSN and effectively take over the entire system.

The process of establishing keys when nodes are already deployed is similar to the process of key establishment when a new node is added to the network. There are four stages that represent the key establishment of new node U deployed in the network: key predistribution, neighbor discovery, pairwise key establishment, and key erasure. During the initial stage of key predistribution, node U is loaded with the key K_i by the controller and derives the master key K_u using it. For neighbor discovery, node U first initializes a timer to activate at time t_{min}, then starts communicating with its neighbors by broadcasting a HELLO message containing its ID. Node V responds to this message with a reply containing its ID. The ACK of V is then authenticated using its master key K_v derived from K_i. Node U verifies the authentication of V by generating the master key K_v as node V shares K_i with it: $U \rightarrow * : U$ and $V \rightarrow U : V, MAC(K_v, U|V)$.

For the third stage of pairwise key establishment, node U computes the pairwise key K_{uv} with node V using V's identity. Node V can also do the same thing with U. There is no need for authenticating node U to V as any future messages authenticated with K_{uv} will prove node U's identity. In the fourth and final stage, key erasure, node U erases K_i and all the master keys of the other nodes after the time expires. Then node U will not be able to establish pairwise keys with any other nodes in the DSN so that, though an adversary captures a node, the communications between it and another node cannot be decrypted without the key K_i.

Pairwise Shared Keys do have computational overhead since each node U in the network must verify the MACs generated by neighboring nodes and each must reply with a message including its identity and an MAC. Each node must also generate a pairwise key between every other neighboring node in the network. Because a HELLO message includes only a node ID and an *ACK* message has only an ID and an *MAC*, both can be adjusted in a single packet. Also the space required for storing a preloaded is only one key K_i; therefore, the communication and storage overheads are small.

The HELLO message in our scheme is not authenticated, so an adversary may try to attack the network by constantly sending these messages, which will drain a DSN's resources. There are two solutions for this attack: the controller may try to load each new node with the group key of the network so that the nodes can verify the authentication of the message by verifying the group key in the message, or else the controller might try to add some randomness into the IDs of the newly added nodes such that false ones will be detected and dropped. The assumption made here is that the sensor nodes are able to permanently eliminate the master key K_i from their memory, which may not be possible in all cases. One of the unique advantages of the scheme above is that once pairwise keys are established between neighboring nodes in an area of a DSN, they cannot be established again, which protects the network from clone attacks. In clone attacks, an adversary tries to attack the network by installing a number of nodes with keys acquired from compromised nodes, which then establish pairwise keys with other nodes in the network and compromises the entire DSN with just a few nodes. The scheme stated above restricts this kind of attack to a local area as the cloned nodes cannot establish pairwise keys with other nodes in the network that are not neighboring nodes of the one compromised.

The security of the above scheme can be even increased by regularly changing the master key K_i. If an adversary compromises the sensor node before the establishment time, the master key K_i can be obtained and then the whole network can be compromised. By changing the master key regularly, not only is this attack averted, but also attacks caused later by the same adversary are averted, as the master key could still be acquired by compromising a node and deriving the key from its memory. There is one more security threat that must be addressed here: if an adversary attacks the DSN and succeeds in compromising the nodes before key establishment time t_{est}, the network can then be attacked through new nodes added into the network using the correct master key. This problem, however, can be solved. Suppose that the controller wants to add N_i nodes into the network in the time interval T_i, N_i ID's are generated for the nodes based on a random seed S_i and each of the N_i nodes is loaded with a unique ID. The nodes can now establish pairwise keys as stated and when the controller later broadcasts N_i and S_i into the network using a broadcasting scheme like μTESLA, the nodes verify whether those attempting communications are valid or not based on N_i and S_i. The pairwise keys of nodes that are not valid are then deleted from the memory of all the nodes. Compared to other approaches, this method offers greater efficiency and smaller overhead, while also protecting the network from clone attacks and other serious attacks.

Establishing Cluster Keys

The cluster key establishment is based on the pairwise key establishment. If node U wants to establish a cluster key with its neighbors $v_1, v_2, v_3, \ldots, v_n$, first it generates a key K_c and then encrypts that key using the pairwise key which it shares with each neighbor. Node U then transmits this encrypted message to its neighbors. Node v_1 decrypts the key using the pairwise key which it shares with U, and then stores the key in a buffer. Next it sends back its own cluster key to node U. When any of the nodes are revoked, node U generates a new cluster key in the same way and transmits the key to all remaining nodes.

Establishing Group Keys

A group key, which is shared between a base station and all the nodes in a DSN, is needed when the base station wants to send a message or query to all the nodes of that DSN. One way of achieving this is using the hop-by-hop method in which the base station encrypts messages using the cluster key which it has and then broadcasts the message to all the nodes in its neighborhood. The nodes would decrypt the message and then encrypt it using the cluster key which they share with their neighbors. In this way, the message can be received by all the nodes in the network. This is efficient, but has an overhead of encryption and decryption at every node.

A simple method to establish a group key is to preload each node with the group key before deployment, but this is still within the scope for rekeying the group key which will be necessary. Unicast-based group rekeying can also be considered for which the base station needs to send the group key to each node in the network, but this involves much communication overhead. However, Zhu, Setia, and Jajodia [6] proposed an efficient scheme based on cluster keys in which the transmission cost will only be one key. In DSNs, all messages sent by the base station must be authenticated or an adversary may impersonate it. The group key must be updated every time when a node is revoked. Therefore the first issue to consider is how node revocation can be done in this scheme. μTESLA, based on a one-way key chain and delayed disclosure of keys, is an efficient method to broadcast messages into a DSN, as previously discussed. To bootstrap μTESLA, each node should be preloaded with the commitment of the key chain. If K_g is the new group key and U is the node to be revoked, the base station broadcasts the following message:

$$M : Controller \rightarrow * : u, f_{K_g'}(0), MAC(k_i^T, u|f_{K_g'}(0)) \qquad (16)$$

, where $f_{K_g'}(0)$ is the key that enables the node to verify the authentication of the group key. The server then distributes the MAC key k_i^T after one μTESLA interval. After a node V receives the message M, it verifies the authenticity of the message using μTESLA. If node V is neighbor of U, V will remove its pairwise key shared with U and update its cluster key. This process can also be used for updating the group key.

For secure key distribution, this scheme uses a protocol that is the same as the beaconing protocol for which all the nodes are organized in a breadth-first spanning tree where each node not only remembers the parent and children of a spanning tree, but also the other neighbors. The new group key K_g' is distributed to all the nodes in the network using the spanning tree established by the routing protocol. The base station initiates the process by sending the group key to all its neighbors in the network. The nodes that receive the message verify its authentication by calculating $f_{K_g'}$ and by checking whether it is the same as the verification key received earlier in the node revocation message. The algorithm continues recursively down the spanning tree with the help of each node, which transmit the group key to neighbors while encrypting the message using their cluster keys. As discussed earlier, this hop-by-hop scheme does not involve much overhead as only one key is encrypted and decrypted, and as rekeying of the group key is infrequent. However, it is desirable to change the group key more often or an intruder may compromise the entire DSN by obtaining one node and, thereby, deriving the group key.

6.2. Local Broadcast Authentication

Local broadcast is different from global broadcast in that in local broadcast a node generally does not know what packet it is going to generate next and messages generally consist of aggregated sensor readings or routing protocols. μTESLA is not suitable for local broadcasting because μTESLA does not provide authentication immediately, which is needed in some local broadcast cases. Also, in μTESLA nodes need to keep the packets in their buffers until the authenticating key arrives, which increases the storage space required. A packet that has to travel L nodes will at least need L μTESLA intervals, thereby affecting latency of the network. Pairwise keys cannot be used for local broadcast because, if a node has n neighbors, the approach requires the sender node to calculate n *MACs* for each message. Local broadcast needs a method where a node can broadcast a message to all its neighbors using a single *MAC* and cluster keys, with a problem as follows. If an adversary can compromise a node, the cluster key from that node is available and can be used to attack the network by impersonating that node or a neighboring node. If nodes X, U, and V are three vertices of a triangle, X is compromised, and U wants to send messages to X and V, X can use node U's cluster key to impersonate it and send false messages to V.

Fortunately, Zhu, Setia, and Jajodia [6] have designed a scheme called One-Way Key Chain-Based Authentication for defeating this attack totally. This scheme is based on μTESLA in that each node generates a one-way key chain and sends the commitment of it to their neighbors. This transferring is done using the pairwise keys already shared with neighbors. If a node wants to send a message to its neighbors, it attaches the next authorization key from its key chain to the message. The receiving node can verify the validation of the key based on the commitment it has already received. The One-Way Key Chain-Based Authentication is designed based on two observations: a node only needs to authenticate to its neighbors and that a node V will receive a packet before a neighboring X receives it and resends it to V. This observation is true because of the triangular inequality among the distances of nodes involved. An adversary may still try to attack the nodes by shielding node V while U is transmitting a message, and then later send a modified packet to V with the same authorization key; but this attack can be prevented by combining the authorization keys with the cluster keys. When this is done, the adversary does not have the cluster key and so cannot impersonate node U. However, this scheme does not provide a solution for attacks from inside where the adversary knows U's cluster key.

6.3. LEAP Performance Evaluation

Overhead

Since the performance of the pairwise key establishment have already been discussed, here we review other factors of performance like the computational cost, communication cost and storage requirement of this approach. As mentioned, a cluster key is established based upon the pairwise keys of a node. Let us suppose that the number of neighbors to a node is n; if the cluster key has to be updated the scheme must perform

n encryptions, which is computationally expensive. The value of n depends upon the density of the scheme; the computational cost increases as the network is denser. While for securing distribution of a group key, the number of decryptions is equal to the size of the network. The total number of encryptions is also equal to the size of the network; so if the size of the network is M, the total number of symmetric operations will be $2M$. From these derivations, the computational cost of the scheme is dependent upon the density of the network d. Zhu, Setia, and Jajodia [6] stated that the average number of symmetric operations of the scheme is about $2(d-1)^2/(M-1)+2$. If the density of the network is reasonable, the computational cost may not be a bottleneck to the scheme.

Also, the cost decreases with the increases of M. The communication cost of the scheme is the same as the computational cost. The average number of keys a node has to transfer for updating keys due to revocation is $(d-1)^2/(N-1)+2$. Just like computational cost, communication cost increases with a increase in the density of the network and decreases with an increase in the size of the network. The storage requirement of this scheme is a bit high because each node must store four types of keys in it. Considering the degree of node to be d, a node has to store one individual key, d pairwise keys, d cluster keys, and one group key. Also, a node must store a one-way key chain and a commitment for each neighbor for local broadcast. If L is the number of keys stored in a key chain, the total number of keys the node has to store in this scheme will be $3d+2+L$. Again, the storage requirement of LEAP depends upon the density of the network.

Resilience to Attack

An adversary might launch a selective forwarding attack in which a compromised node drops the packets containing the routing information of selected nodes and forwards the other packets normally. LEAP can minimize the affects of the scheme by minimizing this problem to a local area. As LEAP uses local broadcast, the attack's effects will not transfer to more than 2-hops, which will result in defeating the purpose of such an attack. LEAP can also prevent a HELLO attack in which an adversary attacks the network by repeatedly transmitting HELLO messages and thereby depletes the network's resources. This attack is averted since the nodes in a LEAP scheme accept packets only from authenticated neighbors. The sinkhole and wormhole attacks, however, are difficult to solve. In the sinkhole attack, a compromised node attracts packets by advertising information like high battery power, etc., then later drops all the packets. In the wormhole attack an adversary launches two nodes in the network, one near the target of interest and the other near the base station. The adversary then convinces the nodes near the target, which would generally be multiple hops away from the base station, that they are only two hops away thereby creating a sinkhole. Also, nodes that are far away think that they are neighbors because of the wormhole created. In LEAP an adversary cannot launch a wormhole attack after key establishment as at that point every node has knowledge about its neighbors so it is not easy to convince a node that it is near a particular compromised node. An insider node must then succeed in compromising two nodes for creating a wormhole and those nodes must be near the target of interest and the base station after the key establishment phase is complete. Although an adversary may try, it is difficult to create an attractive sinkhole without being detected.

LEAP includes efficient protocols for supporting four types of key schemes for different types of messages broadcasted in DSNs and includes an efficient scheme for local broadcast authentication. LEAP is an efficient scheme for key establishment that resists many types of attacks on the network, including the sybil, sinkhole, wormhole, and so on. LEAP also provides efficient schemes for node revocation and key updating in DSNs.

7. A KEY MANAGEMENT SCHEME FOR WIRELESS DSNS USING DEPLOYMENT KNOWLEDGE

Throughout the discussion in this chapter, a significant piece of information regarding DSNs has not yet been mentioned, i.e., the deployment knowledge of these networks. As sensor nodes are randomly deployed in an area, it is difficult to obtain deployment knowledge. Some information on deployment knowledge is achievable if deployment followed a particular order. For example, if sensor nodes are scattered using an airplane pattern, these nodes might be grouped or placed in a particular order before deployment and, based on this pattern, an approximate knowledge of node position can be acquired.

Deployment knowledge offers numerous advantages when used in DSNs such as achieving better storage, better resilience to node capture and more. In their study of key establishment techniques in sensor networks, Du, Deng, Han, Chen, and Varshney [8] propose a scheme using deployment knowledge that is based on the Basic Scheme, which has already been discussed earlier in this chapter. Deployment knowledge in this scheme is modeled using probability density functions (pdfs). All the schemes discussed until now considered the pdf to be uniform; and when uniform, knowledge about the nodes can not be derived from it. In LEAP, Du *et al.* consider non-uniform pdfs, which means that they assume the position of sensor nodes to be at certain areas. Their method first models node deployment knowledge in a DSN and then develops a key predistribution scheme based on this model.

7.1. Modeling of the Deployment Knowledge

The *deployment point* and the *resident point* are two terms that must be briefly understood when discussing the Deployment Model. *Deployment point* of a sensor node is the point at which the sensor node is actually deployed; i.e., the node is dropped where the deployment is done through an airplane deployment point. *Resident point* is the point at which the sensor actually resides after deployment. Let us assume the deployment area to be a 2-dimensional region $X \times Y$. The pdf for node I, for $I = 1, ..., N$ over the two-dimensional area is found by $f_i(x, y)$, where $x \in [0, X]$ and $y \in [0, Y]$. Generally nodes are deployed in groups, therefore the pdfs of the final resident points of all the sensors in a group is the same as the group of sensors deployed in a single deployment point. The group deployment model is designed as following in Du *et al.* [8]:

- N sensor nodes that are deployed in a place are divided into $t \times n$ equal size groups. Each group $G_{i,j}$ for $i = 1, ..., t$ and $j = 1, ..., n$ is from the deployment point with index (i, j); and (x_i, y_j) is the deployment point for this group.

- The Deployment Model follows a grid-based approach with all deployment points arranged in a grid.

- The pdf of the resident points for node K in group $G_{i,j}$ is $f_K^{ij}(x, y|K \in G_{i,j}) = f(x - x_i, y - y_i)$.

Two groups that are deployed close together share some common keys. The amount of key overlap decreases as the deployment distance between the groups increases. When using the Basic Scheme, keys are drawn from the same key pool S; but, using the Deployment Model, different key pools are allowed for different groups so that the key pool can be divided into sub-key pools of $|S_c|$ keys each. The combination of all the sub-key pools still yields S.

Sensor nodes can be deployed in many different ways such as deployment through an airplane, using a vehicle, etc. In this scheme, deployment is considered as a Gaussian distribution, which is widely studied and practiced. The deployment distribution for any node k in group $G_{i,j}$ follows a two dimensional Gaussian distribution. The pdf of the resident points for the node k in group $G_{i,j}$ is [22]

$$f_k^{ij}(x, y|k \in G_{i,j}) = f(x - x_i, y - y_j) \tag{17}$$

When $f(x, y)$ is uniform, we cannot determine which nodes are close together prior deployment as the resident points of the nodes are uniformly distributed over the region. When $f(x, y)$ is random we can tell which nodes are close together. Though the distribution function is not uniform, the sensor nodes still need to be deployed evenly through the entire region. By selecting an appropriate distance between deployment points, the probability of finding a node in each small region can be made equal.

7.2. Key Predistribution Using Deployment Knowledge

In a key pre-distribution scheme based on the Deployment Model, it is assumed that N sensor nodes are deployed in a place (point) and are divided into $t \times n$ equal size groups, each group $G_{i,j}$ for $i = 1, ..., t$ and $j = 1, ..., n$. It is also assumed that the deployment points are arranged in a grid.

As with the Basic Scheme, key predistribution in the Deployment Model also consists of three phases: key predistribution, shared key discovery, and path key establishment. This scheme differs only in the first stage while the other two stages are similar to that of the basic scheme.

- *Key Predistribution*: The most important step in this phase is to divide the key pool into $t \times n$ key pools. The goal of dividing the key pools is to ensure that neighboring key pools have more keys in common. Two key pools are neighbors if their deployment groups have nearby resident points. The concept of dividing the key pool is discussed briefly in the next section. After the key pool is divided, each node in a group is selected and keys are installed from corresponding subset key pools.

- *Shared Discovery Phase*: In this phase each node must find its common keys shared with neighbors. There are many ways for doing this, but the simplest is to make the nodes broadcast their identifiers list to other nodes. If the nodes discover that they share a common key with other nodes, this key can be used as their communication link. When disclosing the nodes' identities is not desired, Merkle's Challenge Response Technique can be employed [9] in which each node sends a puzzle to neighboring nodes for each stored key and, if those nodes share a key in common with the source node, they will respond with the correct solution creating a key link for secure communication.

- *Path Key Establishment*: When two neighboring nodes do not share a common key, they can discover one using Path Key Establishment. If node U wants to communicate with V and the two do not share a common key, node U must communicate with its neighbor I, saying that it wants to communicate with V. Node U then sends its ID and a secret key to node I and, if I shares a common key with V, it sends the message to V encrypted with that shared key. Through this path U $\rightarrow$ I $\rightarrow$ V or V $\rightarrow$ I $\rightarrow$ U, both nodes U and V can communicate with each other using a secret key.

7.3. Creating Key Pools

Key pools that are deployed nearby should share certain keys in common. To assign keys to each key pool $S_{i,j}$ for $i = 1, ..., t$ and $j = 1, ..., n$, it is assumed that the pools are deployed in a grid: (a) key pools that are horizontal or vertical share $a|S_c|$ keys, where $0 \leq \alpha \leq 0.25$; (b) Key pools that are diagonal share $b|S_c|$ keys, where $0 \leq b \leq 0.25$ and $4a + 4b = 1$; (c)Two non-neighboring key pools share no keys.

Here, (a) and (b) are overlapping steps and to achieve the properties stated, the key pool is divided into eight total partitions, each with keys that are shared by the other nodes. Du *et al.* [8] developed a method to select keys for each subset key pool $S_{i,j}$, considering a grid scheme and given a global key pool S (the subset of key pools for each deployment point). The keys for the first subgroup (the group placed in the first row and first column) $S_{1,1}$ are selected from the global key pool S, and then keys for the second group in the same row are selected from the row left to it and S. This process continues for each row from left to right:

- Select $|S_c|$ keys for the group placed in $S_{1,1}$ and remove those keys from S;
- Select $a|S_c|$ keys for group $S_{1,2}$ from the key pool $S_{1,1}$, and the remaining keys w from the global key pool S, and then remove the selected w keys from S.
- Select $a|S_c|$ keys for group $S_{2,1}$ from the key pools $S_{1,1}, S_{3,1}, S_{2,2}$ and select $b|S_c|$ from the key pools $S_{1,2}$ and $S_{3,2}$. Then select and remove the remaining w keys from the global key pool S.

If w is the remaining keys that are to be selected from S, and $S_{i,j}$ be the group number in a grid, the selection procedure for different groups will be:

$$W = \begin{cases} [1-(a+b)] \times |S_c|, \text{for} j = 1 \\ [1-2\,(a+b)] \times |S_c|, \text{for} 2 < j \leq n-1 \\ [1-(2a+b)] \times |S_c|, \text{for} j = n \end{cases} \tag{18}$$

Selecting the size of the key pool $|S_c|$ from S is also critical in this scheme. To ensure that no key in the network is shared between more than two nodes, a rule must be established: if a group G_1 selects the required keys from its neighbor G_2, no other group is allowed to select those keys. Since each group selects distinct keys from their neighbors, the size of $|S_c|$ is equal to the sum of all those keys and is calculated as

$$|S_c| = \frac{|S|}{tn - (2tn - t - n)a - 2(tn - t - n + 1)b} \tag{19}$$

In their paper, Du *et al.* [8] analyzed the performance of their scheme based upon the connectivity, communication overhead, and resilience against node capture of the network. Two parameters for connectivity called the Global Connectivity and Local Connectivity are defined for analyzing the performance of connectivity in the Deployment Model. Global Connectivity is the ratio between the nodes that are isolated and the total number of nodes in the network. Local Connectivity refers to the probability of any two nodes sharing at least one key. Both Global and Local Connectivity are affected by the key predistribution scheme where $B(n_i, n_j)$ is the event that two nodes share at least one key, and $A(n_i, n_j)$ is the event that both the nodes are neighbors, hence

$$p_{local} = \Pr(B(n_i, n_j)|A(n_i, n_j)) \tag{20}$$

Let λ be the ratio of the shared key pool between two nodes and $|S_c|$. Du *et al.* [8] calculated the probability $p(\lambda)$ that two nodes share at least one key when they have $\lambda|S_c|$ in common as

$$p(\lambda) = 1 - \Pr(\text{two nodes do not share any key}) \tag{21}$$

Global Connectivity is also required for better efficiency. The key sharing in a wireless DSN the using Deployment Model is not uniform, so uniform distribution techniques like Erdos' Random Graph Theory cannot be applied. Shakkottai *et al.* [23] determined the connectivity of a wireless network with unreliable nodes that can be used for random distribution. The authors simulated the results for Global Connectivity using m for the number of nodes in a node's key ring, which showed that when $m = 200$, no nodes are wasted due to a lack of security links while only 0.12 percent of nodes are wasted when $m = 100$. The overlapping factors a and b has to be selected carefully, when $a = 0.25$ and $b = 0$, each group shares keys only with horizontal and vertical neighbors only; when $a = 0$ and $b = 0.25$, each group shares keys only with diagonal neighbors only. The values of a and b depend upon many factors like the size of the nodes key ring m, different types of neighbors, etc.

8. CONCLUSIONS

Key management for Distributed Sensor Networks (DSNs) is a critical issue that has been addressed through several proposed schemes presented in various papers. This chapter provides an overview of these techniques, each of which offers different advantages and disadvantages. A balance between the requirements and resources of a DSN determines which key management scheme should be employed. A DSN used in a battlefield demands more security than one used in places like shopping centers; also, the former can be made more costly but the later needs to be as cheap as possible. Decisions regarding the key management scheme to be used must be based on these requirements for efficiency.

8.1. Summary of Schemes Pros and Cons

The Basic Scheme

- *Pros*: flexible, efficient, and fairly simple to employ, while also offering good scalability.

- *Cons*: cannot be used in circumstances demanding heightened security and node-node authentication.

The Q-Composite Scheme

- *Pros*: provides better security than the Basic Scheme by requiring more keys for two nodes to share one for communication, which makes it difficult for an adversary to compromise a node.

- *Cons*: vulnerable to breakdown under large-scale attacks, does not satisfy scalability requirements.

Multipath Key Reinforcement

- *Pros*: offers better security than the Basic Scheme or the Q-Composite.

- *Cons*: creates communication overhead that can lead to depleted node battery life and to the chance for an adversary to launch DOS attacks.

Random Pairwise Scheme

- *Pros*: offers the best security of all the schemes with perfect resilience against node capture as the keys used by each node are unique, and also provides resistance against node replication.
- *Cons*: does not support networks of large size, and does not satisfy scalability requirements.

Polynomial Pool-Based Key Predistribution

- *Pros*: allows the network to grow to a larger size after deployment.
- *Cons*: t-collision resistant (compromising more than t polynomials leads to network compromise.

Polynomial Grid-Based Key Predistribution

- *Pros*: offers low communication and computation overhead on the network, guarantees a total connected graph when there are no compromised nodes and all the nodes are in transmission range.
- *Cons*: overhead of node storage as each must not only share the polynomial keys but also the ID's of compromised nodes to avoid attack.

SPINS: SNEP and μTESLA

- *Pros*: one of the best memory efficient schemes discussed, provides strong security features with less complexity, universal design allows use in many low-end devices, incurs low communication cost, and offers authentication and strong data freshness with a minimum overhead.
- *Cons*: μTESLA overhead from releasing keys after a certain delay, possible message delay.

LEAP

- *Pros*: offers efficient protocols for supporting four types of key schemes for different types of messages broadcasted, reduces battery usage and communication overhead through In-Network Processing, and uses a variant of μTESLA to provide local broadcast authentication.
- *Cons*: requires excessive storage with each node storing four types of keys and a one-way key chain, computation and communication overhead dependant upon network density (the more dense a network, the more overhead)

Wireless DSNs Using Deployment Model

- *Pros*: only to consider deployment knowledge that can minimize the number of keys and help increase resilience or resistance to node capture and reduce network overhead, increases overall connectivity of the network graph, and offers same pros as the Basic Scheme on which it is based

8.2. Future Work for DSNs

More schemes should be developed to make efficient use of sensor nodes' limited resources. Greater emphasis should be given to the security in key management schemes, particularly as a majority of sensor node deployment is in hostile environments where providing strong security features is a must. Though receiving much attention recently, there are many problems to be addressed in DSNs such as finding the compromised nodes in a network, making good use of deployment knowledge, making nodes tamperproof without much overhead, decreasing the bootstrapping time required for the network, and so on. Future research should especially seek techniques for compromised node discovery and efficient methods to revoke compromised nodes.

Numerous lives can be saved in wars with the data collected by sensors, but DSNs in the near future will offer many surprises for humans as they come to be used in daily household matters like locking doors and switching off electronics, or controlling traffic in high-volume areas. Sensors installed in big malls and shopping centers can guide people to their required products easily while those in hospitals can monitor patient condition and those in forests can provide immediate knowledge about disastrous hazards like wildfire. These advantages are only a small fraction of what DSNs could potentially offer when deployed more commonly. Future studies can prove DSNs to be useful in a wider variety of environments.

9. REFERENCES

1. L. Eschenauer and V. D. Gligor. "A key management scheme for distributed sensor networks", *Proc. of the 9th ACM Conference on Computerand Communication Security.*
2. H. Chan, A. Perrig, and D. Song, "Random Key Predistribution Schemes for Sensor Networks", *Proc. of the 2003 IEEE Symposium on Security and Privacy*, May 11-14, pp. 197 - 213.
3. D. Liu and P. Ning, "Establishing pairwise keys in distributed sensor networks," *Proc. of the 10th ACM Conference on Computer and Communications Security (CCS '03)*, pp. 52-61. 2003.
4. W. Du, J. Deng, Y. S. Han, and P. K. Varshney, "A pairwise key pre-distribution scheme for wireless sensor networks," *Proc. of the 10th ACM Conference on Computer and Communications (SecurityCCS'03)*, pp. 42-51, 2003.
5. A. Perrig, et. al, "SPINS: Security Protocols for Sensor Networks,"*Proc. of ACM MOBICOM, 2001.*
6. S. Zhu, S. Setia, and S. Jajodia, "LEAP: Efficient Security Mechanisms for Large-Scale Distributed Sensor Networks," *Proc. of The 10th ACM Conference on Computer and Communications Security (CCS '03)*, Washington D.C., October, 2003.
7. D. Carman, P. Kruus, and B. Matt, "Constraints and approaches for distributed sensor network security," NAI Labs Technical Report No. 00-010, September 2000.

8. W. Du, J. Deng, Y. S. Han, S. Chen and P. K. Varshney, "A Key Management Scheme for Wireless Sensor Networks Using Deployment Knowledge," *Proc. of IEEE INFOCOM 2004.*

9. R. Merkle, "Secure communication over insecure channels," *Communications of the ACM*, Vol. 21, No. 4, pp.294-299, 1978.

10. R. Anderson and A. Perrig, "Key infection: Smart trust for smart dust," Unpublished Manuscript, November 2001.

11. R. Merkle, "Protocols for public key cryptosystems " *Proc. of 1980 IEEE Symposium on Security and Privacy*, 1980.

12. W. Diffie and M. E. Hellman, "New directions in cryptography," *IEEE Trans. Inform. Theory*, IT-22, pp. 644-654, November 1976.

13. C. Karlof and D. Wagner, "Secure Routing in Sensor Networks: Attacks and Countermeasures," *First IEEE International Workshop on Sensor Network Protocols and Applications*, May 11, 2003.

14. A.D. Wood and J.A. Stankovic, "Denial of service in sensor networks," *Computer*, Vol.35, No. 10, Oct. 2002, pp. 54 - 62.

15. C. Karlof, N. Sastry, and D. Wagner, "TinySec: A Link Layer Security Architecture for Wireless Sensor Networks," *Proc. of the Second ACM Conference on Embedded Networked Sensor Systems (SenSys 2004)*. November 2004.

16. F. Ye, H. Luo, S. Lu, and L. Zhang, "Statistical En-route Detection and Filtering of Injected False Data in Sensor Networks," *Proc. of IEEE INFOCOM 2004.*

17. D.W. Carman, B.J. Matt, and G.H. Cirincione, "Energy-efficient and Low-latency Key Management for Sensor Networks," *Proc. of 23rd Army Science Conference*, 2002.

18. M. Chen, W. Cui, V. Wen, and A. Woo, "Security and Deployment Issues in a Sensor Network," Ninja Project, A Scalable Internet ServicesArchitecture, Berkeley,

19. W. Zhang and Guohong Cao, "Group Rekeying for Filtering False Data in Sensor Networks: A Predistribution and Local Collaboration-Based Approach," *Proc. of INFOCOM 2005.*

20. D. Liu, P. Ning, K. Sun, "Efficient self-healing group key distribution with revocation capability," *Proc. of the 10th ACM conference on Computer and communication security*, 2003.

21. J. Lee and D. R. Stinson, "Deterministic key pre-distribution schemes for distributed sensor networks," *To appear in Lecture Notes in Computer Science (SAC 2004 Proceedings)*, 2004.

22. A. Leon-Garcia, *Probability and Random Processes for Electrical Engineering*, 2nd ed. Reading, MA: Addison-Wesley Publishing Company Inc., 1994.

15

SECURE ROUTING IN AD HOC AND SENSOR NETWORKS

Xu (Kevin) Su
Department of Computer Science
The University of Texas at San Antonio
E-mail: xsu@cs.utsa.edu

Yang Xiao
Computer Science Department
University of Alabama
101 Houser Hall
Box 870290
Tuscaloosa, AL 35487-0290 USA
E-mail: yangxiao@ieee.org

Rajendra V. Boppana
Department of Computer Science
The University of Texas at San Antonio
E-mail: boppana@cs.utsa.edu

Both mobile ad hoc networks and wireless sensor networks are gaining popularity due to the fact that they promise viable solutions to a variety of applications. A Mobile Ad hoc NETwork (MANET) consists of a collection of wireless hosts that are capable of communicating with each other without use of a network infrastructure or any centralized administration. A wireless sensor network (WSN) consists of tiny senor nodes with sensing, computation, and wireless communication capabilities. In sensor networks, hundreds or thousands of nodes with low-power, low-cost, and possibly mobile but more likely at fixed locations, collectively monitor an area. However, both MANETs and WSNs are vulnerable to different attacks due to their fundamental characteristics such as open medium, dynamic topology, absence of central administration, distributed cooperation, and constrained capability. Before mobile ad hoc and sensor networks can be successfully deployed, security issues must be addressed. In this chapter we introduce security attacks on MANETs and WSNs and review the recent approaches of secure network routing protocols in both mobile ad hoc and sensor networks.

1. INTRODUCTION

Both mobile ad hoc networks and wireless sensor networks are gaining popularity due to the fact that they promise viable solutions to a variety of applications. A Mobile Ad hoc NETwork (MANET) consists of a collection of wireless hosts that are capable of communicating with each other without use of a network infrastructure or any centralized administration. In such a network, each mobile node can operate not only as a host but also as a router, forwarding packets for other mobile nodes which may not be within direct wireless transmission range of each other. Since MANETs can be easily and inexpensively set up as needed, they have a wide range of applications, especially in situations with geographical or terrestrial constrains such as battlefields, military applications, and other emergency and disaster situations. A wireless sensor network consists of tiny senor nodes with sensing, computation, and wireless communication capabilities. In sensor networks, hundreds or thousands of nodes with low-power, low-cost, and possibly mobile but more likely at fixed locations, collectively monitor an area. These large sensor networks generate a substantial amount of data and send the data to one or more points of centralized control called *base stations* or *sinks*. A base station is typically a gateway to another network, and it can be a powerful data processing or storage center, or an access point for human interface. Networking unattended sensor nodes are increasingly popular to provide economical solutions to both military and civil applications such as real-time traffic monitoring, wildfire tracking, wildlife monitoring, weather monitoring, security and tactical surveillance, target filed imaging, and disaster management etc.

However, recent wireless research work indicates that the ad hoc networks and sensor networks are more vulnerable than conventional wired and wireless networks due to their fundamental characteristics such as open medium, dynamic topology, absence of central administration, distributed cooperation, and constrained capability (power and computation constraints). Use of wireless links makes ad hoc and sensor networks more prone to physical security threats than wired networks, ranging from passive eavesdropping to active interference. Without proper security, mobile hosts and sensor nodes are easily captured, compromised and hijacked by malicious nodes. The adversary might listen to or/and modify the messages in the wireless communication channel, inject erroneous messages, delete messages, and even impersonate other nodes. Some or all of the nodes in a MANET might rely on batteries and all sensor nodes in a WSN have low battery power. The adversary could force a node to replay packets to exhaust its energy.

Before ad hoc and sensor networks can be successfully deployed, security issues must be addressed. Recently significant progress has been made in securing ad hoc and sensor networks via the development of secure routing protocols.

To assess the security needs of ad hoc and sensor networks effectively and to evaluate and choose various policies, we need some systematic way of defining the requirements for security and characterizing the approaches to satisfying those requirements. Usually the following three aspects of information security are considered:

- **Security attack**: any action to compromise the security of information in the system.
- **Security mechanism**: a mechanism that is designed to detect, prevent, or recover from a security attack.
- **Security service**: a service that enhances the security of the data processing systems and information transfers of an system. These services are used to counter security attacks, and they make use of one or more security mechanisms to provide the service.

Our purpose in this chapter is to review the recent approaches to secure routing protocols in ad hoc and sensor networks. The rest of the chapter is organized as follows. We first give an overview of current routing protocols without security enforcement and describe several often used routing protocols on which most recent secure routing protocols are based in Section 3. In Section 4, different security services are introduced. In Section 5, security attacks on ad hoc and sensor networks are classified and some well-known attacks are given. We then give an review on current security mechanisms to secure routing protocols in ad hoc and sensor networks in Section 6. Finally, Section 7 contains concluding remarks.

2. ROUTING PROTOCOLS

During the last decade, extensive research works have been conducted on designing routing protocols in both mobile ad hoc networks and wireless sensor networks, and many efficient routing protocols have been proposed.

In MANETs, the routing protocols can be categorized by the timing of acquisition of routing information and the methods by which the routes are maintained into three categories: reactive protocol, proactive protocol, and hybrid protocol. In the reactive (on-demand) protocols, such as Ad hoc On-demand Distance Vector (AODV) [39], Dynamic Source Routing (DSR) [22], and Temporally Ordered Routing Algorithm (TORA) [37], a route is created only when desired by the source node. In the proactive protocol, such as Destination Sequence Distance Vector (DSDV) [38] and Adaptive Distance Vector (ADV) [3] etc., each node maintains one or more routing tables to store routing information, and any changes in the network topology need to be reflected by propagating updates throughout the network. There are other hybrid protocols, such as Zone-based Routing Protocol (ZRP) [14], which employ both mechanisms.

In WSNs, routing can be divided into *flat-based* routing, *hierar-chical-based* routing, and *location-based* routing depending on the network structure. In *flat-based* routing protocols, such as Sensor Protocols for Information via Negotiation (SPIN) [15, 27], Directed Diffusion [21], Rumor Routing [4], Minimum Cost Forwarding Algorithm (MCFA) [48] etc., all nodes are typically assigned equal roles or functionality. While in *hierarchical-based* routing protocols, such as LEACH protocol [16], Power-Efficient Gathering in Sensor Information Systems (PEGASIS) [28], and Threshold-sensitive Energy Efficient Sensor Network Protocol (TEEN) [30] etc., nodes play different roles.

For example, higher energy nodes can be used to process and send the information while low energy nodes can be used to perform the sensing in the proximity of the target. In *location-based* routing protocols, such as GPSR [26], GEAR [51], SPAN [9], GAF [45], and CEC [46] etc., sensor nodes' positions are exploited to route data in the network. Furthermore, these protocols can be classified into *multipath-based*, *query-based*, *negotiation-based*, *QoS-based*, or *coherent-based* routing approaches depending on the protocol operation.

Since most recent work to secure ad hoc and sensor network routing protocols are based on the several main routing protocols: AODV, DSR, DSDV, and Directed Diffusion, we describe these routing protocols as follows.

2.1. Destination Sequence Distance Vector – DSDV

In DSDV [38] routing, each router maintains a routing table giving the distance from itself to all possible destinations. Each routing table entry consists of destination IP address, the distance to it and the next node in the path. Each router periodically broadcasts this table information to each of its neighbor routers, and uses similar routing updates received from neighbors to update its table. DSDV solves the looping problem by attaching sequence numbers to routing entries. A node increments its current sequence number and includes it in the updates originated at that node. Along with distance information this sequence number is propagated. Any node that invalidates its entry to a destination because of loss of next hop node, increments the sequence number and uses the new sequence number in its next advertisement of this route. A node invalidates or modifies its routing entry if a neighbor broadcasts a routing entry to the same destination with a higher sequence number. An invalid entry can become valid when the node receives an advertisement of this route with the same sequence number (as the one it has) and better metric or higher sequence number. The routing table entries in all the nodes for a given destination collectively specify a virtual destination-based tree to send packets to that destination. A simplistic view of DSDV is that it maintains one such destination tree for each node in a distributed manner. To keep up with network changes, DSDV algorithms use periodic and triggered routing updates. Periodic updates include full routing table and occur once in 30-90 seconds. Triggered updates occur in between periodic updates if enough number of routing entries has changed and often include only newly modified entries. ADV [3] is another distance vector algorithm, which is similar to DSDV but reduces the routing overhead by varying the size and frequency of routing updates in response to traffic variance and node mobility.

2.2. Ad hoc On-demand Distance Vector – AODV

AODV [39] is based upon the distance vector algorithm. The difference is that AODV is reactive, as opposed to proactive protocols like DSDV, i.e., AODV requests a route only when needed and does not require nodes to maintain routes to destinations that are not actively used in communications.

- **Route discovery:** A node broadcasts a RREQ when it determines that it needs a route to a destination and does not have one available. This can happen if the destination is previously unknown to the node, or if a previously valid route to the destination expires. To prevent unnecessary broadcasts of RREQs the source node uses an expanding ring search technique as an optimization. In the expanding ring search, increasingly larger neighborhoods are searched to find the destination. The search is controlled by the time-to-live (TTL) field in the IP header of the RREQ packets. If the route to a previously known destination is needed, the prior hop-wise distance is used to optimize the search.

- **Route maintenance:** Every routing table entry maintains a route expiry time which indicates the time until which the route is valid. Each time that route is used to forward a data packet, its expiry time is updated to be the current time plus ACTIVE ROUTE TIMEOUT. A routing table entry is invalidated if it is not used within such expiry time. AODV uses an active neighbor node list for each routing entry to keep track of the neighbors that are using the entry to route data packets. These nodes are notified with route error (RERR) packets when the link to the next hop node is broken. Each such neighbor node, in turn, forwards the RERR to active neighbors in its own list, thus invalidating all the routes using the broken link.

2.3. Dynamic Source Routing – DSR

A routing entry in DSR [22] contains all the intermediate nodes to be visited by a packet rather than just the next hop information maintained in DSDV and AODV. A source puts the entire routing path in the data packet, and the packet is sent through the intermediate nodes specified in the path. If the source does not have a routing path to the destination, then it performs a route discovery by flooding the network with a route request (RREQ) packet. The RREQs record route information as they visit intermediate nodes on the way to the destination. Any node that has a path to the destination in question can reply to the RREQ packet by sending a route reply (RREP) packet. The reply is sent using the route recorded in the RREQ packet. A node that receives a RREQ can use the path recorded to improve its path to the source. To reduce the cost of route discovery, each node maintains a cache of source routes which it has learned or overheard. It aggressively uses to limit the frequency and propagation of RREQs. When an intermediate node discovers that a source route is broken, the source node is notified with a route error (RERR) packet. The source node can then attempt to use any other route to destination already in its cache or can invoke route discovery again to find a new route. To limit the need for route discovery, DSR also allows nodes to operate their network interfaces in promiscuous mode and snoop all (including data) packets sent by their neighbors. Since complete paths are indicated in data packets, snooping can be very helpful in keeping the paths in the route cache fresh. To further reduce the cost of route discovery, the RREQs are initially broadcasted to neighbors only (zero-ring search), and then to the entire network if no reply is received. Another optimization feasible with DSR is the gratuitous route replies; when a node overhears a packet containing its address in the unused portion of the path in the packet header, it sends the shorter path information to the source of the packet. Also, an intermediate node may replace a packet's current path specification with an alternate path if it is unable to send the packet to the original next hop node.

2.4. Directed Diffusion

Directed Diffusion protocol is a data-centric routing protocol in which all data generated by sensor nodes is named by attribute-value pairs. This protocol can be divided into four phases as follows,

- **Interest Propagation.** Interest describes a task required to be done by the network. It can be defined by a list of attribute-value pairs such as name of objects, interval, duration, geographical area, etc. For each active task, the sink periodically broadcasts an interest message to each of its neighbors. Every node maintains an interest cache. When a node receives an interest, it first checks to see if the interest exists in the cache. If no matching entry exists, the node creates an interest entry and save the parameters from the received interest. Each sensor that receives the interest setups a gradient toward the sensor nodes from which it receives the interest. This process continues until gradients are setup from the sources back to the sink or Base Station (BS). More generally, a gradient specifies an attribute value and a direction.

- **Low-rate data Propagation and routing setup.** In this phase, when a sensor node detects a target, it searches its interest cache for a matching interest entry. If matching entry exists, the node sends low-rate data to the nodes for whom it has a gradient. A node that receives a data message from its neighbor attempts to find a matching interest entry in its cache. If no match exists, the data message is silently dropped. If a match exists, the node adds the received message to the data cache and resends it to its neighbors.

- **Reinforcement.** After the low-rate data reaches the sink or BS, the sink selects and reinforces one particular (e.g., the best) path in order to draw down higher quality events (e.g., higher-rate data).

- **Data Propagation.** In this phase, the source node computes the highest requested event rate among all its outgoing gradients and sends them to its neighbors. The node which receives the message examines the matching interest entry's gradient list. If there is a lower data rate than the received data rate, it may down-convert the data to the appropriate gradient. And it also does some in-network data aggregation before resending the message in order to make the data aggregation more efficient.

3. SECURITY SERVICE

Highly level security requirements for ad hoc and sensor networks are basically identical to security requirements for any other communication systems. They can be classified as follows:

- **Confidentiality:** Ensures that the information is accessible only to authorized entities. Transmission of sensitive information (data packets) in ad hoc and sensor networks requires confidentiality. Disclosure of such information to enemies could cause devastating consequences. Routing information (control packets) must also remain confidential to certain extent, since such information can be used by the enemies to identify and locate their attacking targets.
- **Authentication:** Ensures that the origin of a message is correctly identified and its identity is not forged. Without authentication, the malicious node could impersonate other node to gain unauthorized access to resource and sensitive information.
- **Integrity:** Ensures that only authorized identities are able to modify system assets and transmit information. A message could be corrupted because of link failure, or malicious attack.
- **Data Freshness:** Ensures the freshness of messages.
- **Nonrepudiation:** Ensures that the origin of a message can not deny having sent the message. It is very useful to detect and isolate compromised nodes.
- **Access control:** Requires that access to information resources may be controlled by or for the target system. To achieve this control, each entity trying to gain access must be first identified, or authenticated, so that access rights can be tailored to the individual.
- **Availability:** Ensures the survivability of network service despite malicious attacks (e.g., denial-of-service attacks).

4. SECURITY ATTACK

Existing routing protocols are subject to a lot of security attacks. The attacks can be classified in different ways. One way is to divide attacks into four categories according to where the attacker deploys the attack in the flow of information from a source to a destination shown in Figure 1:

- **Interruption:** An asset of the network is destroyed or becomes unavailable or unusable. This is an attack on **availability**. Examples include silently discarding control or data packets.

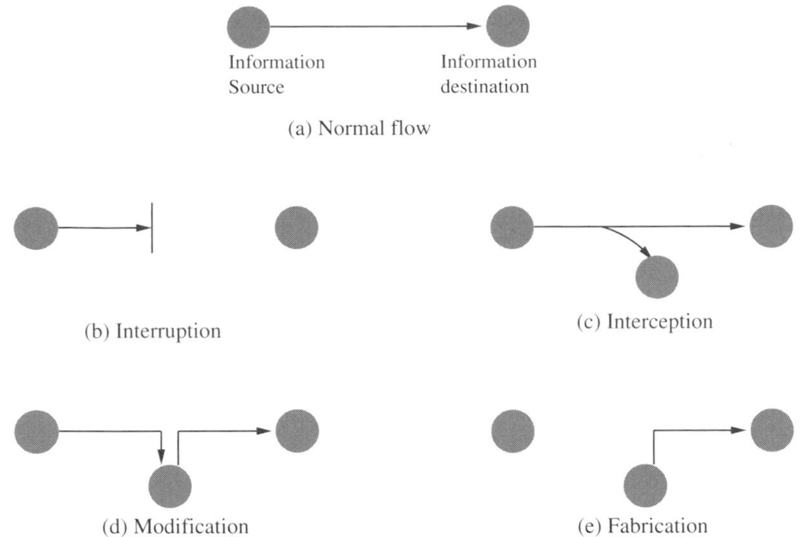

Figure 1. Security Threats.

- **Interception:** An unauthorized node gains access to an asset of the network. This is an attack on **confidentiality**. Examples include eavesdropping control or data packets in the networks.
- **Modification:** An unauthorized node not only gains access to but tampers with an asset. This is an attack on **integrity**. Examples include modifying control or data packets.
- **Fabrication:** An unauthorized node inserts counterfeit objects into the system. This is an attack on **authenticity**. Examples include inserting false routing messages into the network or impersonating other node.

A more useful categorization of these attacks is in terms of *passive attacks* and *active attacks* [43, 10], shown in Figure 2:

- **Passive attacks:** A passive attack does not disrupt the operations of a routing protocol, but only attempts to discover valuable information by eavesdropping, or silently discard messages received. Three types of passive attacks are **release of message contents**, **traffic analysis**, and **message dropping**.
  - The **release of message contents** is easily understood. The malicious node may leak confidential information to unauthorized users in the network, such as routing or location information. We would like to prevent the enemies from learning the contents of the sensitive data.

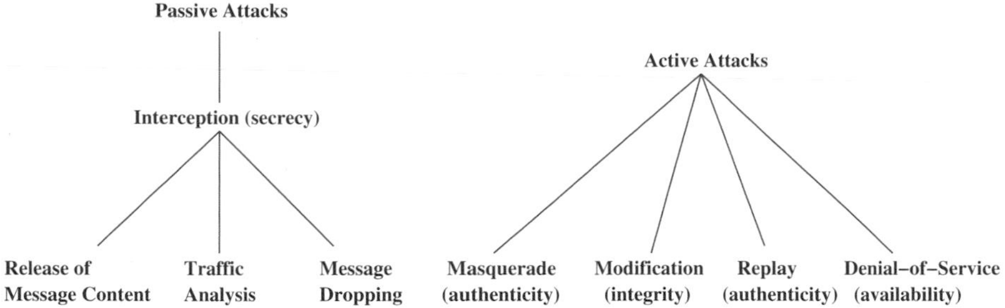

Figure 2. Active and Passive Security Threats.

- The second passive attack, **traffic analysis**, is more subtle. Suppose that we use encryption to mask the contents of messages, the enemies might still be able to observe the pattern of the messages and determine the location and identity of communication hosts.
- The third passive attack, **message dropping**, is to silently discard the message carried on by an intermediate host. For example, in AODV and DSR, when a malicious node receives a control packet (e.g., RREQ, RREP) or a data packet, instead of looking up route table and forwarding a packet, the malicious node drops the packet silently.

It is usually hard to distinguish passive attacks from Byzantine failures [8, 2] in ad hoc and sensor networks because passive attacks do not involve any alteration of contents of the messages. Message loss can occur because of topology change or unreliable wireless media. However, it is feasible to prevent the success of these attacks. Thus, the emphasis in dealing with passive attacks should be put more on prevention rather than detection.

- **Active attacks:** An active attack involves modification of the contents of messages or creation of false messages. It can be subdivided into four categories: **masquerade**, **replay**, **modification of messages**, and **denial of service**.

  - A **masquerade** takes place when one node pretends to be a different node. It is usually combined with other active attacks by the adversary to deploy security attacks. For example, a malicious node may impersonate another node while sending the control packets to create an undesirable update in the routing table.
  - **Replay** involves the passive capture of a valid message and retransmission to create an unauthorized effect.
  - **Modification of message** means that some portion of a legitimate message is altered, or that message is delayed, reordered to produce an unauthorized effect.

- **Denial of service(DoS)** prevents the normal use of communication facilities. In ad hoc and sensor network routing, DoS attacks can be classified into two categories: *DoS attack on routing traffic* and *DoS attack on data traffic*. An attacker can launch DoS attacks against a network by disseminating false routing information so that established routes for data traffic transmissions are invalid. For example, an attacker can create a *routing loop*, causing packets to travel nodes in a cycle without reaching their destinations, or *partition* the network by injecting malicious routing packets to prevent one set of nodes from reaching others. An attacker can also launch DoS attacks on data traffic by ejecting a significant amount of data traffic into the network to clog the network. Both of these two types of attacks might be used to consume valuable network resources such as bandwidth, or to consume node resources such as memory or computation power.

Active attack can be also further divided into *external attacks* and *internal attacks*. An *external attack* is one caused by nodes that do not belong to the network. An *internal attack* is the one launched from compromised or hijacked nodes that belong to the network. Internal attacks are often more severe compared to external attacks.

Next, we describe some well-known active attacks against ad hoc and sensor networks such as *Black Hole and Gray Hole Attacks* , *Wormhole Attacks*, *Rushing Attacks*, *Jellyfish Attacks*, *Sinkhole Attacks*, and *Sybil Attacks*.

- **Black Hole and Gray Hole Attacks:** The malicious node can easily deploy black hole attack [17] by advertising shorter distance (e.g., smaller hop count) or more fresh routes (e.g., a larger destination sequence number) to the destination or both to make the source node think that the route discovery process is complete and the route which is passing the malicious node is the best one. As a result, all the packets through the malicious node are simply consumed or lost. The malicious node could be said to form a black hole in the network, and this is called *black hole attack*. As a special case of black hole attack, the malicious node could create a *gray hole*, in which it selectively drops some, not all of packets that it receives.

- **Wormhole attacks:** In a *wormhole attack* [19], two attacker nodes collude together. One attacker node receives packets at one point and "tunnel" them to another attacker node via a private network connection, and then replays them into the network. The wormhole puts the attacker nodes in a very powerful position compared to other nodes in the network. For instance in reactive (on-demand) routing protocols such as AODV or DSR, the attackers can tunnel each route request RREQ packets to another attacker which is near to destination node of the RREQ. When the neighbors' of the destination hear this RREQ, they will rebroadcast this RREQ and then discard all other received RREQs in the same route discovery process. This type of attack prevents other routes instead of the wormhole from being discovered, and thus creates a permanent Denial-of-Service attack by dropping all the data, or selectively discarding or modifying certain packets as needed. The wormhole attack is very dangerous against ad hoc networking routing protocols and is harder to detect than other attacks since there is collusion between attackers.

- **Rushing attacks:** Rushing attack [20] is another type of DoS attack against all currently reactive (on-demand) routing protocols in MANETs. An attacker can forward route request packets (e.g., RREQs in AODV and DSR) more quickly than legitimate nodes, and thus increase the chance that routes which include the attacker will be discovered rather than other valid routes. After the attacker includes itself into the routes, it can launch different attacks such as dropping the packets that it receives, or modifying the content of the packets etc.

- **Jellyfish attacks:** Jellyfish (JF) attack [1] is targeted against closed-loop flows such as TCP which apply congestion control. In general, there are three kind of Jellyfish attacks: *JF Reorder Attack*, *JF Periodic Dropping Attack*, and *JF Delay Variance Attack*. TCP uses "ACK-N" to indicate that all segements $1, \ldots, N$ has been received. In *JF Reorder Attack*, before delivering all packets, the attacker place the packets received in a re-ordering buffer rather than an FIFO buffer. Such persistent re-ordering of packets will result in near zero goodput, even if all the packets will be delivered. In *JF Periodic Dropping Attack*, the attacker chooses a good dropping pattern to drop packets for only a small fraction of time, it can cause repeated timeout phases for other nodes, which might result in near-zero throughput. In *JF Delay Variance Attack*, the attacker still place packets into an FIFO buffer but wait a random time before delivering each packet, thus significantly increasing delay variance. Such high delay variance can (1) cause TCP to send traffic in bursts and lead to increased collisions and loss, (2) cause incorrect estimation of available bandwidth, and (3) result in an extremely high retransmission timeout (RTO) value.

- **Sinkhole attacks:** In a sinkhole attack for sensor networks, the attacker tries to lure nearly all the traffic from a particular area through a compromised node, creating a metaphorical sinkhole with the attacker at the center. Like black hole attacks in ad hoc networks, sinkhole attacks typically work by making a compromised node look especially attractive to surrounding nodes with respect to the routing algorithm.

- **Sybil attacks:** In a Sybil attack [13], the attacker presents multiple identities to other nodes in the network. The Sybil attack can significantly reduce the effectiveness of fault-tolerant distributed storage systems, routing algorithms, data aggregation, voting, and fair resource allocation etc.

There are also many other attacks. In [24], Jakobsson *et al.* describe several *stealth attacks* against ad hoc networks. Wireless physical layers are often vulnerable to jamming. Media Access Control protocols are also vulnerable to attacks. For example, in IEEE 802.11, an attacker can suppress its neighbor nodes from sending packets by periodically sending Clear-To-Send (CTS) frames in which "Duration" field is set a value greater than or equal to the interval between such frames. We will not discuss those in details since they are beyond the scope of this chapter.

5. SECURITY MECHANISMS

Before mobile ad hoc networks and sensor networks can be successfully deployed, security issues must be addressed first. Recently significant progress has been made in securing mobile ad hoc networks and sensor networks via the development of secure routing protocols. We divide these works into three categories based on the main characteristics of their security mechanisms: *Prevention Mechanisms*, *Detection Mechanisms*, *Other Mechanisms*.

Most of these mechanisms apply either asymmetric cryptography or symmetric cryptography or both. In asymmetric (public key) cryptography, there are RSA, Elliptic Curve Cryptography (ECC) etc. Digital signature is one way using asymmetric cryptography to authenticate the origin of messages. For symmetric cryptography, hash functions are used, such as HMAC, MD5, SHA-1 etc. Hash-chain and hash-tree use one-way hash function to encrypt messages. Asymmetric cryptography is generally much slower and more power-consumable compared to symmetric cryptography, and thus symmetric cryptography is preferred in MANETs, especially for sensor networks with less power and less computational capability.

No matter whether we use asymmetric cryptography or symmetric cryptography, we need to distribute the public key (in asymmetric cryptography) or shared key (in symmetric cryptography). Key distribution and management is another big issue in mobile ad hoc and sensor networks. Since this is quite beyond from our topic, we do not discuss them here.

5.1. Prevention Mechanisms

ARAN

Sanzgiri and Dahil [41] propose ARAN, an on-demand routing protocol for ad hoc networks. In ARAN, each node that forwards a route discovery or a route reply message must sign it. When a source node, S, needs to send data to a destination D, it broadcasts a *route discovery* packet (RDP):

$$S \rightarrow brdcast : [RDP, IP_D, cert_S, N_S, t]_{K_{S-}}$$

The RDP includes a packet type identifier "RDP", the IP address of the destination IP_D, S's certificate $cert_S$, a nonce N_S, and the current time t, all signed with S's private key. When a node B, the neighbor of S, receives the RDP broadcast, it rebroadcasts:

$$B \rightarrow brdcast : \left[[RDP, IP_D, cert_S, N_S, t]_{K_{S-}} \right]_{K_{B-}, cert_B}$$

Upon receiving the RDP, B's neighbor C validates the signature with the given certificate, then C removes B's certifcate and signature, records B as its predecessor, signs the contents of the message originally broadcast by S, appends its own certificate, and broadcasts the message as follows:

$$C \rightarrow brdcast : \left[[RDP, IP_D, cert_S, N_S, t]_{K_{S-}} \right]_{K_{C-}, cert_C}$$

All other nodes follow the same procedure if they receive RDP broadcasts, and they can use (N_S, IP_D) to check if it received the RDP before.

When the first RDP message arrives at the destination, D, the destination unicasts a Reply (REP) packet back along the reverse path to the source using digital signature to sign the packet, the same way as in the route discovery procedure.

During the route maintenance, all the Error (ERR) messages must be signed, and they are forwarded along the path toward the source without modification.

This scheme involves too expensive computation since every message is signed in a point-to-point manner, and it is prone to attacks using fabricated ERR messages.

SEAD

In SEAD [18], hash chains are used to secure Destination-Sequenced Distance-Vector routing protocol (DSDV). A one-way hash chain is built on a one-way hash function H which maps an input of any length to a fixed-length bit string. It can be denoted as: $H : \{0,1\}^* \rightarrow \{0,1\}^\rho$, where ρ is the length in bits of the output of the hash function. For each node, to create a one-way hash chain, it chooses a random number $x \in \{0,1\}^\rho$ and computes a list of values:

$$h_0, h_1, h_2, \ldots, h_n$$

where $h_0 = x$, and $h_i = H(h_{i-1})$, $0 < i \leq n$, for some n. In SEAD, it assumes that there is some mechanism for a node to distribute an authentic element h_n from its hash chain. It assumes that all metrics in any routing update are less than m, and n is divisible by m. Let $k = \frac{n}{m} - i$, where i is the sequence number in some routing update entry. An element from the following group:

$$h_{km}, h_{km+1}, \ldots, h_{km+m-1}$$

is used to authenticate the entry. For example, if the metric value (e.g., hop count) is j, $0 \leq j < m$, then h_{km+j} is used to authenticate the routing update entry. If the sequence number in a route reply message is i and metric value is j, we can check if $H^{n-km-j}(h_{km+j}) = h_n$ to know whether the this route reply message is valid or not.

SEAD can prevent any node from advertising a route to some destination claiming a higher destination sequence number or smaller metric. However, it can not prevent the malicious node from advertising a route with the same metric as in the advertisement it received, and after using all the elements of the hash chain, a new one must be computed and distributed to all other nodes.

Ariadne

Ariadne [17] is proposed to secure DSR protocol by the same authors of SEAD. It authenticates routing messages using one of three schemes: shared secrets between each pair of nodes, shared secrets between end-to-end nodes combined with broadcast authentication TESLA [35], or digital signatures. However, it requires clock synchronization, which would not be realistic requirement for mobile ad hoc networks.

SAODV

Zapata and Asokan [52] propose a security mechanism to protect AODV's routing information: Secure AODV (SAODV). Two mechanisms are used to secure the AODV messages (RREQs, RREPs, and RERRs): (i) digital signatures to authenticate the non-mutable fields of the messages (e.g., destination sequence number) in an end-to-end manner, (ii) hash chains to secure the mutable information of the message (hop count). The RREQ and RREP messages include four new fields Hash_Function, Max_Hop_Count, Top_Hash, and Hash. When a node originates an RREQ or an RREP message, it performs the following operations:

1. Generate a random number *seed* and set Hash field to the *seed* value:

$$Hash = seed.$$

2. Set the Max_Hop_Count field to the TimeToLive value in the IP Header:

$$Max\_Hop\_Count = TimeToLive.$$

3. Set the Hash_Function field to the identifier of the hash function h that will be used. If a fixed hash function is used throughout the network, such as MD5 or SHA1 etc., the Hash_Function field is not needed anymore.

4. Set Top_Hash field value by hashing *seed* MAX_HOP_Count times:

$$Top\_Hash = h^{MAX\_Hop\_Count}(seed).$$

When a node receives an RREP or an RREP message, it can easily verify the hop count by comparing if

$$Top\_Hash = h^{MAX\_Hop\_Count - Hop\_Count}(Hash),$$

where Hop_Count is the hop count value in the message. Since all other fields are signed using digital signature by either source or destination node, current node can verify if other fields have been modified in the middle or if the message is fabricated. Before rebroadcasting an RREQ or relaying an RREP, the node applies the hash function to the value in the Hash field and write the new value back to the Hash field. There is a problem in applying digital signatures on RREP messages because AODV allows intermediate nodes to reply RREQ messages if they have a "fresh enough" route to the destination. To solve this problem, the authors propose two mechanisms. The first one, also the obvious one, is that intermediate nodes can not initiate an RREP even if they have a fresh route to the destination. The second one, aslo more complex one, is that, every time a node S generates an RREQ message, it also includes the RREP flags, the prefix size and the signature that can be used by any intermediate node I that creates a reverse route to the originator of the RREQ, S, to reply other RREQs in which the destination is S. The intermediate node should include two lifetime: the old one which is needed to verify the signature of the route destination and the new lifetime of this route which is signed by itself.

Considering that RERR messages have too much mutable information, in their scheme, every node will use digital signatures to sign the whole message and any node that receives it can verify the signature. Since the destination sequence numbers are not signed by the corresponding node, a node will not update any destination sequence number in its routing table based on RERR messages.

RAP

Hu *et al.* [20] present a new attack, *rushing attack* that results in DoS when used against on-demand ad hoc networking routing protocols. They proposed *Rushing Attack Prevention (RAP)* against rushing attacks for on-demand routing protocols. RAP consists of three generic mechanisms: *secure Neighbor Detection*, *secure Route Delegation*, and *randomized Route Request forwarding*. *Secure Neighbor Detection* allows each node to verify that its neighbor is within a given maximum transmission range. After a node A forwarding a Route Request message determines that the node B is its legitimate neighbor, it signs a *Route Delegation* message, allowing node B to forward the Route Request message. Once node B verify that node A is within its allowable range, it signs an *Accept Delegation* message. *Randomized Route Request forwarding* is used to make sure that arriving Route Requests with low latency are only slightly more likely to be forwarded than others.

SRP

Papadimitratos and Haas [33] propose the Secure Routing Protocol (SRP) which focuses on multiple routes discovery for on-demand source routing protocols such as DSR[22] or ZRP[14]. SRP relies exclusively on the mutual authentication of the end nodes only (source and destination). A *security association (SA)* between the source node S and destination node D is assumed. There is a shared key $K_{S,D}$ between them. When a source node S starts the route discovery, it broadcasts a route request message which includes a query sequence number and a random query identifier and Message Authentication Code (MAC) using $K_{S,D}$. The source, destination, the query sequence number and the random query identifier are inputs for the calculation of MAC. The identifier of the traversed intermediate nodes are accumulated in the route request packet. When a route request arrives at the destination D, D will verify the route request and calculate an MAC covering the contents of the route reply and unicast route reply back to S along the reverse of the route accumulated in the route request. The destination responds to one or more request packets of the same query, so that the source can get multiple paths to the destination. When the source node S receives the route reply packet, it will verify that the packet is originated from the destination.

Since SRP does not require any authentication of the relay nodes, it makes the protocol more light-weight, but also more prone to attacks.

SMT

Papadimitratos and Haas [34] devise a *Secure Message Transmission protocol* (SMT), based on SRP [33], to safeguard the data transmission against malicious behaviors. The principle of SMT is that both source and destination make use of an "Active Path Set" which consists of diverse, preferably link disjoint paths that are assumed valid initially. The source disperses each outgoing packet into a number of pieces. This operation introduces redundancy such that the destination can successfully reconstruct the dispersed message if sufficient number of pieces are received. Each dispersed piece is transmitted across a different route chosen from the "Active Path Set" and it carries a Message Authentication Code, so that the destination can verify its integrity and the authenticity of its origin.

SNEP and μTESLA:

In [36], two secure building blocks: SNEP and μTESLA are proposed for sensor networks. SNEP, which stands for Sensor Network Encryption Protocol, provides data confidentiality, two-party data authentication, integrity, and freshness. In SNEP, data confidentiality is achieved through encryption using the shared key between source and sink. To achieve two-party authentication and data integrity, a message authentication code (MAC) is used. Using secret shared counter between the sender and the receiver, weak freshness is assured. μTESLA is a protocol for efficient, authenticated broadcast and flooding that uses only symmetric key cryptography and requires minimal packet overhead. μTESLA achieves the asymmetric necessary for authenticated broadcast and flooding by using delayed key disclosure and one-way key chains constructed with a publicly computable cryptographically secure hash function. Replay can be prevented since messages authenticated with previously disclosed keys are ignores. Like TESLA, it requires loose time synchronization too.

INSENS

In [11], Deng *et al.* propose an intrusion-tolerant routing protocol for wireless sensor networks (INSENS). INSENS constructs forwarding tables at each node to facilitate communication between sensor nodes and a sink or base station. Route discovery consists of three phases. First, the base station floods a request message to all reachable sensor nodes in the network. Second, sensor nodes send their (local) topology information using a feedback message to the base station. Finally, the base station computes the forwarding tables for each sensor node based on the information received in the second phase and sends them to the corresponding nodes using a routing update message. To counter spurious request message, a scheme similar to μTESLA where one-way key chains are used to identify a message originating from the base station. Moreover, multipath routing is built into INSENS to address the notion of compromised nodes.

SIA

Przydatek *et al.* [40] propose a novel framework for secure information aggregation in large sensor networks. They focus on defending stealthy attacks. In this type of stealthy attacks, the attackers' goal is to make the base station accept false aggregation results, which is significantly different from the true results determined by the measured values. A general approach, called *aggregate-commit-prove*, is proposed. This technique consists of three parts. In the first part, the aggregator collects data from the sensors and computes the aggregation result. In the second part, the aggregator is responsible for committing to the collected data. This commitment ensures that the aggregator actually uses the data collected from the sensors. One way to perform this commitment is to use Merkle hash-tree construction. In the final part, the aggregator uses an interactive proof to prove the correctness of the results. This interactive proof usually contains two steps. In the first step, the base station checks to ensure that the committed data is a good representation of the true data in the sensor networks. The second step, the base station checks if the aggregator is cheating. This can be done by checking whether the aggregation result is close to the committed result.

SDD

In [44], Wang *et al.* propose a new secure directed diffusion protocol (SDD) which provides a simple scheme to securely diffuse data in sensor networks. SDD is also divided into four phases as in directed diffusion protocol. In the first phase of interest propagation, modified authentication TESLA protocol is used to ensure the interest is from the sink. In the low-rate data propagation and secure route setup phase, modified immediate authentication TESLA protocol is used to ensure that data from the source node could be read by the intermediate nodes and have not been modified. Each intermediate node also adds its own identity to a node list in the forwarded message and adds its "signature" to the message by encrypting a nonce using its shared key between itself and the base station. In the third phase of reinforcement, the base station selects a path and reinforces it. This reinforcement is protected in the same way as an interest is protected in the first phase. In the last phase of data propagation, the data is protected the same way as in the second phase for the low-rate data. However, in SDD, in-network aggregation, which can make the data propagation more efficient, is forbidden,

5.2. Detection Mechanisms

Watchdog and Pathrater

In order to mitigate attacks on packet forwarding, Marti *et al.* [31] propose *watchdog* to identfy malicious nodes and *pathrater* to help routing protocol to avoid these nodes. Each node is set to *promiscuous* mode. After a node forwards a packet, the node's *watchdog* verifies if the next node in the path also forwards the packet by *promiscuously* listening to the next node's transmissions. If the next node does not forward the packet as it is supposed to do, it is marked as a malicious node. The *pathrater* uses this knowledge of misbehaving nodes to choose the network path that is most likely to deliver packets. Although it is promising approach, there are three key limitations. First, it cannot be deployed in combination with dynamic power management where each node can reduce its power level while transmitting and thus this transmission might not be heard by its upstream node. Second, *watchdog* assumes that all nodes use omnidirectional antenna. If a malicious node use unidirectional antenna, then it can fool its upstream node by sending packet back toward the upstream node. Third, a node can also uses variable power transmission to cheat the upstream node.

Identification

In some approaches, once a victim has detected a DoS attack, it will try to establish a new route. A more sophisticated approach is trying to identify the malicious node(s) on the route. Awerbuch *et al.* [2] propose a mechanism to detect a malicious link on the route using an adaptive probing technique. This approach is based on using acknowledgements (acks) of the data packets from the destinations. If a source detects that the number of lost packets (without acks) is larger than a given threshold, it performs a binary search on the path to identify the malicious link. In order to realize that, it polls specific nodes via "probes" and waits for replies. The protocol assumes that the malicious node can not distinguish between probing packets and normal ones, and is also unable to know when the source has started a probing process.

Although it is a good approach, how to select the threshold and the performance of this approach still need more investigation.

Examination

Just *et al.* [23] propose a distributed *probing* technique to detect and mitigate malicious packet dropping attacks in ad hoc networks. In this approach, every node periodically monitors the forwarding behavior of other nodes. Instead of using promiscuous listening in [31], a node A sends a probe message (indistinguishable from normal data) to another node C one or more hops away from the node to be checked, B, and waits for an acknowledgement from C within a certain time period. If A can not receive the acknowledgement in the time period, then B is suspected. The best thing of this approach is that it can also be implemented in the application layer and does not require the modification of routing protocols. However, the overhead of this approach is too much and might not work well in mobile ad hoc networks with high mobility.

Packet Leashes

Wormhole attack [19] is very dangerous against ad hoc network routing protocols since it can create a permanent DoS attack. There are some ways to prevent wormhole attacks such as using a secret method for modulating bits over wireless transmission, or using RF watermarking, modulating the RF waveform in a way known only to authorized nodes, to authenticate a wireless transmission, etc. However, these approaches suffer from certain problems as follows. Once a node is compromised, the secret method for modulating bits is likely to fail unless the radio is kept inside a tamper-resistant hardware. When a waveform is exactly captured, RF watermarking fails too. In contrast to these approaches, Hu *et al.* [19] propose a general approach, *packet leash*. The principle is to add information (leash) to a packet designed to restrict the packet's maximum allowed transmission distance. They propose two methods: *geographical leashes* and *temporal leashes*. A *geographical leash* ensures that the recipient of the packet is within a certain distance from the sender. The *temporal leash* ensures that a packet has an upper bound on its lifetime, which restricts its maximum travel distance. Each type of leash can prevent wormhole attacks. However, since in *geographical leash* each node must know its own location, and all nodes must have tightly synchronized clock in *temporal leash*, these requirements restrict their applicability.

Mitigation

Malicious route request flooding attack targets at the overall network performance. These malicious nodes behave like normal nodes except that they do route discoveries much more frequently than normal ones. Desilva and Boppana [12] propose a statistical profiling technique to detect the malicious nodes and an adaptive statistical packet dropping mechanism to mitigate this type of attack, and the proposed scheme reduces the loss of throughput. In this approach, each node monitors the route requests which it receives and maintains a count of RREQs received for each RREQ sender. For node A, when the number of RREQs received from RREQ sender M in a period is larger than some threshold which is related to the average RREQs sent for each node in the whole network, then M is marked as a malicious node. This approach is simple and does not use any additional network bandwidth.

Validation

To defend against the Sybil attack, Newsome *et al.* [29] describe two methods to validate identities: *direct validation* and *indirect validation*. In *direct validation*, a trusted node directly tests whether another node identity is valid. In *indirect validation*, other trusted nodes are allowed to vouch for (or against) the validity of a joining node. Newsome *et al.* present a novel approach, radio test, to direct validation. This approach assumes that any physical device has only one radio and the radio is incapable of simultaneously sending or receiving on more than one channel. In the radio test, a node assigns each of its neighbors a different channel on which to broadcast some messages. The node then randomly chooses a channel and listens. If the neighbor that was assigned that channel is legitimate, it should hear the message. Another technique to defend against the Sybil attack proposed in [29] is to use random key predistribution techniques. The principle of this technique is that with a limited number of keys in a key pool, a node that randomly generates identities will not possess enough keys to take on multiple identities and thus will not be able to exchange messages in the network.

5.3. Other Mechanisms

Researchers have studied approaches to enforce node cooperation in mobile ad hoc networks. Buchegger *et al.* [5] propose a protocol, called CONFIDANT, for making misbehavior unattractive. It aims at not only detecting and avoiding, but also isolating misbehaving nodes, thus making it unattractive to deny cooperation. CONFIDANT consists of four components in each node: a *Monitor* which identifies deviations from the normal routing behavior, a *Reputation System* which rates other nodes according to the observed behavior, a *Path Manager* which maintains path rankings and performs specific actions upon receiving routing messages, and a *Trust Manager* which sends and receives alarm messages to and from other trust managers. Similarly, Michiardi and Molva propose CORE [32], a generic mechanism based on reputation to enforce cooperation among nodes in mobile ad hoc networks.

Buttyan and Hubaux propose the usage of a virtual currency, *nuglet* [6], to stimulate packet forwarding. They propose two payment models: *Packet Purse Model* and *Packet Trade Model*. In the *Packet Purse Model*, the source of a packet pays by loading some nuglets in the packet before sending it, the intermediate nodes can acquire some nuglets after forwarding the packet. In the *Packet Trade Model*, a packet does not carry nuglets. Each intermediate node "buys" the packet from the previous one with some nuglets, and "sells" it to the next one for more nuglets. Later they propose a new approach [7] which is based on a counter, called nuglet counter. Packets do not need to carry nuglets, when the node sends its own packet, the nuglet counter decreases, and the counter increases when the node forwards a packet.

Zhou *et al.* propose *Sprite* [53], a simple, cheat-proof, credit-based system for stimulating cooperation among selfish nodes in mobile ad hoc networks. Sprite provides incentive for mobile nodes to cooperate and report actions honestly, and thus stimulates cooperation among selfish mobile nodes. Moreover, Sirinivasan *et al.* [42] address the issue of user cooperation in ad hoc networks, where each node is constrained by finite energy capacity so that the network has a limited life time.

More recently, Yu *et al.* propose HADOF [50], a set of mechanisms to protect mobile ad hoc networks against routing disruption attacks by insider attackers, which can be implemented upon the existing source routing protocols such as DSR. In HADOF, each node monitors the behavior of the valid routes in its route cache and collect the packet forwarding statistics submitted by the nodes on this route. Each node keeps *cheating records* for other nodes, if a node is detected as dishonest, this node will be excluded from future routes. Moreover, each node tries to build *friendship* with others to speed up malicious node detection. Each source node also discovers multiple routes to the destination, and *adaptive route rediscovery* is used to determine when new routes should be discovered. HADOF can handle various attacks with little overhead, such as black hole attacks, gray hole attacks, rushing attacks, and wormhole attacks, etc.

Yi *et al.* [49] propose a routing protocol, called Security-Aware ad hoc Routing (SAR), which incorporates security attributes as parameters into ad hoc route discoveries, to improve QoS of routing protocols. Yang *et al.* [47] present a new secure enhancement scheme (SES), which incorporates a new metric for routing. The new metric takes into account both the distribution of possible existing attacks and the possibility of future potential attacks.

In [1], Aad *et al.* study two types of attacks: JellyFish attack and Black Hole attack. This is the first paper that provides an analytical model to quantify the scalability of DoS attacks as a function of performance parameters such as mobility, system size, node density, and counter-DoS strategy, etc.

In [25], Karlof *et al.* give a good discussion on attacks on routing protocols in wireless sensor networks. Some countermeasures and design considerations for routing in sensor networks are suggested.

6. CONCLUSION

Both mobile ad hoc networks and wireless sensor networks are gaining popularity due to the fact that they promise viable solutions to a variety of applications. Since MANETs can be easily and inexpensively set up as needed, they have a wide range of applications, especially in situations with geographical or terrestrial constrains such as battlefields, military applications, and other emergency and disaster situations. Networking unattended sensor nodes are increasingly popular to provide economical solutions to both military and civil applications such as real-time traffic monitoring, wildfire tracking, wildlife monitoring, weather monitoring, security and tactical surveillance, target filed imaging, and disaster management, etc. However, both MANETs and WSNs are vulnerable to different attacks due to their fundamental characteristics such as open medium, dynamic topology, absence of central administration, distributed cooperation, and constrained capability (power and computation constraints). In order to make mobile ad hoc and sensor networks work properly, security issues must be addressed first. Recently significant progress has been made in securing mobile ad hoc and sensor networks via the development of secure routing protocols.

In order to assess the security needs of MANETs and WSNs effectively and to evaluate and choose various policies, three aspects of information security are considered: security service, security attack, and security mechanism.

In this chapter, we first go over what kind of security services are needed in ad hoc and sensor networks, then introduce different attacks, and finally review the recent approaches to secure routing protocols in both mobile ad hoc networks and sensor networks. We divide current approaches into three groups: prevention mechanism and detection mechanism, and all other related works as the third group.

To secure routing in ad hoc and sensor networks, both control packets and data packets must be protected from attacking. Although significant progress has been made to detect and prevent known security attacks, there still will be new potential attacks, and more research work must be done to make routing in ad hoc and sensor networks more secure and practically deployable.

Since it is usually hard to distinguish passive attacks from Byzantine failures, the emphasis in dealing with passive attacks should be put more on prevention rather than detection. While for active attacks, it is better to put more efforts on detection mechanisms and reputation-based approaches.

7. REFERENCES

1. I. Aad, J. Hubaux, and E. Knightly, Denial of Service Resilience in Ad Hoc Networks, *Proceedings of ACM MOBICOM Wireless Security Workshop (WiSe)*, 2004.
2. B. Awerbuch, D. Holmer, C. Nita-Rotaru, and H. Rubens, An On-demand Secure Routing Protocol Resilient to Byzantine Failures, *Proceedings of ACM MOBICOM Wireless Security Workshop (WiSe)*, 2002.
3. R. V. Boppana and S. Konduru, An adaptive Distance Vector Routing Algorithm for Mobile, Ad Hoc Networks, *Proceedings of IEEE Infocom*, 2001.
4. D. Braginsky and D. Estrin, Rumor Routing Algorithm for Sensor Networks, *Proceedings of the First Workshop on Sensor Networks and Application (WSNA)*, 2002.
5. S. Buchegger and J. Boudec, Performance Analysis of the CONFIDANT Protocol: Cooperation Of Nodes Fairness In Dynamic Ad-hoc Networks, *Proceedings of IEEE/ACM Symposium on Mobile Ad Hoc Networking and Computing (MobiHOC)*, 2002.
6. L. Buttyan and J. Hubaux, Enforcing Service Availability in Mobile Ad-Hoc WANs, *Proceedings of the First IEEE/ACM Workshop on Mobile Ad Hoc Networking and Computing (MobiHOC)*, 2000.
7. L. Buttyan and J. Hubaux, Stimulating Cooperation in Self-Organizing Mobile Ad Hoc Networks, *Mobile Networks and Applications*, 2003.
8. M. Castro and B. Liskov, Practical Byzantine Fault Tolerance, *Proceedings of the Third Symposium on Operating Systems Design and Implementation*, 1999.
9. B. Chen, K. Jamieson, H. Balakrishnan, and R. Morris, SPAN: An Energy-Efficient Coordination Algorithm for Topology Maintenance in Ad Hoc Wireless Networks, *Wireless Networks*, Vol. 8, No. 5, pp. 481-494, 2002.
10. H. Deng, W. Li, and D. Agrawal, Routing Security in Wireless Ad Hoc Networks, *IEEE Communication Magazine*, (2002) pp. 70-75.
11. J. Deng, R. Han, and S. Mishra, INSENS: Intrusion-tolerant Routing in Wireless Sensor Networks, *Technical Report CU-CS-939-02, Department of Computer Science, University of Colorado*, 2002.
12. S. Desilva and R. V. Boppana, Mitigating Malicious Control Packet Floods in Ad Hoc Networks, *IEEE Wireless Communications and Networking Conference*, March 2005.
13. J. R. Douceur, The Sybil Attack, *The 1st International Workshop on Peer-to-Peer Systems(IPTPS'02)*, 2002.

ABOUT THE EDITORS

Yang Xiao worked at Micro Linear as an MAC (Medium Access Control) architect involving the IEEE 802.11 standard enhancement work before he joined Department of Computer Science at The University of Memphis in 2002. Dr. Xiao is the director of W^4-Net Lab, and was with CEIA (Center for Information Assurance) at The University of Memphis. He is currently with Department of Computer Science at The University of Alabama. He is an IEEE Senior member. He was a voting member of IEEE 802.11 Working Group from 2001 to 2004. He currently serves as Editor-in-Chief for *International Journal of Security and Networks (IJSN)* and for *International Journal of Sensor Networks (IJSNet).* He serves as an associate editor or on editorial boards for the following refereed journals: *(Wiley) International Journal of Communication Systems, (Wiley) Wireless Communications and Mobile Computing (WCMC), EURASIP Journal on Wireless Communications and Networking*, and *International Journal of Wireless and Mobile Computing.* He serves as a guest editor for *IEEE Wireless Communications*, special issue on "Radio Resource Management and Protocol Engineering in Future Broadband and Wireless Networks in 2006, a (lead) guest editor for *International Journal of Security in Networks (IJSN)*, Special Issue on "Security Issues in Sensor Networks in 2005, as a (lead) guest editor for *EURASIP Journal on Wireless Communications and Networking*, Special Issue on "Wireless Network Security" in 2005, as a (sole) guest editor for *(Elsevier) Computer Communications journal*, special Issue on "Energy-Efficient Scheduling and MAC for Sensor Networks, WPANs, WLANs, and WMANs" in 2005, as a (lead) guest editor for *(Wiley) Journal of Wireless Communications and Mobile Computing*, special Issue on "Mobility, Paging and Quality of Service Management for Future Wireless Networks" in 2004, as a (lead) guest editor for *International Journal of Wireless and Mobile Computing*, special Issue on "Medium Access Control for WLANs, WPANs, Ad Hoc Networks, and Sensor Networks" in 2004, and as an associate guest editor for *International Journal of High Performance Computing and Networking*, special issue on "Parallel and Distributed Computing, Applications and Technologies" in 2003. He serves as editor/co-editor for ten edited books: *WiMAX/MobileFi: Advanced Research and Technology, Security in Distributed and Networking Systems, Security in Distributed, Grid, and Pervasive Computing, Security in Sensor Networks, Wireless Network Security,*

Adaptation Techniques in Wireless Multimedia Networks, Wireless LANs and Bluetooth, Security and Routing in Wireless Networks, Ad Hoc and Sensor Networks, and *Design and Analysis of Wireless Networks*. He serves as a referee/reviewer for many funding agencies, as well as a panelist for US NSF and a member of Canada Foundation for Innovation (CFI)'s Telecommunications expert committee. He serves as TPC for more than 70 conferences such as INFOCOM, ICDCS, ICC, GLOBECOM, WCNC, etc. His research areas are wireless networks, mobile computing, and network security. He has published more than 140 papers in major journals and refereed conference proceedings related to these research areas. E-mail: yangxiao@ieee.org

Xuemin (Sherman) Shen received a B.Sc. (1982) degree from Dalian Maritime University, China, and M.Sc. (1987) and Ph.D. (1990) degrees from Rutgers University, New Jersey, USA, all in electrical engineering. Currently, Dr. Shen is with the Department of Electrical and Computer Engineering, University of Waterloo, Canada, where he is a Professor and the Associate Chair for Graduate Studies. His research focuses on mobility and resource management in interconnected wireless/wireline networks, UWB wireless communications systems, wireless security, and ad hoc and sensor networks. He is a co-author of three books, and has published more than 200 papers and book chapters in wireless communications and networks, control, and filtering. He was Technical Co-Chair for the IEEE GLOBECOM'03, ISPAN'04, QShine'05, IEEE Broadnets'05, and WirelessCom'05, and is Special Track Chair of the 2005 IFIP Networking Conference. He serves as an Editor/Associate Editor for IEEE Transactions on Wireless Communications, IEEE Transactions on Vehicular Technology, Computer Networks, ACM/Wireless Networks, Wireless Communications and Mobile Computing (Wiley), and International Journal Computer and Applications. He has also served as Guest Editor for IEEE JSAC, IEEE Wireless Communications, and IEEE Communications Magazine. He received the Outstanding Performance Award in 2004 from the University of Waterloo, the Premier's Research Excellence Award (PREA) in 2003 from the Province of Ontario, Canada, for demonstrated excellence of scientific and academic contributions, and the Distinguished Performance Award in 2002 from the Faculty of Engineering, University of Waterloo, for outstanding contributions in teaching, scholarship, and service.

Ding-Zhu Du received his M.S. degree in 1982 from Institute of Applied Mathematics, Chinese Academy of Sciences, and his Ph.D. degree in 1985 from the University of California at Santa Barbara. He worked at Mathematical Sciences Research Institute, Berkeley in 1985-86,at MIT in 1986-87, and at Princeton University in 1990-91. He was an associate-professor/professor at Department of Computer Science and Engineering, University of Minnesota in 1991-2005, a Program Director for Theory of Computing at National Science Foundation in 2002-2005, a professor at City University of Hong Kong in 1998-1999, and a research professor at Institute of Applied Mathematics, Chinese Academy of Sciences in 1987-2002. Currently, he is a professor at Department of Computer Science, University of Texas at Dallas and a Dean of Science at Xi'an Jiaotong University. His research interests include combinatorial optimization, communication networks, and theory of computation. He has published more than 160 journal papers and 40 books. He is the editor-in-chief of *Journal of Combinatorial Optimization* and book series on *Network Theory and Applications*. He is also in editorial boards of more than 15 journals. He is well-known for proving the Gilbert-Pollak conjecture on the Steiner ratio, the Derman-Leiberman-Ross conjecture on optimal consecutive 2-out-of-n systems in reliability, and the global convergence of Rosen gradient projection method in nonlinear programming. In 1998, he received received CSTS Prize from INFORMS (a merge of American Operations Research Society and Institute of Management Science) for research excellence in the interface between Operations Research and Computer Science. In 1996, he received the 2nd Class National Natural Science Prize in China. In 1993, he received the 1st Class Natural Science Prize from Chinese Academy of Sciences. In 1992 the proof of Gilbet-Pollak conjecture was selected by 1992 Year Book of Encyclopaedia, Britannica, as the first one among six outstanding achievements in mathematics in 1991.

Yang Xiao
Department of Computer Science
The University of Alabama
101 Houser Hall
Box 870290
Tuscaloosa, AL 35487-0290 USA
E-mail: `yangxiao@ieee.org`

Xuemin (Sherman) Shen
Department of Electrical and Computer Engineering
University of Waterloo Waterloo
Waterloo, Ontario, Canada N2L 3G1
E-mail: `xshen@bbcr.uwaterloo.ca`

Ding-Zhu Du
Department of Computer Science
University of Texas at Dallas
Richardson, TX 75083, USA
E-mail: `dzdu@utdallas.edu`

INDEX

(continued from page ii)

SDMA for Multipath Wireless Channels
Limiting Characteristics
and Stochastic Models
I.P. Kovalyov ISBN 3-540-40225-X

Digital Television
A Practical Guide for Engineers
W. Fischer ISBN 3-540-01155-2

Speech Enhancement
J. Benesty (Ed.)
ISBN 3-540-24039-X

Multimedia Communication Technology
Representation, Transmission
and Identification of Multimedia Signals
J.R. Ohm ISBN 3-540-01249-4

Information Measures
Information and its Description in Science
and Engineering
C. Arndt ISBN 3-540-40855-X

Processing of SAR Data
Fundamentals, Signal Processing,
Interferometry
A. Hein ISBN 3-540-05043-4

Chaos-Based Digital Communication Systems
Operating Principles, Analysis Methods, and
Performance Evalutation
F.C.M. Lau and C.K. Tse
ISBN 3-540-00602-8

Adaptive Signal Processing
Application to Real-World Problems
J. Benesty and Y. Huang (Eds.)
ISBN 3-540-00051-8

Multimedia Information Retrieval and
Management Technological
Fundamentals and Applications D. Feng, W.C.
Siu, and H.J. Zhang (Eds.)
ISBN 3-540-00244-8

Structured Cable Systems
A.B. Semenov, S.K. Strizhakov,and I.R.
Suncheley
ISBN 3-540-43000-8

UMTS
The Physical Layer of the Universal Mobile
Telecommunications System
A. Springer and R. Weigel
ISBN 3-540-42162-9

Advanced Theory of Signal Detection
Weak Signal Detection in Generalized
Obeservations
I. Song, J. Bae, and S.Y. Kim
ISBN 3-540-43064-4

Wireless Internet Access over GSMand UMTS
M. Taferner and E. Bonek
ISBN 3-540-42551-9